Multidisciplinary Management of Head and Neck Cancer

Ravi A. Chandra • Ryan J. Li
Editors

Multidisciplinary Management of Head and Neck Cancer

Contemporary Applications and Techniques

Springer

Editors
Ravi A. Chandra
Department of Radiation Medicine
Oregon Health & Science University
Portland, OR
USA

Department of Radiation Oncology
Mid-Atlantic Permanente Medical Group, P.C.
Rockville, MD
USA

Ryan J. Li
Department of Otolaryngology-Head and Neck Surgery
Oregon Health & Science University
Portland, OR
USA

ISBN 978-3-031-05972-8 ISBN 978-3-031-05973-5 (eBook)
https://doi.org/10.1007/978-3-031-05973-5

This Springer imprint is published by the registered company Springer Nature Switzerland AG
The registered company address is: Gewerbestrasse 11, 6330 Cham, Switzerland

Foreword

In order to provide optimal care for patients with often complex neoplasms of the upper aerodigestive tract, it is essential that a comprehensive toolbox be available for providers and patients to embark on a care path that is likely to yield the best evidence-based results. The present textbook has been carefully edited by Drs. Chandra and Li, to the degree that this volume provides an up to date though readily pragmatic approach to guide highest quality patient care. Therein lies the main strength of this contribution of which I will briefly highlight in this Foreword.

First and foremost, the truly interdisciplinary and multidisciplinary approach to establishing an accurate diagnosis and evidence-based treatment is clearly a priority throughout this work. Each of the authors regularly participates in multidisciplinary cancer care as a normative approach to patient-centric care.

Second, the chapter contributors are investigators who can effectively write in a manner that will allow this book to be a worthwhile investment for a rather broad spectrum of learners and practitioners.

Third, while some of the data points for certain chapters will eventually become outdated at varying rates, the editors have ensured that the strong foundation for this contribution will remain relevant for several years.

Fourth, having worked closely with both book editors on multiple projects, I can attest to their collective ability to adhere to high standards to complement the meticulous organization of the chapters.

Finally, I believe that individuals who "go deep" into gaining pragmatic knowledge about best practices for head and neck cancer will be well served by this book. Ultimately, our patients will be the major beneficiaries. Furthermore, institutions involved in training providers across the interprofessional spectrum should provide access to this book in their respective libraries and data repositories. Our own NCI Comprehensive Cancer Center will certainly purchase this book. Drs. Chandra and Li—well done!

Hanover, NH, USA
January 2022

Charles R. Thomas Jr.

Foreword

With this book Dr. Ravi Chandra and Dr. Ryan Li have assembled a cohort of experts in the field of head and neck oncology to create a contemporary and detailed reference on the multidisciplinary care of patients afflicted with head and neck cancer. This book focuses on important topics that pertain to the current state-of-the-art delivery of comprehensive care for head and neck cancers.

The change in the epidemiology of head and neck cancer from primarily a smoking-related disease to one primarily caused by the human papilloma virus (HPV) has impacted the therapy and care of head and neck cancer patients. Chapters provide excellent updates on advances in epidemiology, biomarkers, and imaging. Therapeutic advancements in minimally invasive surgery and reconstruction, radiation, systemic therapy, and immunotherapy are well covered. In addition, organ preservation approaches and deintensification strategies for appropriate patients are explored.

Higher cure rates for HPV-related head and neck cancer translate into patients living longer and burdened with treatment-related sequelae. Dental care, supportive survivorship care, and voice, speech, and swallowing rehabilitation are important aspects of care for these patients, and these topics are well covered in separate chapters.

Drs. Chandra and Li are to be congratulated for their effort in creating an excellent text geared to all members of the multidisciplinary treatment team caring for patients with head and neck cancer. Practitioners and trainees alike will find this book to be a timely and informative resource.

David W. Eisele
Department of Otolaryngology-Head
and Neck Surgery
Johns Hopkins Medicine
Baltimore, MD, USA

Preface

Head and neck cancer represents an evolving area of cancer care, one that is both challenging and rewarding to the treating oncologist. The disease and the treatment pose enormous burdens to patients and their caregivers. The landscape of cancer management continues to evolve rapidly, based in part on the evolution of surgical and radiation techniques, systemic therapy, advanced imaging, changing epidemiology, and improvements in supportive care. Head and neck cancer is the quintessential team sport: Longitudinal close collaboration between treatment and supportive disciplines is essential to excellent patient care.

This book provides an informative and contemporary reference for trainees and practitioners who care for patients with head and neck cancer. Divided into 14 chapters written by leading experts in the field, this text incorporates updates in surgical techniques (minimally invasive and robotic techniques, reconstructive approaches), radiation medicine (new data on dose and fields, oligometastatic disease, retreatment), medical oncology (targeted, immunotherapy and molecular agents), in addition to more overarching topics such as side effects, biomarkers/nanotechnology, imaging/radiomics, and epidemiology. Several phase II and III trials have been published in the last few years and are placed in context with respect to current management.

We would like to thank all of our authors for their efforts in making this a successful endeavor. We would also like to acknowledge Vinodh Thomas, Margaret Moore, Arul Viveaun S, and the team at Springer for their generous contributions of time, talent, expertise, and discipline to ensure an outstanding text.

Portland, OR, USA — Ravi A. Chandra
Portland, OR, USA — Ryan J. Li

Contents

Contributors

Nishant Agrawal Department of Surgery, Section of Otolaryngology-Head and Neck Surgery, University of Chicago Medical Center, Chicago, IL, USA

Vladimir Avkshtol Department of Radiation Oncology, University of Texas Southwestern Medical Center, Dallas, TX, USA

Ravi A. Chandra Department of Radiation Medicine, Oregon Health & Science University, Portland, OR, USA

Department of Radiation Oncology, Mid-Atlantic Permanente Medical Group, P.C., Rockville, MD, USA

Bhishamjit S. Chera Department of Radiation Oncology, University of North Carolina, Chapel Hill, NC, USA

Michael J. Dienberg Oregon Health & Science University, Portland, OR, USA

Prashanth Girdhar Department of Radiation Oncology, Tata Memorial Centre, Varanasi, India

Zhen Gooi Department of Surgery, Section of Otolaryngology-Head and Neck Surgery, University of Chicago Medical Center, Chicago, IL, USA

Gaorav P. Gupta Department of Radiation Oncology, University of North Carolina, Chapel Hill, NC, USA

Patrick K. Ha Department of Otolaryngology Head and Neck Surgery, University of California, San Francisco, San Francisco, CA, USA

Jocelen Hamilton Head and Neck Cancer Care Program, Stanford Cancer Center, Palo Alto, CA, USA

Adam Howard Department of Surgery, Section of Otolaryngology-Head and Neck Surgery, University of Chicago Medical Center, Chicago, IL, USA

Amy Juliano Department of Radiology, Harvard Medical School/ Massachusetts Eye and Ear, Boston, MA, USA

Hyunseok Kang Department of Medicine, University of California, San Francisco, CA, USA

Z. A. Kohutek Department of Radiation Oncology, Vanderbilt University Medical Center, Nashville, TN, USA

Nancy Lee Department of Radiation Oncology, Memorial Sloan Kettering Cancer Center, New York, NY, USA

Ryan J. Li Department of Otolaryngology-Head and Neck Surgery, Oregon Health & Science University, Portland, OR, USA

Garren M. I. Low Department of Otolaryngology-Head and Neck Surgery, Oregon Health & Science University, Portland, OR, USA

William M. Mendenhall Department of Radiation Oncology, University of Florida College of Medicine, Gainesville, FL, USA

Margaret B. Mitchell Department of Otolaryngology-Head & Neck Surgery, Harvard Medical School/Massachusetts Eye and Ear, Boston, MA, USA

B. A. Murphy Ingram Cancer Center, Vanderbilt University Medical Center, Nashville, TN, USA

Zachary A. Oaks Department of Radiation Oncology, University of North Carolina, Chapel Hill, NC, USA

Henry S. Park Department of Therapeutic Radiology, Yale School of Medicine, Smilow Cancer Hospital, New Haven, CT, USA

Ramesh Paudyal Memorial Sloan Kettering Cancer Center, New York, NY, USA

Jack M. Qian Harvard Radiation Oncology Program, Massachusetts General Hospital/Brigham and Women's Hospital/Dana-Farber Cancer Institute, Boston, MA, USA

Jeremy Richmon Department of Otolaryngology-Head & Neck Surgery, Harvard Medical School/Massachusetts Eye and Ear, Boston, MA, USA

Cristina P. Rodriguez Division of Medical Oncology, University of Washington, Seattle, WA, USA

Perrin E. Romine Division of Medical Oncology, University of Washington, Seattle, WA, USA

Jonathan D. Schoenfeld Department of Radiation Oncology, Dana-Farber/Brigham and Women's Cancer Center, Harvard Medical School, Boston, MA, USA

Akash Deelip Shah Memorial Sloan Kettering Cancer Center, New York, NY, USA

Suparna Shah Department of Otolaryngology-Head and Neck Surgery, Oregon Health & Science University, Portland, OR, USA

Colette J. Shen Department of Radiation Oncology, University of North Carolina, Chapel Hill, NC, USA

David J. Sher Department of Radiation Oncology, University of Texas Southwestern Medical Center, Dallas, TX, USA

Siddharth H. Sheth Division of Oncology, Department of Medicine, University of North Carolina, Chapel Hill, NC, USA

Amita Shukla-Dave Memorial Sloan Kettering Cancer Center, New York, NY, USA

Heather M. Starmer Department of Otolaryngology, Head and Neck Surgery, Stanford University, Palo Alto, CA, USA

Madeleine P. Strohl Department of Otolaryngology Head and Neck Surgery, University of Miami, Miami, FL, USA

Kyaw Z. Thein Division of Hematology/Medical Oncology, Oregon Health & Science University, Portland, OR, USA

Bhanu P. Venkatesulu Department of Radiation Oncology, Stritch School of Medicine, Loyola University, Chicago, Illinois and Edward Hines Veterans Affairs Hospital, Chicago, IL, USA

Vivek Verma Department of Radiation Oncology, University of Texas M.D. Anderson Cancer Center, Houston, TX, USA

Katherine Wai Department of Otolaryngology and Head and Neck Surgery, University of California, San Francisco, CA, USA

Kaveh Zakeri Department of Radiation Oncology, Memorial Sloan Kettering Cancer Center, New York, NY, USA

1 De-intensification Strategies for Head and Neck Cancer

Madeleine P. Strohl, Patrick K. Ha, Kaveh Zakeri, and Nancy Lee

Introduction

The incidence of HPV-positive OPSCC has risen over the past three decades, particularly among younger individuals (<60 years of age), Caucasian men, and those of higher socioeconomic status [1–3]. The median patient age is 50 years with a 4:1 male to female ratio [4]. This is a distinct epidemiological difference for those with HPV-negative disease, who are typically older and have a significant smoking and/or alcohol history. Rather, risk factors are related to sexual behavior as HPV-OPSCC is caused by high-risk HPV. Oral sex is an established risk factor for oral HPV infection. Individuals with HPV+ OPSCC are less likely than HPV-negative patients to have a large smoking history.

M. P. Strohl (✉)
Department of Otolaryngology Head and Neck Surgery, University of Miami, Miami, FL, USA
e-mail: mps193@miami.edu

P. K. Ha
Department of Otolaryngology Head and Neck Surgery, University of California, San Francisco, San Francisco, CA, USA
e-mail: Patrick.Ha@ucsf.edu

K. Zakeri · N. Lee
Department of Radiation Oncology, Memorial Sloan Kettering Cancer Center, New York, NY, USA
e-mail: Zakerik@mskcc.org; Leen2@mskcc.org

Prognosis

HPV+ OPSCC is associated with a significantly better survival than is HPV-negative OPSCC [5, 6]. In a landmark study in 2010, Ang et al. published a retrospective analysis of the association between tumor HPV status and survival among patients with stage III to IV OPSCC undergoing chemoradiation treatment. With a median follow-up period of 4.8 years, the 3-year overall survival rate for those with HPV-related disease was 82.4% compared to 57.1% among patients with HPV-negative disease. They had a 58% reduction in the risk of death even after adjusting for age, race, stage, smoking history, and treatment [6].

According to a recent 2020 report on the trends in incidence of cancers of the oral cavity and pharynx in the United States from 2007 to 2016, there has been a relatively large increase in oropharynx cancers that is driven by HPV infection, while there is a decreasing incidence of alcohol- and tobacco-related cancers of the oral cavity (Fig. 1.1). It is expected that this trend will continue until at least 2060 [7].

A unique, separate TNM staging system through the American Joint Committee on Cancer (AJCC) 8th Edition for HPV + OPSCC has been developed as a result of these stark epidemiological and prognostic differences (Table 1.1). Many patients have nodal involvement on presentation, with a small primary tumor. The new system accounts for this presen-

R. A. Chandra, R. J. Li (eds.), *Multidisciplinary Management of Head and Neck Cancer*,
https://doi.org/10.1007/978-3-031-05973-5_1

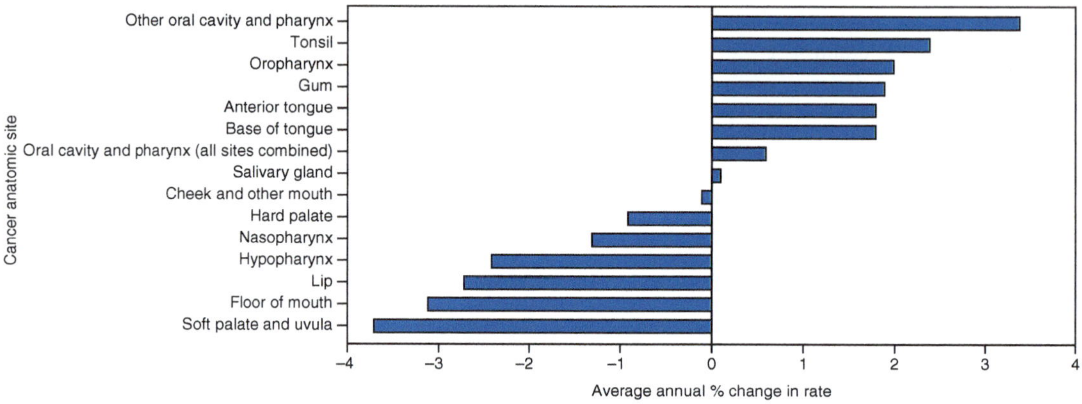

Fig. 1.1 Trends in incidence of cancers of the oral cavity and pharynx. (From: Ellington TD et al. MMWR 2020; 69 (15):433–8)

Table 1.1 A. AJCC (8th edition) TNM categories and definitions for HPV-associated oropharyngeal squamous cell carcinoma [8]

T category	Definition
T0	No primary tumor identified
T1	Tumor size ≤2 cm in greatest dimension
T2	Tumor size >2 cm but ≤4 cm in greatest dimension
T3	Tumor size >4 cm in greatest dimension or extension to lingual surface of epiglottis
T4	Moderately advanced tumor invading larynx, extrinsic tongue muscles, medial pterygoid, hard palate, or mandible or beyond
Clinical N category	
Nx	Regional nodes cannot be assessed
N0	No regional nodal metastases
N1	Metastasis to one or more ipsilateral nodes, ≤6 cm
N2	Metastasis to contralateral or bilateral lymph nodes, ≤6 cm
N3	Metastasis in any cervical lymph node, >6 cm
Pathologic N category	
Nx	Regional nodes cannot be assessed
pN0	No regional nodal metastasis identified
pN1	Metastasis to 4 or fewer lymph nodes
pN2	Metastasis to 5 or more lymph nodes
M category	
M0	Absence of distant metastasis
M1	Presence of distant metastasis

B. AJCC (8th edition) prognostic staging for HPV-associated oropharyngeal squamous cell carcinoma

T category	N category	M category	Stage group
T0, T1, or T2	N0, N1	M0	I
T0, T2, or T2	N2	M0	II
T3 or T4	N0, N1	M0	II
T3 or T4	N2	M0	III
Any T	Any N	M1	IV

tation and has resulted in downstaging of patients with HPV+ OPSCC largely by decreasing the influence of neck disease [9]. The new staging system increases the proportion of patients presenting with stage 1 disease from 3.2% to 63.5% [10]. 93.9% of patients presenting with HPV disease are down-staged with the 8th edition. The 8th edition staging system also differentiates between clinical and pathological nodal (N) staging. Inclusion of pathologic nodal count in the staging definitions allowed for more precision on prognostic elements. Haughey et al. first published on the significance of the number of nodes in HPV-related cancer in 2016 [11]. They found a significant decrease in overall survival between groups of patients with less than four pathologic nodes and those with five or more. There was progressively worse overall survival among three groups of patients: those with pT1–T2 and ≤4 nodes, those with pT1–T2 and ≥5 nodes or pT3–T4 and ≤4 nodes, and those with pT3–T4 and ≥5 nodes (90%, 84% and 48%, respectively). While the number of nodes does seem to affect survival, the role of adjuvant therapy in these groups remains to be seen.

Traditional Treatment Paradigm

Currently, the National Comprehensive Cancer Network does not distinguish between HPV-positive and HPV-negative disease in treatment recommendations for oropharyngeal cancer. As such, the current standard of care for all OPSCC includes surgery, radiation therapy (RT), and chemoradiation (CRT). Traditionally, open surgical techniques were associated with high morbidity and complications, and therefore, RT or CRT has been favored as less morbid. HPV or p16 testing is routinely recommended, and the clinical staging definitions take into consideration the new AJCC 8th edition staging. However, the treatment recommendations regardless of HPV-status are based on randomized studies involving patients for whom the HPV status of their tumor was not specified.

Surgery or definitive radiation therapy is recommended for small, lower stage tumors (cT1-T2, cN0-N1), whereas larger primary tumors and/or nodal stages are recommended for definitive chemoradiation or surgery with adjuvant treatment. For upfront radiation, doses of 66–70 Gy to high-risk sites over a 6- to 7-week course is recommended. In the adjuvant setting for high-risk features, doses of 60–66 Gy over a 6-week course are recommended [12–14]. A platinum-based regimen is standard of care for systemic chemotherapy.

Rationale for De-escalation

While standard of care RT and CRT are evidence based and effective, they are also associated with severe acute and late toxicities affecting patient quality of life, such as severe mucositis, dry mouth, and dysphagia [15]. In a 2008 study, Machtay et al. showed 43% of patients who underwent chemoradiation for locally advanced head and neck cancer developed severe late toxicity. There is a well-established correlation between cumulative radiation therapy (RT) dose delivered and short- and long-term morbidity. Radiation doses below 60 Gy are associated with improved functional swallowing outcomes [16], whereas each 10 Gy delivered above 55 Gy results in increasing rates of dysphagia [17]. Additionally, transoral minimally invasive surgical techniques have pioneered a new era of surgical management for this disease, leading to a re-thinking of treatment paradigm. Considering the combination of better prognosis and younger age of HPV+ OPSCC patients, there has been a focus on reducing treatment toxicities and optimizing quality of life while maintaining favorable oncologic outcomes. It is thought that many patients with HPV+ OPSCC may not require the aggressive intensified traditional RT or CRT regimens and may still have excellent survival, while avoiding the severe toxicities associated with treatment.

Mechanisms of Improved Prognosis

The mechanisms of the improved prognosis of patients with HPV+ OPSCC are not yet well understood. There are a few patient- and tumor-related factors that have been hypothesized to be related. The first is that patients with HPV+ OPSCC have an overall reduced risk of cancer due to a reduced lifestyle risk. Patients are younger at diagnosis and less likely to have significant alcohol or tobacco histories. HPV+ OPSCC patients may also be healthier overall, as the prevalence of comorbidities such as diabetes, chronic obstructive pulmonary disease, and depression is lower in this population [18].

Second, it is thought that there is an immune component to the observed improved survival in this population. It is hypothesized that there is an adaptive immune response directed against the viral antigens expressed by tumor cells in HPV+ OPSCC. In studies to support this notion, higher levels of tumor infiltrating lymphocytes have been noted in HPV+ disease, as well as enrichment of genes associated with CD8 T-cell effector functions in patients with improved survival [19–21].

Finally, HPV+ OPSCC tumors are thought to have a higher sensitivity to both RT and chemotherapy compared to HPV-negative disease. It is not yet completely understood why this is true. DNA damage in HPV-related disease occurs through a different mechanism than HPV-negative disease, and HPV-related tumors on a whole have fewer genetic alterations. DNA damage repair mechanisms may be impaired and therefore lead to more cell cycle dysregulation and cell death [22, 23]. Thus, HPV+ tumors may be more sensitive to DNA damage as a whole, and thus more effectively treated by RT or CRT.

Overview of De-intensification Strategies

There are two main paradigms for de-intensification. The first is surgery followed by risk-based adjuvant therapy. Proponents of this approach feel that there may be less long-term toxicity with surgery compared to definitive CRT. The second paradigm is modification of the traditional definitive CRT approach. Various approaches are under investigation, including using RT alone, substitution of platinum chemotherapy with cetuximab or immunotherapy, and using response-stratified RT following induction chemotherapy or according to hypoxia imaging.

Surgery Alone or with Risk-Based Adjuvant Therapy

Ongoing advances in surgical techniques permit transoral resection of oropharyngeal tumors while minimizing morbidity and preserving function as much as possible. Traditional open surgical approaches to the oropharynx are associated with substantial morbidity as a result of poor access to the tumors. Transoral surgical techniques including transoral laser microsurgery (TLM) and transoral robotic surgery (TORS) have revolutionized the treatment of HPV+ OPSCC [24]. Weinstein et al. established that TORS was a safe and effective technique for surgical removal of oropharyngeal tumors [25]. The ability to surgically remove the tumor with oncologically sound margins is at least as good as seen with traditional open techniques.

Proponents of a surgical de-intensification approach stress than there may be less long-term toxicity with surgery followed by risk-based adjuvant therapy compared to traditional definitive CRT. Many studies have looked retrospectively at quality of life and functional outcomes following transoral surgery for oropharyngeal squamous cell carcinoma [26–29]. When compared to traditional open surgical approaches to the oropharynx, transoral surgical approaches have shown superior functional swallowing outcomes [30, 31].

De-intensification of postoperative RT and CRT in conjunction with transoral surgical techniques is an area of intense interest. While surgical resection allows for increased local control at the primary site and neck, it also allows for risk-based stratification of adjuvant radiation and/or chemotherapy based on pathologic risk factors. A number of clinical trials that are ongoing or have been recently completed are exploring the role of de-escalation of adjuvant treatment after surgery.

For tumors amenable to minimally invasive surgical approaches, the goal is to spare or diminish the need for adjuvant treatment based on the final pathology (Fig. 1.2). Transoral surgical treatment combined with neck dissection without adjuvant treatment can be possible for pathologically low-risk, small tumors. This has been established in a series of small retrospective reviews. In a retrospective study published in 2013, Olsen et al. reported that 18 patients underwent TORS and neck dissection alone for their disease with acceptable functional swallowing outcomes and 100% 3-year disease-specific survival and 90.9% local control rate at 3 years [32]. In another study published in 2012 by Weinstein et al., 30 patients with HPV+ OPSCC underwent TORS and neck dissection alone for their disease. Tumors were staged from T1 to T4a. At a mean follow-up of 2.7 years, there was a 3% local failure rate [33].

The Eastern Cooperative Oncology Group (ECOG) 3311 is a phase II clinical trial that enrolled patients to investigate reduced doses of adjuvant RT following transoral resection of intermediate-risk HPV+ tumors [34]. A total of 519 patients with clinical T1–T2 stage III/IV AJCC 7th edition without matted neck nodes underwent TORS and neck dissection and were subsequently assigned to study arms based on pathologically assessed risk. Patients with clear margins, 0–1 lymph nodes and no extranodal extension were observed (arm A). Patients with clear/close margins, 2–4 pathologically positive nodes, or extranodal extension less than 1 mm were randomized to receive either 50 Gy (arm B) or 60 Gy (arm C) postoperatively. Patients with a positive margin, greater than 4 pathological lymph nodes and greater than 1 mm of extranodal extension were assigned to standard dose CRT with cisplatin (arm D). Two-year progression-

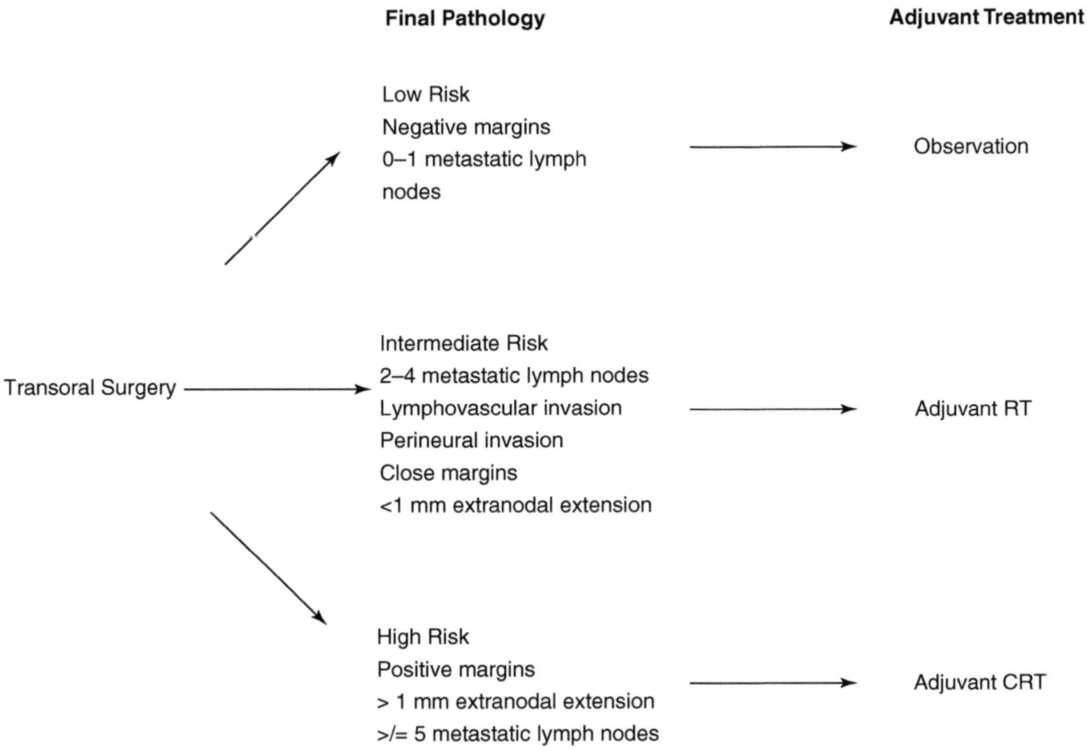

Fig. 1.2 Proposed algorithm for a surgical approach to de-intensification

free survival for arms A, B, C, and D were 93.9%, 95.0%, 95.9%, and 90.5%, respectively. Study authors concluded that for intermediate-risk patients, transoral surgery followed by reduced dose postoperative RT without chemotherapy appears to be sufficient. These findings suggest that a significant number of patients who undergo transoral surgery could have elimination of chemotherapy from their adjuvant regimen, as well as a reduction in their postoperative RT dosing.

The Post-Operative Adjuvant Treatment of HPV-Positive Tumors (PATHOS) is another phase II trial examining the impact of transoral laser resection of tumors followed by reduced dose adjuvant RT on functional outcomes and survival. Patients undergo TLM resection of tumors and are then randomized to reduced dose RT or standard dose RT. PATHOS includes two randomizations: 50 vs. 60 Gy for intermediate post-op risk factors and 60 Gy alone vs. 60 Gy + cisplatin for high-risk factors (positive margins or ECS). Results are still pending [35].

The Adjuvant De-escalation, Extracapsular Spread, p16 Positive, Transoral (ADEPT) trial is a phase II trial investigating whether chemotherapy can be held for patients with extracapsular spread (ECS). Patients are included in the trial if they have negative margins at the surgical site and must have ECS in nodal metastases on final pathology. Radiation volume is also not delivered to the primary bed in patients with completely resected T1 or T2 disease, and reduced dose adjuvant RT to the neck is being investigated [36, 37].

The De-Escalated Adjuvant Radiation Therapy for HPV-Associated Oropharynx Cancer (DART-HPV) study is currently randomizing patients with HPV+ OPSCC who have undergone transoral resection and meet criteria for adjuvant treatment to receive standard adjuvant CRT or docetaxel with reduced adjuvant RT (30 Gy delivered over 2 weeks). The primary endpoint is treatment toxicity [38].

The AVOID trial was a single-arm study of 60 patients with HPV-related oropharynx cancer treated with TORS followed by adjuvant radiation with omission of the primary site for tumors without close surgical margins, perineural invasion, or lymphovascular invasion. Patients received 60–66 Gy to the neck alone with the addition of chemotherapy in cases of extranodal extension. With median follow up of 2.4 years, the 2-year local control, progression-free survival, and overall survival were 98.3%, 92.1%, and 100%, respectively. Only one patient had a primary site recurrence. Further investigation and long-term follow-up is warranted before this strategy can be routinely adopted.

The ORATOR trial was a phase II randomized trial performed across six hospitals in Canada and Australia comparing differences in quality of life outcomes between those who received surgery and those who received definitive chemoradiation. Patients with T1-T2, N0-N2 HPV+ OPSCC were randomized to receive either definitive RT or CRT (if N1-N2) with 70 Gy or TORS plus neck dissection followed by adjuvant RT or CRT based on surgical pathology. Patients treated with upfront RT or CRT showed superior swallowing-related quality of life scores 1 year after treatment compared to the surgery group; however, the difference was not felt to be clinically meaningful [39]. This is the first randomized study to compare quality of life outcomes between surgery and radiation. The results were somewhat surprising given prior retrospective studies favoring functional and quality of life outcomes following transoral surgery compared to radiation. One criticism of the ORATOR trial was that a high percentage of early stage patients that underwent surgery received multimodality treatment. A follow-up clinical trial, ORATOR2, is currently underway. Further studies are needed to investigate the two modalities head-to-head, and better understand which patients should be selected for surgery in order to avoid or reduce multimodality treatment.

Controversies exist in the surgical de-escalation of HPV+ OPSCC. The definition of negative margin at the primary site is a point of debate [40]. The NCCN guidelines suggest that a negative margin is defined as greater than 5 mm; however, in practice, institutions use different cut offs, as do various clinical trials. For example, ECOG 331 specified 3 mm as a negative margin, whereas the ORATOR trial specified 2 mm as a negative margin. Another area of controversy is the prognostic impact of extracapsular spread. In the NCCN guidelines, this is an indication for adjuvant chemotherapy; however, as stated previously, this is based on evidence from trials performed among patients in which HPV status was unknown. Further studies are needed to better understand these pathologic factors and their role in de-intensification therapy after surgery.

Transoral Surgery for Work-Up of Unknown Primary

Transoral surgical techniques have also become a useful adjunct in the work-up of HPV+ cancer of unknown primary in the head and neck and has important implications for de-intensification. Transoral tonsillectomy and base of tongue mucosectomy has resulted in increased detection rates for unknown primary. In a systematic review of the literature, van Weert et al. showed a 72% detection rate for occult tumors for TORS compared to a 41% detection rate for conventional exam under anesthesia with direct biopsies [41]. Successful detection of the primary site has the benefit of possible de-intensification of adjuvant treatment or eliminate the need for adjuvant treatment altogether. In a 2015 study by Graboyes et al., of 47 patients who received adjuvant therapy, 36 patients were spared radiation to the primary site as the primary site had been identified and cleared with negative margins. Seventeen patients presenting with unknown primary disease had identification of their primary site and neck dissection and were spared the need for any adjuvant radiation [42].

Definitive Radiotherapy/ Chemoradiotherapy De-intensification

Various de-intensification approaches are being investigated to limit the toxicities associated with definitive RT and CRT treatment for early stage and advanced stage HPV+ OPSCC, respectively. As discussed previously, HPV+ OPSCC are hypothesized to be more responsive to both radiation and chemotherapy than their HPV-negative counterparts for a variety of hypothesized reasons. Therefore, a reduction in the standard dose of definitive RT and systemic therapy is an attractive way to decrease treatment associated toxicity. Alternatively, response-based de-escalation approaches, using induction chemotherapy, mid-treatment response, or hypoxia imaging, are also being investigated.

In a prospective, multi-institutional phase II study published in 2015, Chera et al. reported on 43 patients treated with reduced dose radiation (60 Gy) along with weekly low-dose cisplatin (30 mg/m^2). Patients were included if they had T0 to T3 and N0 to N2c HPV-related disease with minimal smoking history. Following completion of treatment, patients were evaluated for complete clinical response based on physical and radiologic examination, and pathological clinical response based on planned surgical intervention. Patients underwent directed biopsies at the primary site and those with nodal disease underwent posttreatment neck dissection. The complete pathological response rate was 86%, with decreased incidence of severe and very severe treatment toxicities [43]. Long-term follow-up demonstrated 3-year local control, regional control, cause-specific survival, distant metastasis-free survival, and overall survival rates were 100%, 100%, 100%, 100%, and 95%, respectively, with favorable quality of life outcomes [39, 44].

Another de-intensification strategy involved substituting cisplatin with cetuximab in conjunction with standard dose radiotherapy. The hypothesis was that cetuximab would have noninferior tumor control and survival with improved toxicity compared to cisplatin in combination with radiotherapy. Three phase III clinical trials (RTOG 1016, De-ESCALaTE HPV, and TROG 12.01) randomized patients to cisplatin or cetuximab in combination with standard dose radiotherapy [45, 46]. All three trials observed inferior overall survival or progression-free survival with cetuximab-radiotherapy without any improvement in toxicity or symptom burden.

Omission of systemic therapy altogether with reduced dose radiotherapy for patients clinical low-risk disease was investigated in the NRG HN002 randomized phase II clinical trial. Nonsmoking patients with low-risk HPV-related oropharynx cancer were randomized to receive reduced dose 60 Gy radiation in 5 weeks without any systemic therapy or 60 Gy radiation in 6 weeks with weekly cisplatin [47]. Two-year progression-free survival was 87.6% for RT alone and 90.5% for RT with cisplatin, without statistically different swallowing quality of life outcome measures. Omission of cisplatin did not meet the prespecified PFS threshold for further investigation. Collectively, HN002, RTOG 1016, De-ESCALaTE HPV, and TROG 12.01 demonstrate the importance of concurrent cisplatin chemotherapy with radiation therapy.

Another strategy involves reduction in the radiotherapy field size and/or dose of radiation to elective nodal basins in the definitive treatment setting. EVADER is a phase II clinical trial investigating the omission of low-risk elective lymph node basins from standard dose chemoradiotherapy treatment volumes [48]. At Memorial Sloan Kettering Cancer Center, reduced dose elective nodal irradiation from the traditional 50–54 Gy to 30 Gy has been successfully implemented with concurrent chemotherapy [49]. Elective nodal irradiation with 30 Gy resulted in favorable acute toxicities [50].

Immune checkpoint inhibitors as a replacement for platinum-based chemotherapy are an area of interest in reducing the side effects commonly associated with systemic platinum therapy. While their efficacy has been primarily shown in end stage or nonresectable head and neck cancers, there has been increasing interest in their use in the definitive treatment setting. Patients with HPV-related OPSCC are thought to be ideal candidates for immunotherapy as oncogenesis is driven by blockade of the innate immune system, and immune checkpoint inhibitors aid in unleashing the immune antitumor response. Various clinical trials investigating the use of immunotherapy for HPV+ disease are underway [31]. NRG-HN005, CCTG HN.9, and KEYCHAIN are randomized phase II and III clinical trials that compare radiotherapy with concurrent cisplatin to radiation with concurrent immune checkpoint inhibitors.

Response-Based De-escalation of Definitive Chemoradiotherapy

Response-based strategies involve modifying the intensity of treatment according to individual tumor treatment response, which may better account for heterogeneity in tumor biology. Among HPV-associated tumors, there are significant differences in treatment response and biology [51–53]. Response-based strategies include use of induction chemotherapy, mid-treatment response assessment, and hypoxia imaging.

Induction chemotherapy followed by reduced doses or volumes of chemoradiotherapy has been the focus of several phase II clinical trials. ECOG 1308 randomized 80 patients with stage III and IV (AJCC 7th edition) disease to receive an induction chemotherapy regimen followed by either low (54 Gy) or standard dose IMRT with cetuximab, depending on the response to induction chemotherapy [54]. Two-year progression-free survival and overall survival were 80% and 94%, respectively, for patients with primary site complete response treated with 54 Gy of radiation. At 1 year, significantly fewer patients who were treated with the reduced radiation dose reported difficulty swallowing (40% versus 89%, $P = 0.011$) or impaired nutrition (10% versus 44%, $P = 0.025$). In a similar phase II clinical trial, Chen et al. reported on 44 patients with stage III and IV HPV-related disease (AJCC 7th edition) who received induction chemotherapy followed by concurrent paclitaxel and 54 Gy in 27 fractions for partial or complete responders or 60 Gy in 30 fractions for poor responders [55]. Progression-free survival at 2 years was 92% (95% CI 77–92). The OPTIMA trial was a phase II radiation dose and volume de-escalation trial of 62 patients with HPV-related OPSCC. After induction chemotherapy, low-risk patients (those with ≤T3, ≤N2B, and ≤10 pack-year history) with ≥50% response received 50 Gy RT (RT50), while low-risk patients with 30–50% response or high-risk patients with ≥50% response received 45 Gy CRT (CRT45). Those with lesser responses received a standard chemoradiation regimen (CRT75). Two-year progression-free survival and overall survival were 95% and 100% for low-risk patients and 94% and 97% for high-risk patients, respectively. Grade 3+ mucositis occurred in 30%, 63%, and 91% of the RT50, CRT45, and CRT75 groups, respectively ($P = 0.004$), while G-tube rates were 0%, 31%, and 82% for RT50, CRT45, and CRT75 groups, respectively ($P < 0.0001$) [56].

Mid-treatment response during chemoradiotherapy can potentially guide de-escalation of therapy. Two prospective clinical trials (NCT03215719 and NCT03416153) are investigating de-escalated chemoradiotherapy for patients with good response on mid-treatment CT or PET/CT imaging. These strategies avoid the use of induction chemotherapy and its associated toxicity, though further investigation is needed to compare the effectiveness patient selection using mid-treatment response to post-induction chemotherapy response.

Hypoxia imaging with 18F-fluoromisonidazole (18F-FMISO) positron emission tomography (PET) is a novel method of quantifying tumor oxygenation [57]. Since hypoxia is associated with radioresistance, 18F-FMISO can be used to measure tumor response to treatment and select patients for de-escalated therapy [58, 59]. The 30 ROC trial enrolled 19 patients at Memorial Sloan Kettering Cancer Center and investigated whether mid-treatment 18F-FMISO could select patients for dramatic reduction in radiation dose to 30 Gy with standard platinum chemotherapy. The primary oropharynx tumor was resected in all patients prior to radiation and planned neck dissection at 4 months post chemoradiation. Among the 15 patients with no tumor hypoxia or mid-treatment resolution of hypoxia, the 2-year locoregional control and overall survival with 30 Gy were 94.4% and 94.7%, respectively [60].

The subsequent 30 ROC phase II trial accrued 158 patients following resection of the primary tumor and planned neck dissections were omitted. The rates of 1-year locoregional control, distant metastasis-free survival, and overall survival were 94%, 100%, and 100%, respectively [61]. There were eight locoregional recurrences, and all were in undissected lymph nodes and were successfully salvaged with neck dissection. No patients required a feeding tube. A follow-up to the 30 ROC trial is ongoing with omission of both primary tumor resection and planned neck dissection.

Limitations of De-intensification

Despite growing evidence supporting the role of de-intensification, there are limitations to this new approach. Not all trial results have shown favorable results. Three recent phase III de-intensification trials investigating the replacement of cisplatin with cetuximab in combination with radiation therapy for HPV+ disease reported statistically inferior progression-free survival or overall survival with cetuximab [45, 46]. Additionally, cetuximab did not reduce treatment toxicity compared to cisplatin in any of the three trials.

More importantly, HPV+ OPSCC is not a uniformly favorable disease. There is a lack of critical information about the role of confounding variables such as smoking and comorbidities in de-escalation regimens. In Ang et al.'s landmark paper in 2010, while overall survival for HPV+ OPSCC was high, patients with a >10 pack year smoking history or with multiple nodes or large nodal disease had a significantly lower overall 3-year survival at 71% [6]. The favorable prognosis of HPV+ OPSCC may be tempered by the adverse effects of smoking.

Additionally, while HPV+ OPSCC have an initial high response rate to treatment, recurrence rates remain at 13–15% at 2 years and up to 36% at 8 years [5, 6]. Currently, many of the time points in de-escalation studies are too short to evaluate the long-term effects on survival and recurrence, particularly on distant recurrence that may occur up to 5 years after treatment. This is particularly concerning when considering trials evaluating withholding systemic therapy. Long-term follow-up is necessary to best understand the effect of de-escalation on long-term oncologic control.

While many trials investigating de-intensification are ongoing or have been recently completed, it should be remembered that the best level of evidence currently available still supports the use of either definitive RT or CRT for a total dose of 70 Gy over 6–7 weeks with concurrent platinum based chemotherapy or surgery with adjuvant therapy as indicated [62, 63]. Clinicians should not offer de-escalation of treatment outside of clinical trials [64].

Conclusion

Overall, patients with HPV+ OPSCC have a favorable prognosis. Reducing treatment-related toxicities without compromising oncologic outcomes is an important goal for this generally younger, healthier patient population. As there is currently a lack of consistent randomized phase III clinical trial results, treatment de-intensification is not recommended outside of ongoing clinical trials. It is important that the numerous de-intensification strategies are studied rigorously. We will eagerly await the results of these trials to guide the future of treatment of HPV+ OPSCC, in an attempt to maintain the excellent treatment outcomes and minimize the significant side effects.

References

1. Gillison ML, Chaturvedi AK, Anderson WF, Fakhry C. Epidemiology of human papillomavirus-positive head and neck squamous cell carcinoma. J Clin Oncol. 2015;33:3235–42.
2. Chaturvedi AK, Anderson WF, Lortet-Tieulent J, et al. Worldwide trends in incidence rates for oral cavity and oropharyngeal cancers. J Clin Oncol. 2013;31:4550–9.
3. Chaturvedi AK, Engels EA, Pfeiffer RM, et al. Human papillomavirus and rising oropharyngeal cancer incidence in the United States. J Clin Oncol. 2011;29:4294–301.
4. Gillison ML, D'Souza G, Westra W, et al. Distinct risk factor profiles for human papillomavirus type 16-positive and human papillomavirus type 16-negative head and neck cancers. J Natl Cancer Inst. 2008;100:407–20.
5. Fakhry C, Westra WH, Li S, et al. Improved survival of patients with human papillomavirus-positive head and neck squamous cell carcinoma in a prospective clinical trial. J Natl Cancer Inst. 2008;100:261–9.
6. Ang KK, Harris J, Wheeler R, et al. Human papillomavirus and survival of patients with oropharyngeal cancer. N Engl J Med. 2010;363:24–35.
7. Ellington TD, Henley SJ, Senkomago V, et al. Trends in incidence of cancers of the oral cavity and pharynx — United States 2007–2016. MMWR Morb Mortal Wkly Rep. 2020;69:433–38. https://doi.org/10.15585/mmwr.mm6915a1externalicon.
8. Lydiatt WM, Patel SG, O'Sullivan B, et al. Head and neck cancers-major changes in the american joint committee on cancer eighth edition cancer staging manual. CA Cancer J Clin. 2017;67(2):122–37. https://doi.org/10.3322/caac.21389. Epub 2017 Jan 27. PMID: 28128848.
9. Glastonbury CM. Critical changes in the staging of head and neck cancer. Radiol Imaging Cancer. 2020;2:e190022.
10. Cramer JD, Hicks KE, Rademaker AW, Patel UA, Samant S. Validation of the eighth edition american joint committee on cancer staging system for human papillomavirus-associated oropharyngeal cancer. Head Neck. 2018;40(3):457–66. https://doi.org/10.1002/hed.24974. Epub 2017 Oct 9. PMID: 28990257.
11. Haughey BH, Sinha P, Kallogjeri D, et al. Pathology-based staging for HPV-positive squamous carcinoma of the oropharynx. Oral Oncol. 2016;62:11–19. https://doi.org/10.1016/j.oraloncology.2016.09.004. Epub 2016 Sep 23. PMID: 27865363.
12. Bernier J, Domenge C, Ozsahin M, et al. Postoperative irradiation with or without concomitant chemotherapy for locally advanced head and neck cancer. N Engl J Med. 2004;350:1945–52.
13. Cooper JS, Pajak TF, Forastiere AA, et al. Postoperative concurrent radiotherapy and chemotherapy for high-risk squamous-cell carcinoma of the head and neck. N Engl J Med. 2004;350:1937–44.
14. Eisbruch A, Harris J, Garden AS, et al. Multi-institutional trial of accelerated hypofractionated intensity-modulated radiation therapy for early-stage oropharyngeal cancer (RTOG 00-22). Int J Radiat Oncol Biol Phys. 2010;76:1333–8.
15. Machtay M, Moughan J, Trotti A, et al. Factors associated with severe late toxicity after concurrent chemoradiation for locally advanced head and neck cancer: an RTOG analysis. J Clin Oncol. 2008:20;26(21):3582–9. https://doi.org/10.1200/JCO.2007.14.8841. Epub 2008 Jun 16. PMID: 18559875.
16. Mortensen HR, Jensen K, Aksglæde K, Behrens M, Grau C. Late dysphagia after IMRT for head and neck cancer and correlation with dose-volume parameters. Radiother Oncol. 2013;107:288–94.
17. Levendag PC, Teguh DN, Voet P, et al. Dysphagia disorders in patients with cancer of the oropharynx are significantly affected by the radiation therapy dose to the superior and middle constrictor muscle: a dose-effect relationship. Radiother Oncol. 2007;85:64–73.
18. Hess CB, Rash DL, Daly ME, et al. Competing causes of death and medical comorbidities among patients

with human papillomavirus-positive vs human papillomavirus-negative oropharyngeal carcinoma and impact on adherence to radiotherapy. JAMA Otolaryngol Head Neck Surg. 2014;140:312–6.
19. Balermpas P, Rödel F, Weiss C, Rödel C, Fokas E. Tumor-infiltrating lymphocytes favor the response to chemoradiotherapy of head and neck cancer. Onco Targets Ther. 2014;3:–e27403.
20. Russell S, Angell T, Lechner M, et al. Immune cell infiltration patterns and survival in head and neck squamous cell carcinoma. Head Neck Oncol. 2013;5:24.
21. King EV, Ottensmeier CH, Thomas GJ. The immune response in HPV(+) oropharyngeal cancer. Onco Targets Ther. 2014;3:–e27254.
22. Arenz A, Ziemann F, Mayer C, et al. Increased radio-sensitivity of HPV-positive head and neck cancer cell lines due to cell cycle dysregulation and induction of apoptosis. Strahlenther Onkol. 2014;190:839–46.
23. Braakhuis BJ, Senft A, de Bree R, et al. Expression profiling and prediction of distant metastases in head and neck squamous cell carcinoma. J Clin Pathol. 2006;59:1254–60.
24. Ridge JA. Surgery in the HPV era: the role of robotics and microsurgical techniques. Am Soc Clin Oncol Educ Book. 2014:154–9.
25. Weinstein GS, O'Malley BW Jr, Magnuson JS, et al. Transoral robotic surgery: a multicenter study to assess feasibility, safety, and surgical margins. Laryngoscope. 2012;122:1701–7.
26. Dziegielewski 2013. https://doi.org/10.1001/jamaoto.2013.2747. PMID: 23576186. PMCID: PMC4274181.
27. Choby GW, Kim J, Ling DC, et al. Transoral robotic surgery alone for oropharyngeal cancer: quality-of-life outcomes. JAMA Otolaryngol Head Neck Surg. 2015;141(6):499–504. https://doi.org/10.1001/jamaoto.2015.0347. PMID: 25834991.
28. Sethia R, Yumusakhuylu CA, Ozbay I, et al. Quality of life outcomes of transoral robotic surgery with or without adjuvant therapy for oropharyngeal cancer. https://doi.org/10.1002/lary.26796. PMID: 28771728.
29. Hutcheson KA, Holsinger FC, Kupferman ME, et al. Functional outcomes after TORS for oropharyngeal cancer: a systematic review. Eur Arch Otorhinolaryngol. 2015;272(2):463–71. https://doi.org/10.1007/s00405-014-2985-7. Epub 2014 Mar 19. PMID: 24643851.
30. Moore EJ, Olsen SM, Laborde RR, et al. Long-term functional and oncologic results of transoral robotic surgery for oropharyngeal squamous cell carcinoma. Mayo Clin Proc. 2012;87:219–25.
31. Genden EM, Kotz T, Tong CC, et al. Transoral robotic resection and reconstruction for head and neck cancer. Laryngoscope. 2011;121:1668–74.
32. Olsen SM, Moore EJ, Laborde RR, et al. Transoral surgery alone for human-papillomavirus-associated oropharyngeal squamous cell carcinoma. Ear Nose Throat J. 2013;92:76–83.
33. Weinstein GS, Quon H, Newman HJ, et al. Transoral robotic surgery alone for oropharyngeal cancer: an analysis of local control. Arch Otolaryngol Head Neck Surg. 2012;138:628–34.
34. Ferris RL, Flamand Y, Weinstein GS, et al. Transoral robotic surgical resection followed by randomization to low- or standard-dose IMRT in resectable p16+ locally advanced oropharynx cancer: a trial of the ECOG-ACRIN Cancer Research Group (E3311). J Clin Oncol. 2020;38:6500.
35. Evans M, Knott S, Hurt C, et al. PATHOS: a phase II/III trial of risk-stratified, reduced intensity adjuvant treatment in patients undergoing transoral surgery for human papillomavirus (HPV)-positive oropharyngeal cancer. J Clin Oncol. 2018;36:TPS6097.
36. Post operative adjuvant therapy de-intensification trial for human papillomavirus-related, p16+ oropharynx cancer (ADEPT). https://clinicaltrials.gov/ct2/show/NCT01687413. Accessed 15 Oct 2019.
37. Wirth LJ, Burtness B, Nathan C-AO, Grégoire V, Richmon J. Point/counterpoint: do we de-escalate treatment of HPV-associated oropharynx cancer now? And how? Am Soc Clin Oncol Educ Book. 2019:364–72.
38. Evaluation of de-escalated adjuvant radiation therapy for human papillomavirus (HPV)-associated oropharynx cancer. https://clinicaltrials.gov/ct2/show/NCT02908477. Accessed 15 Oct 2019.
39. Nichols AC, Theurer J, Prisman E, et al. Radiotherapy versus transoral robotic surgery and neck dissection for oropharyngeal squamous cell carcinoma (ORATOR): an open-label, phase 2, randomised trial. Lancet Oncol. 2019;20:1349–59.
40. Holcomb AJ, Herberg M, Strohl M, et al. Impact of surgical margins on local control in patients undergoing single-modality transoral robotic surgery for HPV-related oropharyngeal squamous cell carcinoma. Head Neck. 2021;43(8):2434–44.
41. van Weert S, Rijken JA, Plantone F, et al. A systematic review on transoral robotic surgery (TORS) for carcinoma of unknown primary origin: has tongue base mucosectomy become indispensable? Clin Otolaryngol. 2020;45:732–8.
42. Graboyes EM, Sinha P, Thorstad WL, Rich JT, Haughey BH. Management of human papillomavirus-related unknown primaries of the head and neck with a transoral surgical approach. Head Neck. 2015;37:1603–11.
43. Chera BS, Amdur RJ, Tepper J, et al. Phase 2 trial of de-intensified chemoradiation therapy for favorable-risk human papillomavirus-associated oropharyngeal squamous cell carcinoma. Int J Radiat Oncol Biol Phys. 2015;93:976–85.
44. Chera BS, Amdur RJ, Tepper JE, et al. Mature results of a prospective study of deintensified chemoradiotherapy for low-risk human papillomavirus-associated oropharyngeal squamous cell carcinoma. Cancer. 2018;124:2347–54.
45. Gillison ML, Trotti AM, Harris J, et al. Radiotherapy plus cetuximab or cisplatin in human papillomavirus-

positive oropharyngeal cancer (NRG Oncology RTOG 1016): a randomised, multicentre, non-inferiority trial. Lancet. 2019;393:40–50.
46. Mehanna H, Robinson M, Hartley A, et al. Radiotherapy plus cisplatin or cetuximab in low-risk human papillomavirus-positive oropharyngeal cancer (De-ESCALaTE HPV): an open-label randomised controlled phase 3 trial. Lancet. 2019;393:51–60.
47. Yom SS, Torres-Saavedra P, Caudell JJ, et al. Reduced-dose radiation therapy for HPV-associated oropharyngeal carcinoma (NRG Oncology HN002). J Clin Oncol. 2021;39:956–65.
48. Bratman SV, Berthelet E, Butler JB, et al. CCTG HN.10: a phase II single-arm trial of elective volume adjusted de-escalation radiotherapy (EVADER) in patients with low-risk HPV-related oropharyngeal squamous cell carcinoma (NCT03822897). J Clin Oncol. 2020;38(15_suppl):TPS6592.
49. Tsai CJ, McBride SM, Riaz N, et al. Reducing the radiation therapy dose prescription for elective treatment areas in human papillomavirus-associated oropharyngeal carcinoma being treated with primary chemoradiotherapy at Memorial Sloan Kettering Cancer Center. Pract Radiat Oncol. 2019;9(2):98–101.
50. Tsai CJ, Kim JK, McBride S, et al. Prospective personalized elective nodal dose de-escalation in patients with human papillomavirus (HPV)-positive oropharyngeal cancer (OPC) undergoing definitive concurrent chemoradiation (CCRT): acute toxicities and tolerance. Int J Radiat Oncol Biol Phys. 2018;102(3_suppl):E362.
51. Kimple RJ, Smith MA, Blitzer GC, et al. Enhanced radiation sensitivity in HPV-positive head and neck cancer. Cancer Res. 2013;73(15):4791–800.
52. Vainshtein J, McHugh JB, Spector ME, et al. Human papillomavirus-related oropharyngeal cancer: HPV and p16 status in the recurrent versus parent tumor. Head Neck. 2015;37(1):8–11.
53. Rieckmann T, Tribius S, Grob TJ, et al. HNSCC cell lines positive for HPV and p16 possess higher cellular radiosensitivity due to an impaired DSB repair capacity. Radiother Oncol. 2013;107(2):242–6.
54. Marur S, Li S, Cmelak AJ, et al. E1308: phase II trial of induction chemotherapy followed by reduced-dose radiation and weekly cetuximab in patients with HPV-associated resectable squamous cell carcinoma of the oropharynx-ECOG-ACRIN Cancer Research Group. J Clin Oncol. 2017;35:490–7.
55. Chen AM, Felix C, Wang PC, et al. Reduced-dose radiotherapy for human papillomavirus-associated squamous-cell carcinoma of the oropharynx: a single-arm, phase 2 study. Lancet Oncol. 2017;18:803–11.
56. Seiwert TY, Foster CC, Blair EA, et al. OPTIMA: a phase II dose and volume de-escalation trial for human papillomavirus-positive oropharyngeal cancer. Ann Oncol. 2019;30:297–302.
57. Rasey JS, Koh WJ, Evans ML, et al. Quantifying regional hypoxia in human tumors with positron emission tomography of [18F]fluoromisonidazole: a pretherapy study of 37 patients. Int J Radiat Oncol Biol Phys. 1996;36(2):417–28.
58. Terris DJ. Head and neck cancer: the importance of oxygen. Laryngoscope. 2000;110(5 Pt 1):697–707.
59. Brizel DM, Dodge RK, Clough RW, et al. Oxygenation of head and neck cancer: changes during radiotherapy and impact on treatment outcome. Radiother Oncol. 1999;53(2):113–7.
60. Riaz N, Sherman E, Pei X, et al. Precision radiotherapy: reduction in radiation for oropharyngeal cancer in the 30 ROC trial. J Natl Cancer Inst. 2021;113(6):742–51.
61. Lee NY, Sherman EJ, Schöder H, et al. The 30 ROC trial: precision intra-treatment imaging guiding major radiation reduction in human papillomavirus related oropharyngeal cancer. J Clin Oncol. 2021;39(15_suppl):6019.
62. Pignon JP, le Maître A, Maillard E, Bourhis J. Meta-analysis of chemotherapy in head and neck cancer (MACH-NC): an update on 93 randomised trials and 17,346 patients. Radiother Oncol. 2009;92:4–14.
63. Blanchard P, Baujat B, Holostenco V, et al. Meta-analysis of chemotherapy in head and neck cancer (MACH-NC): a comprehensive analysis by tumour site. Radiother Oncol. 2011;100:33–40.
64. Mehanna H. Update on de-intensification and intensification studies in HPV. Recent Results Cancer Res. 2017;206:251–6.

Novel Multidisciplinary Paradigms: Surgery/Radiation, Immunotherapy, Organ Preservation

2

Garren M. I. Low, Kyaw Z. Thein, Suparna Shah, Ravi A. Chandra, and Ryan J. Li

Background

Head and neck cancer (HNC) treatment has evolved with advances in surgery, radiotherapy, and systemic therapies aimed to maximize oncologic control and organ preservation. In addition to chemotherapy, immunotherapy and targeted agents have grown a larger role in systemic therapeutics. All treatment paradigms must carefully consider functional consequences. Given the complex interactions between multimodality therapies, multidisciplinary care is always evolving. The oncologist has the responsibility to provide patients with treatment options, both curative and palliative, and provide understanding of the advantages and disadvantages inherent to each approach.

G. M. I. Low (✉) · S. Shah · R. J. Li
Department of Otolaryngology-Head and Neck Surgery, Oregon Health & Science University, Portland, OR, USA
e-mail: shahsu@ohsu.edu; lry@ohsu.edu

K. Z. Thein
Division of Hematology/Medical Oncology, Oregon Health & Science University, Portland, OR, USA
e-mail: theink@ohsu.edu

R. A. Chandra
Department of Radiation Medicine, Oregon Health & Science University, Portland, OR, USA

Department of Radiation Oncology, Mid-Atlantic Permanente Medical Group, P.C., Rockville, MD, USA
e-mail: chandrav@ohsu.edu

Over the last 30 years, there has been an overall decrease in the incidence of HNC, attributable to a decline in tobacco consumption. Over the same period, there has been a substantial increase in high-risk human papillomavirus (HPV)-associated oropharyngeal squamous cell cancers (OPSCC) [1]. In the last three decades, survival rates of HNC overall have improved from 54.7% in 1992–1996 to 65.9% in 2002–2006. This statistic has been driven largely by the high rate of cure in HPV-associated oropharyngeal squamous cell carcinoma (HPV-OPSCC).

The various sites of head and neck squamous cell carcinomas (HNSCC) are labeled in Fig. 2.1. Each site within the head and neck confers different anatomic and physiologic constraints. A squamous cell carcinoma (SCC) in the nasopharynx, for example, behaves in a very different fashion than one in the adjacent oropharynx. Treatment algorithms, therefore, vary highly dependent on site.

Given the complexity of head and neck anatomy and tumor biology, multidisciplinary approaches to treatment confer enormous benefit to patients. Surgical, systemic, and radiotherapeutic technologies and techniques have made large improvements of late. Some of the most important advances are detailed in this chapter.

In contrast to 20 years ago, HPV-OPSCC now makes up the majority of oropharyngeal cancer diagnoses [1]. Combination chemotherapy and radiation (CRT) demonstrated excellent oncologic control for HPV-OPSCC and became gold

R. A. Chandra, R. J. Li (eds.), *Multidisciplinary Management of Head and Neck Cancer*,
https://doi.org/10.1007/978-3-031-05973-5_2

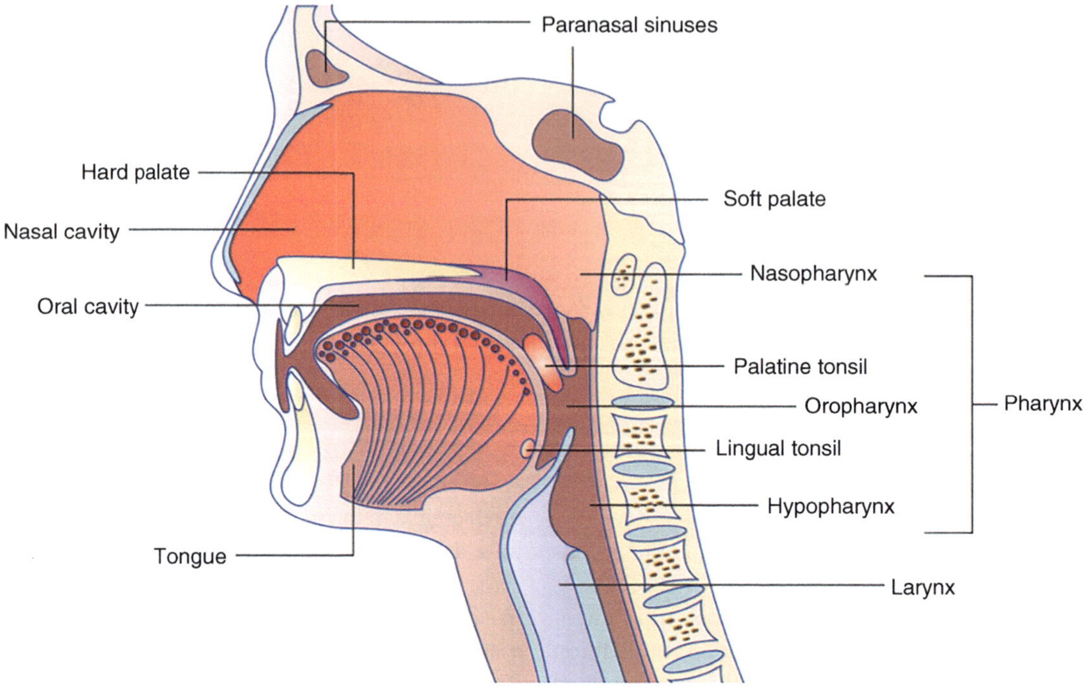

Fig. 2.1 Anatomic regions of the head and neck [2]

standard primary treatment for advanced disease [3, 4]. However, the short- and long-term adverse effects of CRT for OPSCC are significant [4, 5]. These concerns in part led to two separate and intersecting advances within the multidisciplinary treatment of OPSCC: clinical trials aimed at dose de-escalation for CRT protocols (detailed in the Chap. 1) and the reemergence of primary surgical treatment, detailed below.

The treatment of oligometastatic disease provides a new setting through which we can effectively intervene in a patient's cancer progression. Oligometastatic is disease distant from the primary site but in limited quantity. The nascent literature on this currently describes oligometastases as less than or equal to 5 metastatic lesions [6]. This differs from oligoprogression wherein a majority of disseminated disease sites respond to systemic therapy; however, specific lesions progress through treatment [7]. New data demonstrates the value of stereotactic body radiation therapy (SBRT) in select cases to slow or halt progression in select sites of distant metastatic disease. In head and neck cancer, this provides a modality for potential control of a small number of distant metastases when locoregional control has been achieved. Furthermore, this modality can be useful in the setting of recurrent disease.

Many improvements continue to be made at the outer reaches of what is currently considered treatable disease. Multimodality organ preservation techniques continue to be explored in areas where traditional methods would cause intolerable morbidity. Examples include sparing the eye during sinonasal tumor treatment and maximizing the preservation of function for oral cavity cancers [8–10].

Oral cavity squamous cell carcinomas (OSCC) today are predominantly treated with surgery, with adjuvant radiation and systemic therapy as indicated. Unfortunately, the morbidity of surgical treatment for advanced T-stage tongue tumors is debilitating especially for speech and swallow function. In addition, the poor prognosis for very advanced OSCC patients invited reinvestigation of neoadjuvant systemic therapy, with or without radiotherapy, for further prognostication and consideration for definitive nonsurgical management. While definitive CRT protocols have been widely adopted for the treatment of laryngeal, hypopharyngeal, nasopharyngeal, and oropharyngeal primaries, only a limited number of institutions have updated CRT treatment protocols with the intent of improving organ preservation in the treatment of OSCC [10, 11].

Induction chemotherapy has also become an important approach in sinonasal malignancies. We discuss below the implications of this treatment modality on surgical morbidity, mainly orbital preservation [8, 10].

Finally, this chapter will explore the quickly advancing field of immunotherapy and targeted therapies. Early research on targeted therapies in head and neck cancer focused primarily on using cetuximab in place of platinum-based chemotherapy as a de-escalation strategy. While cetuximab ultimately did not demonstrate equivalent oncologic efficacy [5, 12], other promising targeted therapies are emerging. Newer generations of targeted therapies have focused on combined regimens, treating metastatic and locally advanced cancers in the palliative setting. Several of these agents have had good results and factor heavily into decision-making in planning multidisciplinary care for HNC patients. While immunotherapy will be more thoroughly covered in another chapter, we will explore its value in multimodal care here.

Advances in Surgery and Radiation

Oropharynx Cancer

Open surgical management of oropharyngeal cancer historically carried high treatment morbidity with difficult operative exposure. Taking this into account, a majority of the initial treatment algorithms for oropharynx cancers preferentially favored concurrent chemotherapy and radiation as the primary modality for treatment. Early pioneering work to develop transoral robotic surgery (TORS) for the treatment of oropharyngeal cancer catalyzed present-day adoption in many head and neck cancer programs [13, 14]. TORS offered a new technique with the potential lower functional morbidity in select HNC cases due to reduced disruption of neural and muscular soft tissues relative to open pharyngotomy approaches. The surgical robot delivered visualization and wristed instrumentation to tumor sites where the human hand was unable to reach. An endoscope provided a high-definition three-dimensional view of the tumor for resection. The newest iteration of the surgical robot by Intuitive Surgical Inc. (Sunnyvale), the da Vinci Single Port (SP) surgical system, is depicted in Fig. 2.2. The system includes the patient cart with the robotic arms and endoscope, the vision cart, and the surgeon console. The SP was originally

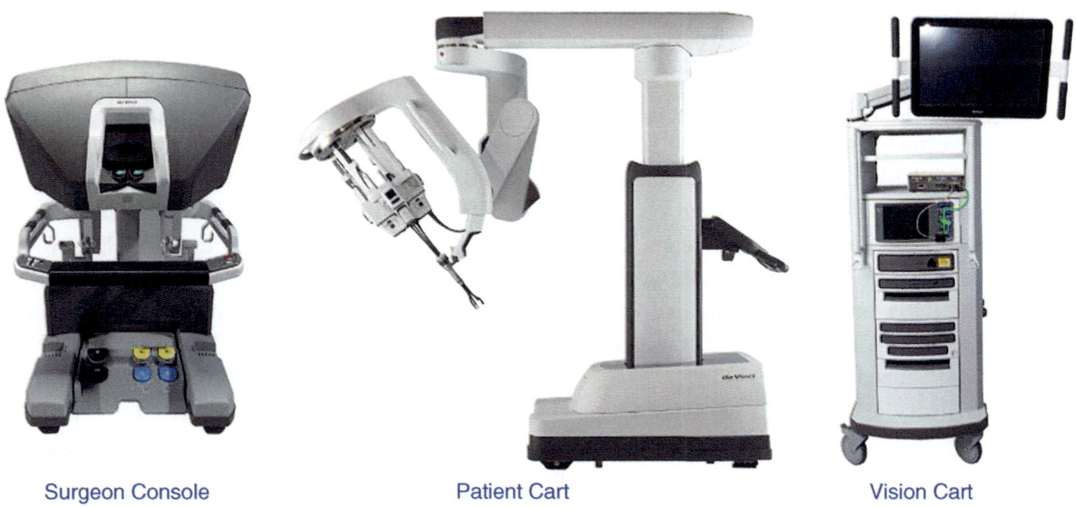

Fig. 2.2 The Intuitive da Vinci Surgical System [15]

developed to be able to introduce the camera and all instruments into the abdomen using a single trocar port, and these same mechanics benefit the head and neck surgeon operating in the narrow corridor of the oropharynx.

For carefully selected tumors, high local control rates are achievable when utilizing TORS. The original description of the TORS radical tonsillectomy reported a series of 27 patients with 93% achieving negative margins. A subsequent study reported a similar rate of negative margins (98%) and observed that positive margins were predictive of poorer local control and survival [16]. A later retrospective multi-institutional study of 410 patients undergoing TORS resection of oropharyngeal malignancies reported 2-year and 3-year locoregional control rates of 91.8% and 88.8%, and overall survival rates of 91% and 87.1%, respectively [17]. The proportions of HPV-positive disease are often not known in these study populations, complicating the evaluation of results. This obscures the expected natural history of the disease in the study populations and clouds the utility of recorded markers of surgical success, such as close margins.

While robotic oropharyngectomy was being established, concurrent chemotherapy and radiotherapy (CRT) continued to be a mainstay of OPSCC treatment with excellent oncologic outcomes in the HPV-OPSCC group. CRT was an effective means of treating local disease and provided a high overall survival in HPV-OPSCC disease [3]. A subset analysis of Radiation Therapy Oncology Group (RTOG) 0129 evaluated and detailed the effect of CRT on HPV-OPSCC. The study evaluated 743 patients with Stage III or IV OPSCC (American Joint Committee on Cancer [AJCC] version 7), enrolled over 3 years, and followed for a median of 4.8 years. Three-year overall survival (OS) in the HPV-positive patients receiving standard CRT was 82.4%, compared to 57.1% for HPV-negative patients [3].

The discovery of HPV as a driver of OPSCC introduced a new tumor biology in a site with anatomic and functional restrictions for both radiotherapy and surgery. The rationale for surgery in part includes obtaining surgical pathology to guide indications for or against adjuvant therapy. In fact, a subset of patients who undergo transoral surgery for OPSCC with neck dissection avoid radiotherapy and systemic therapy altogether [18]. The decision whether to start with surgery followed by adjuvant treatment as necessary, versus definitive CRT, varies highly based upon institutional and patient preference. Ongoing trials are aimed at better defining approaches to deintensification of different treatment modalities (e.g., DART: ClincialTrials.gov identifier NCT02908477, ORATOR 2: NCT01590355, PATHOS: NCT02215265) [19, 20]. The topic of treatment deintensification is discussed at length elsewhere in this text.

Radiotherapy Options for Oligometastatic Disease

The treatment of oligometastastic disease is an area of emerging interest in head and neck cancer. Oligometastases are qualitatively described in the literature as having a few distantly metastatic lesions, rather than widely disseminated disease. This is often described as five or fewer metastatic lesions. The goal in defining oligometastases is the identification of a disease state that is controllable for longer term than a comparison of widely metastatic disease state. Initial research in head and neck cancer has focused on patients with curable local disease that was accompanied with a small number of distant metastases [21].

A randomized phase 2 clinical trial comparing stereotactic ablative radiotherapy (SABR) versus standard of care palliative treatment (SABR-COMET) trial evaluated the use of SABR as compared to standard palliative radiotherapy for oligometastatic disease. This was a broad trial, and less than 30% of patients enrolled had head and neck primary tumors. The majority of metastases treated were in the lungs, but the study also included metastases to bone, adrenal glands, and liver. They evaluated 99 patients between 2012 and 2016, 67% of whom were randomized to the intervention arm. These patients had between one and five metastatic lesions. Median survival in the control arm was

28 months (95% CI 19–33) and 41 months (95% CI 26-not reached) in the SABR group. There were higher rates of grade 2 or higher adverse effects in the intervention group. There were also three (4.5%) treatment-related deaths in the SABR group (one from radiation pneumonitis, one from pulmonary abscess, and one from a subdural hemorrhage following surgery to repair a SABR-related perforated gastric ulcer) [22]. Despite the treatment toxicity, the improved survival as compared to standard of care palliative treatment does suggest a role for SABR as treatment of oligometastatic disease. A Phase 3 trial was recommended for further definition of possible benefit from treatment and identification of subgroups that may benefit.

Immunotherapy

Despite innovation and advances in conventional treatment paradigms for head and neck cancer, the overall 5-year survival remains nearly 50% excluding HPV-associated oropharyngeal cancers. Novel treatment methods seek to improve disease control while mitigating toxicity and functional morbidities. The concept of directed molecular marker therapy is gaining ground in all cancer types. This is especially true for high-risk tumors, recurrent disease, and tumors with known mutations. Immunotherapy is a cancer treatment that potentiates a component of a patient's own immune system to target cancer cells. This approach utilizes molecular inhibitors or monoclonal antibodies that are aimed at targeting certain vital processes necessary for tumor survival. This includes affecting the tumor microenvironment or blood supply resulting in tumor death and preventing the spread of metastatic disease. Within the last two decades, improved understanding of the molecular mechanisms behind tumor growth and metastasis has led to the development of targeted agents with specificity for tumor cells. Tumor immunotherapy is in its infancy in the field of head and neck cancers, and there is hope that continued progress can positively impact patient survival with tolerable morbidity.

Multidisciplinary Regimens Including Epidermal Growth Factor Receptor Blockers

The epidermal growth factor receptor (EGFR) has been found to be overly expressed in many head and neck cancers [23]. Disease recurrence and worse patient outcomes have been associated with increased expression of EGFR and its ligand, transforming growth factor-alpha (TGF-α). Cetuximab, an immunoglobulin-G1 chimeric monoclonal antibody, targets the extracellular ligand-binding domain of the EGFR protein. This was the first molecular targeting agent to demonstrate improved survival in patients with HNSCC, and the initial hope was that it could replace platinum-based CRT when given in conjunction with radiotherapy. Unfortunately, two large multicenter trials failed to prove noninferiority of cetuximab with radiotherapy as compared to cisplatin with radiotherapy in oropharynx cancer [5, 12, 24]. Since De-ESCALaTE HPV and NRG Oncology RTOG 1016 trials were published, enthusiasm has waned in the literature for cetuximab except for use in de-escalation for patients unable to tolerate current standard systemic options.

Multidisciplinary Regimens Including Immune Checkpoint Blockers

Some of the most exciting data of late has come with programmed cell death 1 (PD-1) and programmed death ligand 1 (PD-L1) modulators. Immune checkpoint blockade, targeting the cytotoxic T-lymphocyte antigen 4 (CTLA-4), PD-1, and PD-L1 proteins, has proven to be a fruitful site of oncologic research. PD-1 is an inhibitory receptor expressed on T cells, B cells, regulatory T cells, natural killer cells, and macrophages (Fig. 2.3). Binding to either the PD-1 receptor or the ligand inhibits immune function, and thus inhibitors of these proteins have been targeted as possible immune stimulants to target malignancy [25].

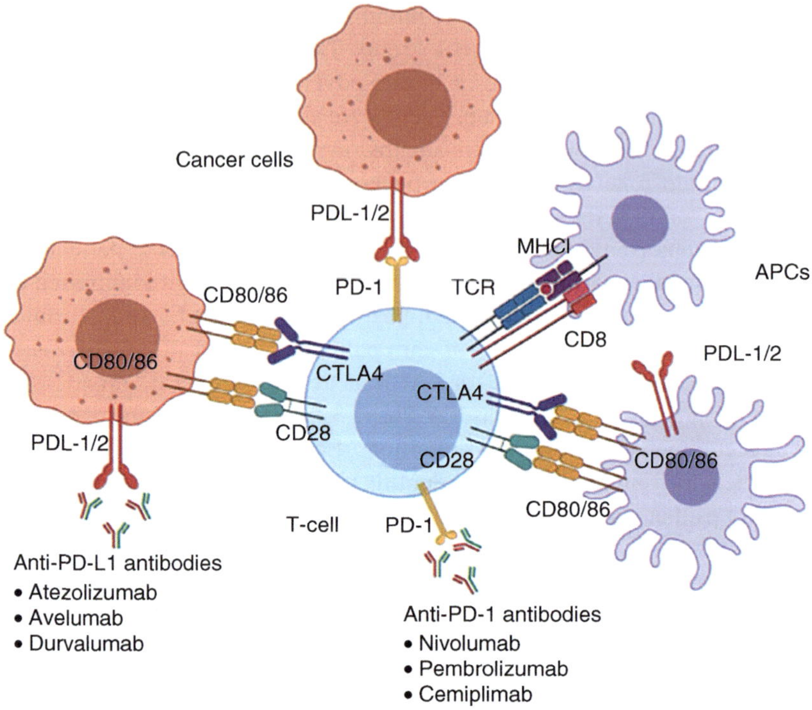

Fig. 2.3 Mechanism of blockade of PD-1 and PDL-1 by targeted therapies [25]

Nivolumab (anti-PD-1 antibody) and ipilimumab (anti-CTLA-4 antibody) have found success in patients with advanced melanoma and other various solid tumors. The multicenter, phase III, randomized clinical trial (CheckMate 067) found that in patients with untreated unresectable stage III or IV melanoma, nivolumab plus ipilimumab or nivolumab alone provided superior survival than ipilimumab alone [26, 27]. CheckMate 141 evaluated nivolumab in unresectable or metastatic HNSCC who progressed on platinum chemotherapy. Nivolumab was found to provide superior OS as compared to the investigator's choice of palliative treatment (16.9% vs. 6.0% alive at 2 years, respectively) with significantly fewer grade 3 or higher adverse effects. Compared to control, the nivolumab intervention group had double the response rate (13% vs. 6%) and double the 1-year overall survival (36% vs. 16.6%) [28, 29]. Importantly, quality of life (QOL) data from this same study showed that the nivolumab group had significantly better QOL outcomes in the first 15 weeks after initiation of treatment [30].

Equal enthusiasm has focused on the PD-1 receptor inhibitor, pembrolizumab. Pembrolizumab has been investigated in recurrent and metastatic HNSCC with good results. KEYNOTE-040, a phase III trial comparing pembrolizumab to methotrexate, docetaxel or cetuximab in patients with recurrent or metastatic HNSCC after a previous platinum containing regimen, showed a median OS of 8.4 months (95% CI 6.4–9.4) compared to 6.9 months (95% CI 5.9–8.0) with standard of care regimens. This was achieved with a significantly better profile of grade III or higher adverse events [31, 32]. With the success of nivolumab and pembrolizumab as second-line treatments, the EAGLE study, a phase III trial investigated employing durvalumab (anti-PD-L1 antibody) with or without tremelimumab (anti-CTLA-4 antibody) as second-line treatment in patients with metastatic or recurrent HNSCC who progressed on prior platinum-based chemotherapy. The intervention groups did not have improved response compared to standard of care chemotherapy (17.9% vs 18.2% vs 17.3%). Survival was also not different between groups [33].

Following these studies, a pivotal phase III comparative study was performed comparing pembrolizumab alone or with chemotherapy to standard of care EXTREME regimen containing cetuximab plus chemotherapy in patients with untreated recurrent or metastatic HNSCC [34]. This study, KEYNOTE-048, is most often cited for providing data to support the use of pembrolizumab monotherapy or in combination with chemotherapy. Pembrolizumab plus cisplatin or carboplatin, and 5-flurouracil, or pembrolizumab monotherapy were all found in the trial to be appropriate first-line treatment for recurrent or metastatic HNSCC. The combination therapy did impart higher levels of adverse effects, but all were found to be noninferior to the control EXTREME regimen [34].

With the landmark KEYNOTE-048 study showing a survival benefit, subsequent trials evaluated immunotherapy as a first-line treatment: CheckMate 651 (nivolumab plus ipilimumab versus EXTREME regimen, presented in ESMO 2021) and KESTREL [35] (durvalumab with or without tremelimumab versus EXTREME regimen, with data announced and awaiting presentation) did not meet the primary endpoint of improving survival. Responses to immunotherapy in patients with recurrent or metastatic HNSCC varied, and different immunotherapies as well as different combination strategies seem to play pivotal roles in augmenting the efficacy of immunotherapy. Mechanistic insight and detailed analysis of the immune contexture in patients with metastatic HNSCC receiving immune checkpoint inhibitors are crucial.

While the checkpoint inhibitors have been used primarily in combination with other forms of systemic therapy, combination radiotherapy and immunotherapy regimens have also been investigated. The safety and tolerability of addition of pembrolizumab to standard chemoradiation with cisplatin were demonstrated in an early phase 1b study [36]. About 85% of HPV-positive and 78% of HPV-negative locally advanced HNSCC achieved complete response at the end of the treatment, and no additional safety concerns were reported. In addition, another phase 2 trial supported the use of pembrolizumab in combination with radiotherapy in patients with locally advanced HNSCC who were platinum ineligible [37]. While preserving a favorable safety profile, 1-year progression free survival (PFS) was 76% (86% in HPV-positive and 67% in HPV-negative) and OS was 86% (93% in HPV-positive versus 80% in HPV-negative). There are many trials employing immunotherapy with radiotherapy that have been launched and are underway; however, results have not been universally positive. A recent phase III study, JAVELIN Head and Neck 100, failed to demonstrate the survival benefit of avelumab plus chemoradiotherapy followed by avelumab maintenance compared to standard chemoradiation in patients with locally advanced HNSCC [38].

Recently, immune checkpoint inhibitors have been explored in the neoadjuvant setting. In a multicenter phase 2 study, the feasibility of neoadjuvant and adjuvant pembrolizumab was studied in patients with locally advanced HPV-negative HNSCC. Although no patients obtained pathological complete response (pCR), 44% of patients had a partial pathologic tumor response (pTR; tumor necrosis, giant cells, and debris in tumor surgery bed). Half of those achieved pTR ≥50% after a single dose of neoadjuvant pembrolizumab, without adversely impacting the surgery or adjuvant therapy [39]. Compared to historical data, patients with high-risk pathology had notably lower 1-year relapse rate (16.7%). Another phase 1b trial explored neoadjuvant PD-1 inhibition with concurrent radiotherapy. This study included 21 patients with locally advanced HNSCC, primarily of the oropharynx, between 2018 and 2019. Patients were treated with 40 Gy in five fractions or 24 Gy in three fractions, with or without concurrent nivolumab. After definitive surgical resection, a major pathological response (mPR) rate of 86% and pCR of 67% was reported. Clinical to pathologic downstaging occurred in 90% of patients. Surgery was followed by adjuvant nivolumab for 3 months. Most of the primary tumors were HPV+ OPSCC (16 of 21, 76.2%). The early results demonstrated proof of concept for immunoradiotherapy strategies, and the literature continues to grow.

Organ Preservation

Neoadjuvant Chemotherapy and Radiation for Advanced Oral Cavity Squamous Cell Carcinoma (OSCC)

Advanced OSCC of the oral tongue and adjacent subsites may necessitate a total or near-total glossectomy for local oncologic control. With a critical role in swallow and speech, loss of oral tongue function is severely debilitating. Free tissue transfer for reconstruction cannot restore the dynamic oral-motor control that is lost by such resection. Quality of life scores have been evaluated at multiple points over the last 30 years for patients who have undergone total glossectomy with laryngeal preservation, and significant deficits remain despite reconstructive surgery [40]. The literature supports what patients know intuitively that the tongue is highly valued for quality of life in ways that are difficult to quantify.

The significant morbidity of total glossectomy has led to the exploration of nonsurgical treatments for OSCC. This is analogous to treatment paradigms for laryngeal cancer. A landmark study demonstrated induction chemotherapy given to patients with advanced laryngeal cancer could select for good responders that may be able to undergo definitive chemoradiotherapy with anatomical preservation of the larynx. Poor responders proceeded with total laryngectomy. Using induction chemotherapy in this way led to exploration of the technique to spare other high-value organs. In one study of advanced oral cavity cancers with induction chemotherapy followed by surgery or definitive chemoradiotherapy based upon response, the surgical group showed a significant survival advantage. The 5-year overall survival (OS) was found to be 32% in the induction chemotherapy cohort and 65% in the surgical cohort [11]. Despite these early findings, there remain significant areas where nonsurgical management can play an important role.

Since that study, several institutions have reported good success with nonsurgical treatment for locally advanced OSCC. A retrospective, single-institution review of definitive chemoradiotherapy management for stage III and IV OSCC patients highlighted both the oncologic outcomes and major morbidity of this approach [41]. Several protocols were utilized including 5-fluorouracil and hydroxyurea along with a third systemic agent, and radiotherapy. Radiotherapy was delivered either once or twice daily, and total dose to the primary site was between 70 and 75 Gy. In total 140 patients with advanced OSCC were treated with these various CRT protocols. Seventy-five percent of these patients had clinical T3 or T4 disease, 68% had N2 or greater disease, and 91.4% were stage IV. Forty-seven percent of the patients had oral tongue primary sites. Median follow-up for this patient group was 5.7 years with a 5-year overall survival of 63.7% and locoregional control survival of 78.6%. They had a 20.7% rate of osteoradionecrosis and a 10.0% rate of long-term feeding tube placement [41].

Another large retrospective series of advanced OSCC patients treated primarily with CRT was reported in 2020 [42]. This cohort included 63% T3–4 patients, 54% stage IV disease. Notably this was a less advanced tumor stage cohort overall as compared to the prior study. In total 1316 OSCC patients were treated with curative intent, and 108 of these were selected for nonsurgical management. They noted that in 49% of these cases, the explicit reason for pursuing nonsurgical treatment was to attempt preservation of oral structure/function. In all nonsurgical cases, definitive radiotherapy with concurrent chemotherapy was planned. Median follow-up was 52 months (range 3–136 months). Their 5-year local control rate was 78%, regional control 92%, disease-free survival (DFS) was 42%, OS was 50%. They reported a Grade 3 or higher rate of osteoradionecrosis of 8%. The studies above are both single arm, retrospective, nonrandomized cohort studies. However, these data show in principle the value of nonsurgical management of advanced OSCC cases that would otherwise require debilitating surgical management. These reports also transparently describe the considerable morbidity inherent to nonsurgical treatment.

Induction Chemotherapy

There are two main roles that induction chemotherapy (IC) is currently used in head and neck cancer therapy. The first is to guide treatment, testing the tumor biology to determine its susceptibility to definitive treatment. The second and overlapping value is reduction of tumor burden that may allow for reduced morbidity of definitive treatment.

Sinonasal undifferentiated carcinoma (SNUC) is a fast-growing sinonasal malignancy that has a well-established treatment algorithm using CRT. A single-arm prospective trial treated 95 patients with SNUC from 2001 to 2018. All underwent IC followed by either consolidation with CRT or surgical extirpation, based on tumor and patient specific factors. The induction chemotherapy regimen in this study was cisplatin (60–80 mg/m^2 on day 1) and etoposide (100–120 mg/m^2 docetaxel (75 mg/m^2 [n = 21]) on days 1–3, administered intravenously every 21 days. When consolidation CRT was performed, it began 4 weeks after the initial dose of IC. In patients who had a partial response (PR) or complete response (CR) to induction chemotherapy, 5-year disease-specific survival (DSS) was 81% in patients who underwent CRT, and 54% in patients who underwent surgery with postoperative radiotherapy or CRT. In patients who did not have a response to induction chemotherapy, the 5-year DSS was 0% in patients who underwent CRT, and 39% in patients who underwent surgery with adjuvant treatment [10].

The same group adopted this protocol for SCC of the sinonasal cavity with promising results. From 1988 to 2017, 123 patients with SCC of the sinonasal cavity were treated with IC followed by definitive CRT or surgery. These were patients with advanced local disease (89% with T4 disease), and 29.3% had regional metastatic disease at the time of presentation. The chemotherapy regimen consisted of platinum-based chemotherapy and taxanes for 88% of patients, but there were some that received additional 5-fluorouracil and cetuximab. After IC, 57.8% of these patients achieved a partial response. Half of these patients underwent definitive surgery, and the other half underwent definitive CRT. The 2-year OS was 61.4% and 5-year OS was 44.2% for the whole population. OS and DFS were significantly worse in the patients who did not achieve partial response; however, many of these patients were censored when lost to follow-up, before the end of the second year. All patients who achieved partial response or better (81% of all patients) were able to preserve their orbital contents. This study showed proof of concept again that IC could be used to assist in organ preservation.

The same group recently published data looking at IC as a possible method of organ preservation in oral cavity SCC (OSCC). They included 120 patients, treated for OSCC from 1995 and 2018. Fifty percent of these patients had T3 or T4 disease. After two cycles of IC, 63.3% of their patients showed at least partial response. Sixty of these patients underwent definitive surgical management. A subgroup of 15 patients were able to undergo a less extensive local resection after tumor shrinkage from IC. Each was able to avoid resection of an oral cavity subsite or avoid a total or subtotal glossectomy. One patient was able to avoid surgery at the primary site entirely and underwent only neck dissection for residual lymph node disease. After surgery was completed, they found that nine patients had achieved pathology-confirmed complete response. They noted that recurrence occurred in 48% of the total patients in the study, but there is no description as to which patients suffered recurrence. Like the above studies, patients who achieved at least a partial response had an improved 5-year DSS (78% vs. 66.9%) and OS (60.1% vs. 51.4%) compared to the whole study population [8]. It should be noted that the current standard of care is that post-IC resection should include the margins of the original tumor regardless of tumor response to the IC therapy. The theory is that the tumor responds nonuniformly to IC at a microscopic level, and thus, clinical examination may not define the true tumor margin post-IC. These data provide support for using IC as a method of preserving more oral-motor function in advanced OSCC.

Conclusion

Much progress has been made in understanding the epidemiology, pathogenesis, and management of head and neck cancers, with the resources of the head and neck oncologist continuing to broaden. Treatment algorithms are modified as scientific advances develop, such as in the implementation of TORS in the primary treatment of oropharynx cancer, or SABR for oligometastatic disease. Even in cases where survival improvements are not currently achievable, efforts to improve the quality of remaining life have value, especially in the context of organ preservation. Finally, the roles of immunotherapy and targeted therapies are rapidly evolving. These have already proven to have substantial value in multidisciplinary cancer care. The combination of various therapies is a nuanced and important resource for patients that require multispecialty collaboration.

References

1. Xu L, Dahlstrom KR, Lairson DR, Sturgis EM. Projected oropharyngeal carcinoma incidence among middle-aged US men. Head Neck. 2019;41(9):3226–34.
2. Sabatini ME, Chiocca S. Human papillomavirus as a driver of head and neck cancers. Br J Cancer. 2020;122(3):306–14.
3. Ang KK, Harris J, Wheeler R, et al. Human papillomavirus and survival of patients with oropharyngeal cancer. N Engl J Med. 2010;363(1):24–35.
4. Maxwell JH, Mehta V, Wang H, et al. Quality of life in head and neck cancer patients: impact of HPV and primary treatment modality. Laryngoscope. 2014;124(7):1592–7.
5. Gillison ML, Trotti AM, Harris J, et al. Radiotherapy plus cetuximab or cisplatin in human papillomavirus-positive oropharyngeal cancer (NRG Oncology RTOG 1016): a randomised, multicentre, non-inferiority trial. Lancet. 2019;393(10166):40–50.
6. Huang F, Wu G, Yang K. Oligometastasis and oligo-recurrence: more than a mirage. Radiat Oncol. 2014;9:230.
7. Patel PH, Palma D, McDonald F, Tree AC. The dandelion dilemma revisited for oligoprogression: treat the whole lawn or weed selectively? Clin Oncol (R Coll Radiol). 2019;31(12):824–33.
8. Abdelmeguid AS, Silver NL, Boonsripitayanon M, et al. Role of induction chemotherapy for oral cavity squamous cell carcinoma. Cancer. 2021;127(17):3107–12.
9. Abdelmeguid AS, Teeramatwanich W, Roberts DB, et al. Neoadjuvant chemotherapy for locoregionally advanced squamous cell carcinoma of the paranasal sinuses. Cancer. 2021;127(11):1788–95.
10. Amit M, Abdelmeguid AS, Watcherporn T, et al. Induction chemotherapy response as a guide for treatment optimization in sinonasal undifferentiated carcinoma. J Clin Oncol. 2019;37(6):504–12.
11. Chinn SB, Spector ME, Bellile EL, et al. Efficacy of induction selection chemotherapy vs primary surgery for patients with advanced oral cavity carcinoma. JAMA Otolaryngol Head Neck Surg. 2014;140(2):134–42.
12. Mehanna H, Robinson M, Hartley A, et al. Radiotherapy plus cisplatin or cetuximab in low-risk human papillomavirus-positive oropharyngeal cancer (De-ESCALaTE HPV): an open-label randomised controlled phase 3 trial. Lancet. 2019;393(10166):51–60.
13. O'Malley BW Jr, Weinstein GS, Snyder W, Hockstein NG. Transoral robotic surgery (TORS) for base of tongue neoplasms. Laryngoscope. 2006;116(8):1465–72.
14. Weinstein GS, O'Malley BW Jr, Snyder W, Sherman E, Quon H. Transoral robotic surgery: radical tonsillectomy. Arch Otolaryngol Head Neck Surg. 2007;133(12):1220–6.
15. Surgical I. DaVinci SP system user manual. 2019.
16. Moore EJ, Van Abel KM, Price DL, et al. Transoral robotic surgery for oropharyngeal carcinoma: surgical margins and oncologic outcomes. Head Neck. 2018;40(4):747–55.
17. de Almeida JR, Byrd JK, Wu R, et al. A systematic review of transoral robotic surgery and radiotherapy for early oropharynx cancer: a systematic review. Laryngoscope. 2014;124(9):2096–102.
18. Wirth LJ, Burtness B, Nathan CO, Gregoire V, Richmon J. Point/counterpoint: do we de-escalate treatment of HPV-associated oropharynx cancer now? And how? Am Soc Clin Oncol Educ Book. 2019;39:364–72.
19. Nichols AC, Theurer J, Prisman E, et al. Radiotherapy versus transoral robotic surgery and neck dissection for oropharyngeal squamous cell carcinoma (ORATOR): an open-label, phase 2, randomised trial. Lancet Oncol. 2019;20(10):1349–59.
20. Ferris RL, Flamand Y, Weinstein GS, et al. Transoral robotic surgical resection followed by randomization to low- or standard-dose IMRT in resectable p16+ locally advanced oropharynx cancer: a trial of the ECOG-ACRIN Cancer Research Group (E3311). J Clin Oncol. 2020;38(15_suppl):6500.
21. Guckenberger M, Lievens Y, Bouma AB, et al. Characterisation and classification of oligometastatic disease: a European Society for Radiotherapy and Oncology and European Organisation for Research and Treatment of Cancer consensus recommendation. Lancet Oncol. 2020;21(1):e18–28.

22. Palma DA, Olson R, Harrow S, et al. Stereotactic ablative radiotherapy versus standard of care palliative treatment in patients with oligometastatic cancers (SABR-COMET): a randomised, phase 2, open-label trial. Lancet. 2019;393(10185):2051–8.
23. Bonner JA, Harari PM, Giralt J, et al. Radiotherapy plus cetuximab for squamous-cell carcinoma of the head and neck. N Engl J Med. 2006;354(6):567–78.
24. Maubec E, Petrow P, Scheer-Senyarich I, et al. Phase II study of cetuximab as first-line single-drug therapy in patients with unresectable squamous cell carcinoma of the skin. J Clin Oncol. 2011;29(25):3419–26.
25. Makuku R, Khalili N, Razi S, Keshavarz-Fathi M, Rezaei N. Current and future perspectives of PD-1/PDL-1 blockade in cancer immunotherapy. J Immunol Res. 2021;2021:6661406.
26. Hodi FS, Chiarion-Sileni V, Gonzalez R, et al. Nivolumab plus ipilimumab or nivolumab alone versus ipilimumab alone in advanced melanoma (CheckMate 067): 4-year outcomes of a multicentre, randomised, phase 3 trial. Lancet Oncol. 2018;19(11):1480–92.
27. Wolchok JD, Chiarion-Sileni V, Gonzalez R, et al. Overall survival with combined nivolumab and ipilimumab in advanced melanoma. N Engl J Med. 2017;377(14):1345–56.
28. Ferris RL, Blumenschein G Jr, Fayette J, et al. Nivolumab for recurrent squamous-cell carcinoma of the head and neck. N Engl J Med. 2016;375(19):1856–67.
29. Ferris RL, Blumenschein G Jr, Fayette J, et al. Nivolumab vs investigator's choice in recurrent or metastatic squamous cell carcinoma of the head and neck: 2-year long-term survival update of CheckMate 141 with analyses by tumor PD-L1 expression. Oral Oncol. 2018;81:45–51.
30. Harrington KJ, Ferris RL, Blumenschein G, et al. Nivolumab versus standard, single-agent therapy of investigator's choice in recurrent or metastatic squamous cell carcinoma of the head and neck (CheckMate 141): health-related quality-of-life results from a randomised, phase 3 trial. Lancet Oncol. 2017;18(8):1104–15.
31. Chow LQM, Haddad R, Gupta S, et al. Antitumor activity of pembrolizumab in biomarker-unselected patients with recurrent and/or metastatic head and neck squamous cell carcinoma: results from the phase Ib KEYNOTE-012 expansion cohort. J Clin Oncol. 2016;34(32):3838–45.
32. Cohen EEW, Soulières D, Le Tourneau C, et al. Pembrolizumab versus methotrexate, docetaxel, or cetuximab for recurrent or metastatic head-and-neck squamous cell carcinoma (KEYNOTE-040): a randomised, open-label, phase 3 study. Lancet. 2019;393(10167):156–67.
33. Ferris RL, Haddad R, Even C, et al. Durvalumab with or without tremelimumab in patients with recurrent or metastatic head and neck squamous cell carcinoma: EAGLE, a randomized, open-label phase III study. Ann Oncol. 2020;31(7):942–50.
34. Burtness B, Harrington KJ, Greil R, et al. Pembrolizumab alone or with chemotherapy versus cetuximab with chemotherapy for recurrent or metastatic squamous cell carcinoma of the head and neck (KEYNOTE-048): a randomised, open-label, phase 3 study. Lancet. 2019;394(10212):1915–28.
35. Argiris A, Harrington K, Tahara M, et al. LBA36 Nivolumab (N) + ipilimumab (I) vs EXTREME as first-line (1L) treatment (tx) for recurrent/metastatic squamous cell carcinoma of the head and neck (R/M SCCHN): final results of CheckMate 651. Ann Oncol. 2021;32:S1310–1.
36. Powell SF, Gold KA, Gitau MM, et al. Safety and efficacy of pembrolizumab with chemoradiotherapy in locally advanced head and neck squamous cell carcinoma: a phase IB study. J Clin Oncol. 2020;38(21):2427–37.
37. Weiss J, Sheth S, Deal AM, et al. Concurrent definitive immunoradiotherapy for patients with stage III-IV head and neck cancer and cisplatin contraindication. Clin Cancer Res. 2020;26(16):4260–7.
38. Lee NY, Ferris RL, Psyrri A, et al. Avelumab plus standard-of-care chemoradiotherapy versus chemoradiotherapy alone in patients with locally advanced squamous cell carcinoma of the head and neck: a randomised, double-blind, placebo-controlled, multicentre, phase 3 trial. Lancet Oncol. 2021;22(4):450–62.
39. Uppaluri R, Campbell KM, Egloff AM, et al. Neoadjuvant and adjuvant pembrolizumab in resectable locally advanced, human papillomavirus-unrelated head and neck cancer: a multicenter, phase II trial. Clin Cancer Res. 2020;26(19):5140–52.
40. Ruhl CM, Gleich LL, Gluckman JL. Survival, function, and quality of life after total glossectomy. Laryngoscope. 1997;107(10):1316–21.
41. Foster CC, Melotek JM, Brisson RJ, et al. Definitive chemoradiation for locally-advanced oral cavity cancer: a 20-year experience. Oral Oncol. 2018;80:16–22.
42. Hosni A, Chiu K, Huang SH, et al. Non-operative management for oral cavity carcinoma: definitive radiation therapy as a potential alternative treatment approach. Radiother Oncol. 2021;154:70–5.

3 Advances in Surgery and Reconstruction: TORS, TLM

Adam Howard, Nishant Agrawal, and Zhen Gooi

Introduction

Robot-assisted surgery for oropharyngeal squamous cell carcinoma (OPSCC) was first utilized around 2006, given its promising result in minimally invasive cardiac and urologic surgery. O'Malley et al. published their experience with the da Vinci robot for transoral resection of base of tongue neoplasms. They described excellent visualization and instrument access to the oropharynx with this technique [1]. Over time transoral robotic surgery (TORS) gained increasing popularity and usage for resectable oropharyngeal tumors [2]. It is now a widely accepted surgical approach to select oropharyngeal neoplasms and was approved by the Food and Drug Administration (FDA) in 2009 for the treatment of early tumor stage OPSCC. Transoral laser microsurgery (TLM) is a technique that dates back decades. While there were reports of its use as early as the 1950s [3], the first major publication regarding TLM technique and efficacy came in 2005 by Holsinger et al. [4] This study described the authors' 20-year experience of using TLM to treat 191 patients with tonsillar cancer. Around the same time, Steiner et al. published experience utilizing TLM to treat base of tongue and pyriform sinus tumors, using a novel laryngoscope, and a laser-mounted microscope [5, 6]. Together, these studies provided early data on the feasibility and efficacy of surgical resection of oropharyngeal tumors using TLM.

There are many reasons why surgical resection of OPSCC by TORS or TLM has gained popularity over the years. Open surgical approaches for tongue base tumors historically required mandibulotomy or pharyngotomy, both of which carry significant morbidity [7]. These approaches provide the benefit of improved exposure and ability to reconstruct the defect with free tissue transfer [7]. However, disadvantages include prolonged hospital stay, higher rates of gastrostomy and tracheostomy dependence, hardware extrusion and malocclusion, impaired speech and swallowing, pharyngocutaneous fistula, and cosmetic deformity [8, 9]. Furthermore, once organ-sparing strategies were identified as a possible alternative to radical surgery, open approaches for oropharyngeal cancer began to decline in favor.

In 1991 the Veterans Administration Laryngeal trial was published. This was a landmark paper that established organ-sparing chemoradiotherapy as a viable treatment for laryngeal and hypopharyngeal cancers [10]. Extrapolation of organ-sparing approaches for use in oropharyngeal cancers demonstrated equivalent oncologic

A. Howard · N. Agrawal · Z. Gooi (✉)
Department of Surgery, Section of Otolaryngology-Head and Neck Surgery, University of Chicago Medical Center, Chicago, IL, USA
e-mail: adam.howard@uchospitals.edu; nagrawal@surgery.bsd.uchicago.edu; zgooi@surgery.bsd.uchicago.edu

R. A. Chandra, R. J. Li (eds.), *Multidisciplinary Management of Head and Neck Cancer*, https://doi.org/10.1007/978-3-031-05973-5_3

outcomes (locoregional control, 5-year overall survival, 5-year disease-specific survival) compared to surgery followed by adjuvant treatment [11]. However, nonsurgical treatment also came with treatment side effects including dysphagia, mucositis, stricture and tissue fibrosis, neuropathy, and gastrostomy tube dependence [12, 13]. For these reasons, surgical approaches such as TORS and TLM, which provide sufficient exposure and allow for complete resection in select tumors, are of interest to the head and neck oncologic community. This chapter will describe the technique of TORS and TLM, discuss outcomes of minimally invasive oropharyngeal surgery, and review completed and in-progress clinical trials using TORS for treatment of oropharyngeal cancer.

Epidemiology

Cancers of the oral cavity and pharynx are the seventh most frequent cancer in the world and the ninth cause of cancer death worldwide [14]. Together, there are approximately 710,000 new cases and 359,000 deaths per year from oral cavity and pharyngeal cancers [14]. The incidence of oropharyngeal squamous cell carcinoma in particular has steadily increased in the United States and other regions of the world over the last few decades [15, 16]. Concurrently, the incidence of cancers of most other head and neck subsites have been declining over the last few decades in the United States and many other regions of the world [17, 18]. This is believed to be in part due to a decrease in tobacco use in the United States and other regions of the world [19, 20]. It is well established that the incidence of human papilloma virus (HPV)-related OPSCC is increasing in the United States and many other developed countries in Europe and Asia based on cancer registry data [21]. In addition, incidence of HPV-unrelated OPSCC in the United States has declined over the last three decades [21].

It is now understood that HPV-associated and HPV-unrelated OPSCCs have different epidemiological characteristics. Patients with HPV-associated OPSCC are more likely to be younger male and never smokers [17, 22]. This is in contrast to HPV-unrelated disease, which is more likely to present in older males with a history of smoking [21]. It is well-established that HPV-associated OPSCCs also have superior prognosis compared to HPV-unrelated OPSCC [17, 23–26]. Despite the overwhelming evidence that HPV-associated OPSCC pertains a more favorable prognosis, treatment guidelines by the National Comprehensive Cancer Network (NCCN) have yet to establish de-escalation for these patients as a recommended approach outside of clinical trials [27]. The NCCN recommends identifying the HPV status of all oropharyngeal tumors for better prognostication and patient counseling, even if the patient will not receive de-escalated treatment [27].

Technique

The oropharynx is a region located posterior to the oral cavity, inferior to the nasopharynx, and superior to the hypopharynx and larynx. It includes the tonsillar region (tonsil, tonsillar fossae, anterior and posterior tonsillar pillars), base of tongue, soft palate, posterior pharyngeal wall, and lateral pharyngeal walls. Transoral resection has been a common treatment for select tonsil cancers for many decades [4]. Holsinger et al. published their technique of transoral lateral oropharyngectomy in 2005 using monopolar and bipolar electrocautery and a Crowe–Davis mouth retractor for exposure [4]. While variations of this approach have developed, the principles of the resection hold true using TORS and TLM. They describe the first step as creating a lateral mucosal incision at the retromolar trigone that incorporates the entire anterior tonsillar pillar, buccopharyngeal fascia, and superior constrictor muscle. These structures are grasped and retracted medially [4]. Medial retraction of these contents exposes the deep margin: the prevertebral fascia, parapharyngeal fat, and their contents. The incisions are carried inferiorly to the glossotonsillar sulcus and superiorly to incorporate a portion of the soft palate. The medial mucosal incision is made along the posterior pharyngeal wall down to the prevertebral fascia. Continued resection results in an en bloc excision of the tonsil, anterior and posterior tonsillar pillars, lateral pharyngeal wall mucosa, and superior constrictor musculature [4].

Transoral laser microsurgery for oropharyngeal tumors was introduced in the early 2000s by Steiner [5, 28]. One of the controversial components of this approach was the concept of piecemeal resection. Historically, oncologic surgeons have practiced en bloc resection of tumors. Steiner demonstrated how tumors of the oropharynx, including the base of tongue, may be safely resected via a piecemeal approach with TLM [5]. The arguments for piecemeal resection include the ability to inspect the depth of tumor infiltration by cutting through the tumor until normal appearing tissue is reached. This enables adjustments to the depth of resection in order to strike a balance of achieving negative margins without over-resection and resulting morbidity [28]. Based on data from this group's experience with TLM, transecting the tumor did not result in increased rate of subsequent regional or distant metastasis [28], as one animal model at the time had demonstrated [29]. Favorable outcomes using this approach were duplicated by Grant et al., where 57 of 59 achieved negative frozen and permanent margins after TLM resection for tongue base SCC of various stages [30]. This group reported 3- and 5-year overall survival (OS) rates of 91% and 69%, respectively, with only seven treatment failures (four local, one regional, and two distant). A smaller retrospective review of BOT tumors resected with TLM was published by Henstrom et al., which demonstrated a 5-year OS of 87%, and only three out of 20 treatment failures [31].

However, the largest study to duplicate the favorable outcomes by Steiner was published by Haughey et al. in 2010 [32]. This was a multi-institutional prospective study that included 204 patients across the United States with advanced stage oropharyngeal SCC treated with TLM resection of primary tumor, neck dissections, and adjuvant therapy based on the presence of high-risk features. This study showed very promising outcomes, with 3-year locoregional control (LRC) rate of 93%, 3-year OS of 86%, and 3-year disease-free survival (DFS) of 82% [32]. Patients who achieved clear margins had significantly higher 3-year overall survival, as did those with p16-positive tumors compared to those with p16-negative tumors. Not only did this study provide more evidence that piecemeal resection of tumors does not confer high local or regional recurrence rates, it also showed that TLM is feasible across institutions and multiple surgeons [32]. Despite the growing popularity of TLM at this time, limitations remained—particularly for base of tongue tumors. TLM requires direct line of sight for adequate resection, and the surgeons need to use one hand to grasp the specimen for proper traction and exposure [3]. Furthermore, laryngoscopes designed for BOT resection at the time were less optimized compared to more modern exposure devices.

In 2009 the FDA approved TORS for resection of T1 and T2 oropharyngeal tumors. Since then, the number of early stage OPSCC treated with primary surgery has steadily increased, from 56% in 2004 to 82% in 2013 based on NCDB analysis [2]. It is critical to understand the goals and limitations of TORS to achieve optimal oncologic outcomes and patient safety. Over years of experience with TORS, absolute contraindications to the procedure have been elucidated. Weinstein et al. published guidelines on contraindications for TORS and classified them into one of three categories [33]. These three categories include vascular, functional, and oncologic contraindications for TORS (Fig. 3.1) [33]. Some examples of contraindications include inability to expose the tumor, invasion of the mandible or skull base, involvement of deep extrinsic tongue musculature, or a retropharyngeal internal carotid artery [3, 33]. Exophytic tumors, early T stage tumors, and well-lateralized tumors are more favorable for TORS compared to endophytic, advanced T stage, and tumors approaching or crossing midline [9]. The goal of TORS is en bloc resection of the primary tumor to achieve negative margins. Achieving negative margins is crucial, as positive margins have consistently been shown to reduce overall survival and locoregional control compared to negative

Vascular	Functional	Oncologic
• Tonsil primary and retropharyngeal carotid artery • Tumor located adjacent to carotid bulb or internal carotid artery • Carotid artery encased by tumor or metastatic lymph nodes	• Tumor located at the midline of the tongue base or vallecula • Tumor resection requiring ≥50% of the tongue base musculature • Resection of tongue base and entire epiglottis	• Stage T4b • Unresectable neck disease • Distant metastases • Neoplastic trismus • Prevertebral fascia involvement • Mandible or hyoid involvement • Tumor extension to lateral neck soft tissue • Involvement of eustachian tube

Fig. 3.1 Contraindications to TORS

margins [32, 34]. Factors that increase the risk of positive margins have been analyzed based on data from NCDB and include higher T stage, low-volume surgical centers, and lymphovascular invasion on final histopathologic analysis [35]. Patient selection is critical for successful resection with TORS. Understanding the strengths and limitation of this tool and correlating to patient- and tumor-specific characteristics allows for a successful transoral surgery.

Currently multiple robots are available for TORS. The da Vinci surgical system was approved by the FDA in 2009, and the multiport variant (da Vinci Si) was the first to achieve widespread use [36, 37]. The FLEX Robotic System is an alternative to the da Vinci Si and was approved for use in head and neck surgeries by the FDA in 2015 [36]. Its advantages over the da Vinci Si included haptic feedback to its controls, flexibility in its operating arms, and improved visualization of the larynx [36]. One of the more recent additions to the head and neck robotics arsenal is the da Vinci singly port (Sp) robot. This was first used in genitourinary surgery and was quickly adapted for use in head and neck cancer [37]. Since 2019, there have been multiple prospective trials analyzing the efficacy, feasibility, and safety of the da Vinci Sp. [37–40] Three of these studies reported on using the robot and complications from surgery. These studies found no increase in postoperative complication rates using the Sp compared to the Si, and no conversions to open surgery [37–40].

One of the most important tenets of transoral surgery for the oropharynx is adequate exposure. This is accomplished using a variety of retractors. Some are designed specifically for BOT or tonsil tumors, whereas others have more universal uses. Once the appropriate retractor based on the patient's anatomy and tumor characteristics is in place, the robot is brought in and positioned. The robotic arms and telescope are inserted into the patient's mouth, and there should be confirmation that the surgeon has visualization and access to the entire surgical field. The surgical steps to a base of tongue tumor resection are tailored based on tumor characteristics. Generally, an anterior incision delineating the anterior margin is created first, allowing for posterior retraction of the specimen during resection [1, 41]. Next, medial and lateral mucosal incisions are created, guided by the extent of gross tumor. The lateral incision is often made through the glossotonsillar sulcus, but may include the inferior portion of the tonsil or the entire tonsil if needed based on tumor size and location. Continued anterior to posterior dissection through normal appearing soft tissue deep to the tumor is aided by constant posterior retraction of the specimen. The dorsal lingual artery is sometimes encountered laterally in the tongue base, and should be ligated. This branch can be traced laterally to the lingual artery if needed for proximal hemostasis. The posterior margin is then made, connecting the medial and lateral incisions, often at the level of the vallecula [1, 41]. The specimen is then removed and inspected for gross margins. Margin sampling for frozen section analysis can be performed from either the specimen or the wound bed. If needed, re-resection of one or multiple margins should be completed to ensure negative margins.

The principles of TORS radical tonsillectomy are similar to radical tonsillectomy using handheld electrocautery, with the exception that TORS allows for superior exposure and access to the base of tongue. The tonsil tumor is resected en bloc with appropriate circumferential margins, using the superior constrictor muscles as the deep margins. This leaves the parapharyngeal fat and prevertebral fascia exposed, which protects the parapharyngeal space contents including the internal carotid artery. A lateral incision is made through the buccal mucosa at the pterygomandibular raphe [42]. A plane between the pharyngeal constrictor and the pterygoid musculature is created and deepened to the levels of the styloglossus and stylopharyngeus muscles. The superior cuts are then made through the soft palate. The extent of included soft palate is determined by tumor size and location. The soft palate mucosal incision is brought down through the soft palate musculature to the level of the prevertebral fascia. Using medial retraction of the specimen, the constrictor muscles are elevated off the prevertebral fascia. A cuff of styloglossus and stylopharyngeus muscles are typically incorporated into the resection specimen . A caudal margin incision is then planned at the level of the tongue base. The amount of tongue base mucosa included is directed by the tumor size and location. The specimen should be inspected for adequate gross margins. Frozen section margins should be sent, and re-resection performed if there are positive margins on frozen histopathologic analysis. Hemostasis is achieved via cautery and/or clipping of vessels transorally. When there is satisfactory hemostasis, a nasogastric feeding tube may inserted and secured. Attention may then be turned to the neck dissections if performed concurrently.

Management of the Neck

For oropharyngeal SCC, the ipsilateral neck must be addressed regardless of site, stage, and HPV status [27]. Oropharyngeal cancer has a high propensity for regional metastasis. It is estimated that around 65–75% of all cases will present with cervical lymph node metastases [43, 44]. Select early T-stage tonsil cancers, such as those without tongue base or soft palate involvement, and only single-level ipsilateral nodal metastases require just the ipsilateral neck to be addressed [27, 45, 46]. The risk of contralateral nodal metastasis for T1 and T2 tonsil cancers is up to 10% [46–48], but this risk is much higher for more advanced tonsil primaries and for other oropharynx tumors [48, 49]. Therefore, for all base of tongue, soft palate, and posterior pharyngeal wall tumors (regardless of T-stage) and for advance T-stage tonsil cancers (or those with tongue base or soft palate involvement, or multilevel ipsilateral neck disease), both sides of the neck must be treated [27]. Neck dissections for oropharyngeal SCC must include levels II–IV [27]. Neck dissection may be performed at the same time as TORS or TLM resection, or the procedures can be staged. Both the approaches have advantages and disadvantages. The advantages of concurrent neck dissection after TORS include: a single general anesthetic, reduced hospital stay and cost, minimizing time to potential adjuvant treatment if necessary, and the option to perform arterial ligation transcervically to reduce the risk of postoperative hemorrhage [50].

Risks and benefits of both the approaches to addressing the neck continue to be investigated. The possibility of pharyngocutaneous fistula (PCF) is of particular interest, as it intuitively seems to be a significant risk when performing oropharyngectomy concurrently with a lateral neck dissection. However, single-institution and database studies have found no increase in rates of postoperative PCF or need for reconstruction between concurrent or staged neck dissections after TORS [51–53]. Critical evaluation of the location and extent of the primary tumor and nearby lateral neck disease is imperative prior to subjecting a patient to TORS with concurrent neck dissection. Multiple single-institution studies have found higher rates of intra-operative PCF in patients with advanced stage disease [53, 54]. However, a 2017 systematic review failed to show an association of PCF rates with timing of neck dissection or stage due to a paucity of complication data reporting among articles [55]. This study also analyzed survival outcomes based on timing of neck dissection after TORS. It found no difference in 5-year OS or LRC between the concurrent and staged neck dissection groups [55].

Postoperative hemorrhage after TORS is a feared complication based on its severity. Meta-analyses and retrospective reviews place the risk of any bleeding event around 5–10%, and serious bleeding events at 2–5% [56–59]. Serious bleeding has been defined using the Mayo Clinic classification for postoperative hemorrhage and includes both the major and severe bleeding categories [56]. This includes copious bleeding requiring operative intervention with vessel ligation or embolization and any bleed that is life-threatening. One advantage of concurrent neck dissection with TORS is the ability to transcervically ligate arterial contributions to the lateral oropharyngeal region. One of the earlier studies on this technique was a retrospective review by Mandal et al. in 2016, which analyzed 224 patients who underwent TORS for OPSCC. The group found the overall rate of postoperative hemorrhage to be 9.82% [56]. The authors could not identify a correlation between tumor size and risk of bleeding. Additionally, while there was a trend toward less severe bleeding in the prophylactic transcervical arterial ligation (TAL) cohort compared to controls, there was not a significant difference in hemorrhage rates between the two groups [56]. A meta-analysis by Bollig et al. in 2020 found a significant reduction in major and severe postoperative hemorrhage rates following transcervical arterial ligation compared to controls, with a risk ratio of 0.28 (95% CI 0.08–0.92) favoring TAL [57]. There was no difference in overall rates of postoperative hemorrhage between the two groups [57]. Another meta-analysis by Stokes et al. in 2021 found that patients with history of prior radiation and those requiring anticoagulation had higher rates of postoperative hemorrhage [59]. When comparing patients who underwent TAL to a control group, there was no significant difference in relative risk of overall hemorrhage following TAL. However, the authors did see a large, but insignificant, relative risk reduction in severe postoperative hemorrhage in the TAL group [59]. Finally, a 2021 meta-analysis from Sharbel et al. found that history of radiation and advanced T-stage tumors had significantly higher rates of postoperative hemorrhage [58]. Patients with history of radiation had an odds ratio (OR) of 2.26 (95% CI 1.12–4.56) predisposing them to hemorrhage, and patients with T3–T4 had an OR of 1.93 (95% CI 1.14–3.29) predisposing them to hemorrhage [58]. The study also found that TAL significantly reduces the rate of major and severe bleeding, but not overall bleeding rates [58]. Taken together, TAL appears to be protective from major and severe postoperative hemorrhage but does not seem to reduce the overall rate of bleeding.

Adjuvant Therapy

There are many factors to consider when discussing adjuvant radiotherapy (RT) or chemoradiotherapy (CRT) after TORS/TLM for OPSCC. For years the standard of care for the treatment of OPSCC, regardless of HPV status, has been definitive radiation or concurrent chemoradiotherapy [27, 60]. It is understood that full-dose definitive RT or CRT for OPSCC causes both early and late toxicities [61–63]. One of the most important and exciting possibilities utilizing TORS for the treatment is OPSCC, is the potential to spare the patient the long- and short-term side effects of RT or CRT. This can be achieved through risk stratifying and careful patient selection. Proper patient selection for TORS is critical and requires consideration of the patient's functional status, patient anatomy, the tumor site and stage, and the stage of the neck. Risk stratification of patients based on HPV status, smoking history, T-stage, and N-stage is useful for prognostic purposes [64, 65]. Based on these aforementioned factors, a patient with oropharyngeal cancer may be amenable to surgical resection of the primary and neck without needing adjuvant RT or CRT. The need for adjuvant RT or CRT following surgical resection is dependent upon the presence of adverse histopathologic features (Fig. 3.2) [27]. Alternatively, a patient who would otherwise be treated with definitive CRT may be a candidate for surgical resection followed by adjuvant RT at a lower radiation dose [66].

To justify treatment with TORS and appropriate adjuvant therapy as opposed to concurrent definitive CRT, the survival and functional outcomes after TORS must be equivalent to definitive CRT. Based on the available data, it appears that TORS can result in improved functional outcomes, including lower rates of gastrostomy tube dependence, compared to definitive CRT [67–70]. However, studies indicate that this functional benefit may decrease, but is not eliminated, if the patient requires adjuvant therapy [69, 71]. Nonetheless, a decrease of radiation dose by 10 Gy to the neck, achievable with TORS and neck dissection and based on histopathologic features, may justify surgical resection with adjuvant RT compared to definitive CRT [66]. In this circumstance, the patient would be spared chemotherapy, which may in and of itself justify upfront surgery compared to definitive CRT for some patients. Understanding adverse histopathologic features and how they influence adjuvant treatment is essential. Guidelines provided by the NCCN recommend adjuvant RT for ≥2 cervical nodal metastases, perineural invasion, lymphovascular invasion, perineural invasion, close margins, nodal metastases in levels IV or V, and advanced T-stage tumors [27]. Adjuvant CRT is recommended for positive margins that are not amenable to re-resection, and for pathologic extra-nodal extension (ENE) [27]. The NCCN defines close margin as microscopic disease within 5 mm of the healthy tissue edge [27]. A positive margin is tumor involving the resected tissue edge [27].

The ideal TORS candidate is one with a small and well-lateralized primary tumor (T1 or T2) with no nodal disease. In this case, the advantages of upfront surgical resection with TORS or TLM with ipsilateral or bilateral neck dissection over definitive CRT is relatively clear. These patients can potentially be spared RT/CRT, assuming resection margins are clear, there are no adverse histopathologic features of the primary tumor, and there are none or one positive cervical lymph nodes without ENE [27]. For patients with two or more cervical metastases and/or other adverse pathologic features but without positive/close margins or ENE, adjuvant RT after surgery is still indicated but can be dose-

Adjuvant radiotherapy	Adjuvant chemoradiotherapy
• Primary tumor ≥T3	
• Close margins (1-5mm)	• Positive margins (<1mm)
• Two or more pathologic lymph nodes	
• Lymphovascular invasion	• Extranodal extension
• Perineural invasion	
• Levels IV or V positive lymph nodes	
• Single lymph node ≥3cm	

Fig. 3.2 Indications for adjuvant treatment after TORS

reduced by up to 10 Gy [27, 66, 72]. The target location of adjuvant radiation is important as well. If a tumor is completely resected via TORS with negative margins and no adverse features, the primary site can be spared radiation altogether, even if adjuvant therapy to the neck is indicated based on neck dissection pathology. This would relatively spare the pharyngeal and laryngeal mucosa and musculature of radiation with the goal of preserving swallowing function [66]. In general, when recommending TORS and neck dissection for treatment of OPSCC, one must expect that adjuvant therapy will either be spared completely or that the dose requirements will be reduced compared to definitive CRT.

This decision becomes even more complex when dealing with advanced stage disease. Patients with multiple clinically evident cervical metastases will require radiation to the neck, either in the form of definitive CRT or neck dissection followed by RT (assuming no ENE) [27]. However, it has been shown that the risk of ENE increases with increasing number of positive lymph nodes [73]. This must be considered prior to surgical treatment, as a patient with many clinically evident cervical lymph node metastases is at a much higher risk of requiring full-dose adjuvant CRT for pathologic ENE . Therefore, patients with numerous cervical metastases or bulky nodal disease (≥6 cm) are at higher risk of requiring adjuvant CRT after neck dissection and are not optimal candidates for surgical treatment.

Outcomes

Minimally invasive transoral surgery for oropharyngeal cancer demonstrated promising oncologic outcomes from some of the first reported data sets. By 2009, the Mayo clinic had published several reports on their experience with transoral resection of oropharyngeal cancers of all stages followed by adjuvant therapy when indicated [30, 31, 42, 74]. These retrospective studies demonstrated 2-year OS of >89% and 5-year OS between 58 and 83%. The 5-year locoregional control rates were 83–92%, and the disease-free survival rates were 84–94% [30, 31, 42, 74]. A multi-institutional retrospective review on TLM for advanced stage OPSCC was published in 2011, which included 214 patients with stage III or IV OPSCC treated with TLM followed by adjuvant therapy when indicated [32]. This article reports a 2-year OS of 89% and 5-year OS of 78%. Disease-specific survival (DSS) at 2 and 5 years were 91% and 84%, respectively. Disease-free survival at 2 and 5 years were 85% and 74%, respectively [32]. The authors found several prognostic variables based on multivariate analysis. Advanced T stage, positive margins, HPV/p16 negative, and lack of adjuvant RT were all correlated with significantly lower 5-year OS [32]. These early results of transoral surgery followed by adjuvant therapy already showed equivalent survival outcomes compared to traditional nonsurgical approaches [11, 64, 75].

Publication of further single-institution retrospective reviews demonstrated excellent oncologic outcomes, with 2-year OS being >90% and 5-year OS ranging from 75% to 95% [72, 76–78]. These studies included all stages of OPSCC, both HPV-associated and HPV-negative. Additional findings include a consistent 2-year disease-free survival rate of 88–92% [72, 76]. Dalton et al. found 5-year disease-specific survival to be 86% and a 5-year LRC rate of 89% [77]. De Almeida et al. published data on a large multi-institutional review of oncologic outcomes after TORS in 2015 [34]. Combined analysis of the included studies demonstrated a 89% 3-year LRC rate, 87% 3-year OS, and 92.5% 3-year DSS [34]. In 2017 Mahmoud et al. published results comparing outcomes of TORS (with adjuvant therapy when indicated) versus nonsurgical treatment for all stages and HPV status of OPSCC based on information from the National Cancer Database (NCDB) [79]. They found a total of 1873 patients with OPSCC who were treated with TORS and adjuvant RT/CRT when indicated. Interestingly, treatment with TORS showed a significantly higher 3-year OS at 93% compared to 83% for nonsurgical treatment for all patients. When this was investigated further, the authors found a significant improvement in 3-year OS with TORS compared to nonsurgical treatments in patients who were HPV-negative (84% vs. 66%, respectively, $p = 0.01$). However, the apparent survival benefit from TORS was not appreciated in the HPV-positive subset (3-year OS for TORS 95% vs. 91% for RT/CRT) [79]. This population-based analysis demonstrated the efficacy of TORS and revealed that it may be superior to nonsurgical treatment for HPV-unrelated OPSCC. This finding was corroborated by a review of 319 patients from the Alberta Cancer Registry in 2019 [13]. This review found that patients with p16-negative OPSCC had significantly higher 5- and 10-year OS when treated with surgery and adjuvant therapy compared to nonsurgical treatment. Interestingly, patients with a >10 pack-year smoking history also had significantly higher 5- and 10-year OS when treated with an upfront surgical approach [13].

Additional population-level analyses emerged, further supporting the role of TORS in treating OPSCC. Li et al. published another NCDB review of the efficacy of TORS in early T-stage OPSCC in 2019 [80]. This article reports outcomes on 2224 patients treated with TORS, and compared these outcomes to 6697 patients treated with non-robotic surgery, and 333 patients treated with TLM. The authors found promising 5-year overall survival rates for each surgical approach: 77%, 74%, and 62%, for TORS, non-robotic, and TLM, respectively ($p = 0.002$) [80]. However this effect was not apparent after controlling for age, tumor size, lymph node dissection, and HPV status. Additional benefits of TORS compared to other surgical approaches were that TORS significantly reduced the likelihood of postoperative chemotherapy (but not radiotherapy), and TORS had significantly lower rates of positive margins compared to the other two surgical approaches [80]. More recently in 2020, Nguyen et al. reported on 9745 patients from NCDB with early T-stage OPSCC treated with TORS or nonrobotic surgery [81]. The authors found that TORS provided a significantly higher 5-year OS at 85% compared to 80%, both on univariate and multivariate analysis after controlling for age, HPV status, comorbidity score, T stage, postoperative radiotherapy, positive margins, and ENE. Additional important findings included lower positive margin rates and lower use of adjuvant CRT (but not RT alone) with TORS compared to nonrobotic surgery [81].

Meta-analyses and systematic reviews analyzing the outcomes of TORS versus nonsurgical treatment for OPSCC have recently emerged. A recent meta-analysis by De Virgilio et al. published in 2020 compared outcomes of TORS (with or without adjuvant therapy) to nonsurgical treatment [82]. This article analyzed a total of 5624 patients; 4322 patients were treated with RT/CRT and 1302 patients were treated with TORS with or without adjuvant therapy. Patients included in this study were of all stages and HPV status. Pertinent findings of this study include the high rates of concurrent chemotherapy with radiotherapy in the nonsurgical group (81% of patients received CRT). Additionally, 68% of patients treated with TORS also had adjuvant therapy. Overall survival rates were significantly higher in the TORS cohort at 91%, compared to 83.6% in the RT/CRT group. Disease-free survival was also significantly higher in the TORS cohort at 89.4%, compared to 79.6% in the RT/CRT cohort [82]. There was a trend toward lower gastrostomy tube dependence rates in the TORS group (1.3%) compared to the RT/CRT group (4%), but this difference was not significant. There was also no significant difference in tracheostomy dependence rates between the two groups. The authors conclude that, despite the lack of randomized controlled trials comparing these two approaches, the available data supports the use of TORS in treating selected OPSCC [82]. Two earlier studies, one systematic review and one meta-analysis, demonstrated similar findings. Both the studies found no difference in 2-year OS between patients treated with TORS versus nonsurgical management, but with lower rates of g-tube placement in patients treated with TORS [69, 83].

Functional outcome differences between primary transoral surgical resection versus definitive nonsurgical treatment of OPSCC have mostly been reported through retrospective cohort studies. For example, one review of a large cancer registry by Clark et al. revealed that there were no significant difference in aspiration-free survival between the TORS plus RT/CRT versus RT/CRT alone [13]. The authors defined aspiration-free survival as the time to documented aspiration on barium and/or video fluoroscopic swallow study as measured by a speech and language pathologist. Interestingly, when subgroup analysis was performed, patients who were p16 negative did have significantly improved aspiration-free survival with surgery and adjuvant therapy compared to RT/CRT alone. This improvement was not seen in the p16-positive subgroup analysis [13]. There have only been a few systematic reviews looking at functional outcomes and validated quality of life (QOL) measures in patients treated with minimally invasive transoral surgery [68, 84]. One meta-analysis of functional outcomes after TORS with or without adjuvant therapy was published in 2015 by Hutcheson et al. [68] The authors included 12 studies in their review and noted the high degree of heterogeneity of outcomes reported between the studies. The authors were able to compare g-tube rates in TORS versus nonsurgical groups and found that the median g-tube placement rate for the TORS group was 23% versus median g-tube placement rate of 46% in the nonsurgical cohort [68]. The authors do not comment on statistical significance for this difference. For all studies included in the analysis, around 70% of TORS patients began an oral diet by discharge, 83% were on an oral diet by week 2 after surgery, and 89% were on an oral diet by week 4 after surgery. The authors noted that evaluating swallowing function was difficult due to the heterogeneity of methods used to evaluate swallowing. The composite MD Anderson Dysphagia Inventory (MDADI) was the most widely used, but still only reported in three articles. At 1 year, the composite MDADI scores for TORS patients ranged from 65 to 78, indicating reasonable swallowing function. The authors report that long-term tracheostomy dependence was exceedingly rare, with only two patients out of 441 requiring long-term tracheostomy [68].

A second systematic review of functional outcomes after TORS published in 2019 by Castellan et al. included 20 articles with a total of 659 patients [84]. This study once again found a high degree of heterogeneity in functional and QOL reporting between the included studies. Though the authors did find that patients treated with TORS had superior self-reported QOL scores on a number of surveys compared to patients who had primary open surgery or definitive RT/CRT [84]. Additionally, multiple studies reported superior swallowing function in the TORS group compared to the open surgery or RT/CRT groups. Lastly, the authors report that the majority of patients who underwent TORS with or without adjuvant therapy returned to baseline QOL and swallowing function between 6 and 12 months after treatment [84].

Prospective studies, including randomized controlled trials, are currently investigating the oncologic and functional outcomes of TORS versus nonsurgical treatment for OPSCC, particularly in the setting of HPV-related disease and treatment de-escalation. Two prospective studies have reported data on functional outcomes between these two groups [85, 86]. One study is a secondary analysis of a prospectively collected cancer registry of HPV-associated OPSCC patients who were either treated with TORS ± adjuvant therapy or RT/CRT. These patients did not receive de-escalated treatment. The authors included 257 patients who were at low or intermediate risk in their analysis. Low-risk patients had AJCC 7th edition staging of T1–2 N0–2a, and intermediate risk patients had T3N0 or T1–2 N2b staging. The authors found that patients who underwent TORS followed by adjuvant radiotherapy had significantly higher acute pharyngeal dysphagia compared to patients who received nonsurgical treatment. The results demonstrate an improvement in this acute dysphagia by 3–6 months after completion of adjuvant RT or CRT, but swallow function did not return to baseline in the TORS plus adjuvant therapy group. By 6 months however, swallowing function was equivalent between the TORS plus adjuvant therapy and the nonsurgical groups [85]. The authors also note that the patients who received TORS and did not require adjuvant therapy had superior swallowing function than groups receiving radiation (primary or adjuvant) at 3 and 6 months post treatment. The authors conclude that TORS followed by adjuvant RT or CRT may lead to acute worsening of swallowing function beyond what patients may experience with nonsurgical treatment alone. The authors acknowledge this may be a trade-off for better long-term swallowing outcomes from TORS compared to nonsurgical treatment [85].

Lastly, a randomized controlled trial by Nichols et al. published in 2019 analyzed the oncologic and functional outcomes of 68 patients with OPSCC treated with either TORS with neck dissection ± adjuvant therapy or RT/CRT based on clinical nodal stage [86]. Patients included in this study had T1–2 and N0–2 staging p16 positive or negative cancers (based on AJCC 7th edition staging system). The patients were randomized into the TORS arm or RT/CRT arm, and baseline characteristics between the groups were similar. From an oncologic standpoint, the study found no difference in 5-year OS or PFS between the two treatment modalities. The study used MDADI to evaluate overall quality of life and swallow-specific scores. The authors found that the patients in the TORS arm had significantly worse swallowing function at 1 year compared to the RT/CRT arm based on MDADI swallowing scores. However, this difference was small and did not translate to a clinically meaningful difference. They did find a clinically meaningful difference in the global and emotional QOL scores, favoring the RT/CRT arm compared to the TORS arm [86]. The authors conclude that their study demonstrates that RT/CRT provides equivalent oncologic outcomes with slightly superior global and emotional QOL scores compared to TORS. Based on results from these prospective trials and how they differ from the conclusions reached from other studies, it's clear that further randomized controlled trials are needed to better understand the impact of TORS on swallowing and QOL.

Prospective Trials

There are many active and completed prospective trials for the treatment of OPSCC. We now know that HPV-associated OPSCC patients have significantly improved outcomes and prognosis due to multiple large prospective randomized controlled trials [23, 24, 64, 87]. Attention has turned to de-escalating treatment for HPV-associated OPSCC in order to reduce morbidity from treatment while maintaining good oncologic and functional outcomes. Many approaches to de-escalating treatment are being investigated, with numerous studies published to date demonstrating compelling results [88–91]. This section will focus on de-escalation strategies that involve surgery (specifically transoral resection) which have been completed or are ongoing. There have been a number of prospective trials involving the use of TORS for the purpose of de-escalation to date (Fig. 3.3) [92, 93]. In 2019 Ma et al. published their results from the MC1273 trial [92]. This trial analyzed the outcomes of 79 patients with HPV-associated OPSCC. Inclusion criteria were surgically resectable HPV-associated OPSCC with no evidence of distant metastases, <10 pack-year smoking history, and either pathologic high-risk (ENE+) or intermediate-risk factors (LVI, PNI, single node ≥3 cm, multiple positive lymph nodes, ≥pT3 stage). Patients underwent surgery (95% with TORS) with neck dissection, and adjuvant therapy was administered based on the presence of the high-risk factor (patients received 36 Gy with concurrent docetaxel) or intermediate-risk factors (received 30 Gy with concurrent docetaxel). Based on the pathologic findings after surgery, 36 patients had intermediate-risk factors (cohort A) and 43 patients had high-risk factor (ENE+, cohort B). The median follow-up time was 35.7 months. The 2-year LRC for all patients was 96%, with 100% LRC in cohort A and 93% LRC in cohort B. For all patients, 2-year PFS was 91% and distant metastasis-free survival was 95%, with 2-year OS of 98.7% [92]. Swallowing function and overall QOL measures improved slightly between pre-adjuvant therapy evaluation and 12-month posttreatment evaluations. Only one patient required a temporary g-tube. The

MC1273 (N = 80)	AVOID (N = 60)	ECOG 3311 (N = 519)
Pathologic risk features but ENE negative: 30Gy + docetaxel Pathologic ENE positive: 36Gy + docetaxel	Involved neck levels received 60-66Gy, uninvolved levels received 54Gy (+chemotherapy for ENE) Actively avoided RT to primary site	Low risk: pT1-2, pN01 with negative margins and no adverse pathologic features Intermediate risk: 2-4 lymph nodes, ENE ≤1mm, negative margins but other adverse pathologic features positive High risk: positive margins, ENE >1mm, ≥5 positive lymph nodes
2-yr PFS: 91% 2-yr OS: 98.7%	2-yr local control: 98% 2-yr PFS: 96.2 2-yr OS: 100%	Low risk: observed 2-yr PFS: 94% Intermediate risk: 50Gy adjuvant RT 2-yr PFS: 95% 60Gy adjuvant RT 2-yr PFS: 96% High risk: 66Gy RT with cisplatin 2-yr PFS 90.5%

Fig. 3.3 Completed de-escalation clinical trials with TORS. All trials used TORS upfront with neck dissection followed by potentially de-escalated adjuvant therapy

authors conclude that aggressive RT dose de-escalation based on pathologic findings after surgery results in acceptable oncologic outcomes with little impact on swallowing or QOL [92].

A second completed prospective trial utilizing TORS in de-escalation of HPV-associated OPSCC was published by Swisher-McClure et al. in 2020 (the AVOID trial) [93]. This was a single-arm trial, including patients with T1–2, N1–3 (AJCC 7th edition staging system) who were treated with TORS and neck dissection. The included patients had negative margins and no adverse histopathologic features at the primary site, but required adjuvant RT or CRT based on lymph node involvement. Patients received dose-reduced RT (60–66 Gy) to involved neck levels, and 54 Gy to uninvolved but at-risk neck levels based on pathologic findings from neck dissection. Patients with ENE also received concurrent chemotherapy. The primary site was treated as an active avoidance structure during the planning process for adjuvant RT (the mean radiation dose to the primary site was 37 Gy). This study included 60 patients with a median follow-up time of 2.4 years. The authors report a 2-year local control rate of 98.3%, with only one patient having local recurrence. Two-year OS was 100%, with 2-year distant metastasis-free survival being 96% and 2-year PFS being 92% [93]. The authors also report that no patients required g-tube support during adjuvant therapy, and no patients were long-term g-tube dependent. The authors conclude that de-intensified adjuvant therapy after TORS for HPV-associated OPSCC that avoids the primary site can result in acceptable oncologic outcomes while minimizing treatment morbidity and side effects [93].

There are many ongoing clinical trial utilizing transoral surgery for the de-escalation of treatment for HPV-associated OPSCC in some capacity. The way surgery is implemented in the de-escalation pathway varies from trial to trial [87]. This section will discuss a number of ongoing trials using TORS, how the trials are structured, and the role that TORS is serving in the de-escalation approaches (Fig. 3.4). Information about these trials can be found on clinicaltrials.gov. ORATOR II (NCT03210103) is a phase II trial with an estimated accrual of 140 with one study arm that aims to de-escalate adjuvant ther-

Study	Low Risk	Intermediate risk	High Risk
PATHOS (N = 242)	pT1-2, pN0-1 with no adverse pathologic features Observed	pT1-3, pN2a-2b, LVI, PNI, and/or close margins (1-5mm) Randomized to adjuvant 50Gy RT or 60Gy RT	Positive margins (<1mm), ENE positive Randomized to adjuvant 60Gy RT or 60Gy CRT with cisplatin
DART-HPV (N = 227)	No low risk arm	Any pathologic risk factors, ENE negative Randomized to 30Gy CRT with docetaxel or 60Gy RT	ENE positive Randomized to 36Gy CRT with docetaxel or 60Gy CRT with cisplatin
DELPHI (N = 384)	pT1-2, pN0, no adverse features De-escalated RT at 49Gy	pT3-4, 1-3 positive lymph nodes De-escalated RT at 54Gy	Positive margins, ENE positive, ≥4 positive lymph nodes Standard dose concurrent CRT

Fig. 3.4 Ongoing de-escalation clinical trials with TORS. All trials use TORS upfront with neck dissection followed by potentially de-escalated adjuvant therapy

apy after TORS. It includes potentially resectable HPV-associated T1–2, N0–2 (AJCC 8th edition) OPSCC. The patients are randomized to either de-escalated CRT (60 Gy RT with concurrent cisplatin) or TORS with neck dissection with or without adjuvant de-escalated RT (50 Gy, as indicated by pathology). Estimated completion date is 2028. PATHOS (NCT02215265) is a phase III study of an estimated 242 participants with the goal of de-escalating adjuvant therapy after TORS and neck dissection based on risk-stratified pathologic findings. Eligible participants include T1–3, N0–2b (AJCC 7th edition) HPV-associated OPSCC, excluding current smokers. All participants undergo TORS and neck dissection. Low-risk patients (pT1–T2 N0–1 without adverse features) are observed. Intermediate-risk patients (T3, N2a–2b, PNI, LVI, close margins) are randomized to receiving either 50 Gy or 60 Gy adjuvant RT. High-risk patients (positive margins, ENE+) are randomized to either 60 Gy RT or 60 Gy CRT with concurrent cisplatin as adjuvant therapy. This study uses pathologic findings from surgery and risk stratification to direct adjuvant therapy. Risk stratification is a common way to direct de-escalated therapy and can be done post-operatively based on pathologic findings or pre-operatively based on patient and clinical staging characteristics [87].

Preliminary results of a highly anticipated trial, ECOG 3311 (NCT01898494), were recently published. This phase II trial included 353 patients who had T1–2, N1–2b (AJCC 7th edition) HPV-associated OPSCC. All patients underwent TORS with neck dissection and were then separated into one of four treatment arms based on pathology results. The low-risk group (arm A), defined as patients with pT1–T2, pN0–N1, and negative margins did not receive adjuvant therapy and were observed. The intermediate-risk patients (negative margins, microscopic ENE (≤1 mm ENE present), LVI, PNI, or 2–4 positive lymph nodes) were randomized to receive ether 50 Gy (arm B) or 60 Gy (arm C) adjuvant therapy. The high-risk patients (positive margins, ENE >1 mm, or ≥5 positive lymph nodes) had adjuvant CRT with 66 Gy and concurrent cisplatin (arm D). The primary endpoint was 2-year PFS for the patients in arm B versus arm C. The median follow-up for all patients was 31.8 months. Two-year PFS for arm A was 93.9%. Two-year PFS for arms B and C were 95% and 95.9%, respectively. Two-year PFS for arm D was 90.5%. The authors report 17 progression events, 7 were locoregional and 10 were distant recurrences (one in arm A, two in arm B, four in arm C, and three in arm D). The authors also report grade III/IV treatment-related adverse events for each arm. Arm A had 15%/2% grade III/IV adverse events. Arms B and C had 13%/2% and 25%/0%, respectively. The abstract does not report adverse events for arm D. The authors conclude that patients with low-risk disease had favorable 2-year PFS, and those with intermediate-risk disease had equivalent 2-year PFS with 50 Gy and 60 Gy adjuvant RT. [94] They recommend that surgery followed by 50 Gy RT should be compared to optimal nonsurgical therapy in a phase III trial. These preliminary results are indeed very promising, and further data and analysis from this trial will hopefully shed more light on the functional differences between the four treatment arms.

Two other notable ongoing trials that use surgery to risk-stratify and direct de-escalated adjuvant therapy include the DART-HPV and DELPHI studies [87]. The DART-HPV trial is a phase III trial with an anticipated 227 enrolled participants. Eligible subjects include surgically resectable HPV-associated OPSCC with at least one risk factor. Possible risk factors include LVI, PNI, single node ≥3 cm, multiple positive lymph nodes, ≥pT3 stage, or ENE+. This is a follow-up phase II trial to the previously discussed MC1273 study. All patients undergo TORS and neck dissection, then are divided into intermediate- or high-risk arms based on the presence of certain risk factors. Risk factors that place patients into the intermediate- or high-risk arms are not clearly defined on clinicaltrials.gov for this study. However, in the MC1273 study, the high-risk arm was only ENE-positive patients, and all other risk factors placed patients in the intermediate-risk group. This may or may not be similar for the DART-HPV trial. The intermediate-risk patients will be randomized to receive either 60 Gy adjuvant RT alone or 30 Gy RT with concurrent docetaxel. The high-risk patients will be randomized to receive either 60 Gy with concurrent cisplatin or 36 Gy RT with concurrent docetaxel. This trial is expected to be completed in 2024. The DELPHI trial (NCT03396718) is a phase I trial with an estimated enrollment of 384 participants. This study includes all HPV-associated and surgically resectable OPSCC. Patients undergo TORS and neck dissection followed by risk stratification based on pathologic findings. Low-risk patients (pT1–2, pN0, no adverse features) undergo dose-reduced RT at 49 Gy. The intermediate-risk arm (pT3–T4, 1–3 lymph nodes involved, other pathologic risk factors) will also have adjuvant RT dose-reduction, but at 54 Gy. Finally, the high-risk group (ENE, positive margins ≥4 lymph nodes involved) will have 60–66 Gy CRT with concurrent chemotherapy.

There are two ongoing studies that use induction chemotherapy to direct definitive treatment, including TORS, for HPV-associated OPSCC (Fig. 3.5). The first is OPTIMA-II (NCT0310782), a phase II study that is a follow-up on the OPTIMA I trial from the University of Chicago. This trial currently has 76 enrolled patients, eligible subjects include T3–4 or N2–3 (advanced disease, based on AJCC 7th edition) HPV-associated OPSCC. All patients receive induction chemotherapy (IC) with nab-paclitaxel, carboplatin, and nivolumab. Patients are risk-stratified into low risk (T1–2, N2a–b, <10 pack-year smoking history) or high risk (T4, bulky N2b, N2c–N3, ≥10 pack-year smoking). Patients in the low-risk arm that have a >50% response rate to IC (based on RECIST criteria) and are TORS-eligible then undergo TORS with neck dissection and reduced adjuvant RT when indicated based on pathologic risk factors. Patients with PNI and/or LVI on histopathologic analysis will receive 40 Gy adjuvant RT after TORS. Patients with positive surgical margins and/or ENE+ will receive 44 Gy adjuvant RT. Patients with none of the above adverse features will not receive adjuvant RT after TORS.

Lastly, the Quarterback 2b trial (NCT02945631) is a phase I trial with an estimated 65 patients. It includes stage III-IV (AJCC 7th edition) HPV-associated OPSCC with ≤20 pack-year history and not currently smoking. All patients receive induction chemotherapy with docetaxel, cisplatin, and 5-fluorouracil. Patients are stratified into low- or high-risk based on their response to IC. Patients with partial or complete response to IC (low-risk) are then treated with dose-reduced CRT at 56 Gy with concurrent carboplatin. Patients who demonstrate no response

Study	Induction Chemotherapy	Low Risk	High Risk
OPTIMA II (N = 76)	3 cycles of nab-Paclitaxel, carboplatin, nivolumab	T1-2, N2a-b, <10 pack year smoking history >50% response and TORS eligible: TORS plus neck dissection +/- reduced RT >50% response and TORS ineligible: 50Gy RT 30-50% response: 50Gy CRT with cisplatin <30% response: 75Gy CRT with cisplatin	T4, bulky N2b, N2c or N3, ≥10 pack year smoking history >50% response: 50Gy CRT with cisplatin <50% response: 75Gy CRT with cisplatin
Quarterback 2b (N = 65)	Docetaxel, cisplatin, 5-fluorouracil	Complete response or partial response 56Gy CRT with carboplatin	No response or progression of disease Surgery plus adjuvant CRT or standard dose CRT with carboplatin

Fig. 3.5 Ongoing de-escalation clinical trials with induction chemotherapy and TORS. All trials use induction chemotherapy followed by potentially de-escalated nonsurgical treatment and/or surgical excision

to IC or have progression while on IC (high-risk) are treated with either surgical resection or standard dose CRT with concurrent carboplatin. This study is expected to be completed in 2023. It is evident based on the ongoing studies that de-escalation for HPV-associated OPSCC may be achievable through a variety of approaches. Performing surgery instead of standard-strength CRT, de-escalating adjuvant therapy based on pathologic findings, and using RT with dose-reduction instead of CRT are all methods of de-escalation that are actively being researched. Transoral surgery plays a key role in many of these trials and likely will continue to be an instrumental part of the treatment options for both HPV-associated and HPV-unrelated OPSCC.

Conclusions

Oropharyngeal squamous cell carcinoma associated with the human papillomavirus is increasing worldwide. The use of transoral, minimally invasive surgery has offered advances in the treatment of this disease. Transoral surgery can be performed using laser microsurgery or robotic surgery. Both the approaches have over a decade of data to support their use, especially in early stage disease. Oncologic outcomes with TORS and TLM are equivalent to standard of care radiotherapy or chemoradiotherapy. These minimally invasive surgical approaches may have the benefit of fewer long-term adverse effects on swallowing and quality of life compared to traditional nonsurgical treatments. Multiple ongoing prospective trials are utilizing TORS as a means to guide treatment de-escalation approaches for HPV-associated oropharyngeal cancer.

References

1. O'Malley BW, Weinstein GS, Snyder W, Hockstein NG. Transoral robotic surgery (TORS) for base of tongue neoplasms. Laryngoscope. 2006;116(8):1465–72.
2. Cracchiolo JR, Baxi SS, Morris LG, et al. Increase in primary surgical treatment of T1 and T2 oropharyngeal squamous cell carcinoma and rates of adverse pathologic features: National Cancer Data Base. Cancer. 2016;122(10):1523–32.

3. Moore EJ, Hinni ML. Critical review: transoral laser microsurgery and robotic-assisted surgery for oropharynx cancer including human papillomavirus-related cancer. Int J Radiat Oncol Biol Phys. 2013;85(5):1163–7.
4. Holsinger FC, McWhorter AJ, Ménard M, Garcia D, Laccourreye O. Transoral lateral oropharyngectomy for squamous cell carcinoma of the tonsillar region: I. Technique, complications, and functional results. Arch Otolaryngol Head Neck Surg. 2005;131(7):583–91.
5. Steiner W, Fierek O, Ambrosch P, Hommerich CP, Kron M. Transoral laser microsurgery for squamous cell carcinoma of the base of the tongue. Arch Otolaryngol Head Neck Surg. 2003;129(1):36–43.
6. Steiner W, Ambrosch P, Hess CF, Kron M. Organ preservation by transoral laser microsurgery in piriform sinus carcinoma. Otolaryngol Head Neck Surg. 2001;124(1):58–67.
7. Jefferson GD, Frey H. Open versus robotic surgery for oropharyngeal cancer. Otolaryngol Clin N Am. 2020;53(6):995–1003.
8. Dziegielewski PT, Mlynarek AM, Dimitry J, Harris JR, Seikaly H. The mandibulotomy: friend or foe? Safety outcomes and literature review. Laryngoscope. 2009;119(12):2369–75.
9. Golusiński W, Golusińska-Kardach E. Current role of surgery in the management of oropharyngeal cancer. Front Oncol. 2019;9:388.
10. Wolf GT, Fisher SG, Hong WK, et al. Induction chemotherapy plus radiation compared with surgery plus radiation in patients with advanced laryngeal cancer. N Engl J Med. 1991;324(24):1685–90.
11. Parsons JT, Mendenhall WM, Stringer SP, et al. Squamous cell carcinoma of the oropharynx: surgery, radiation therapy, or both. Cancer. 2002;94(11):2967–80.
12. Goepfert RP, Yom SS, Ryan WR, Cheung SW. Development of a chemoradiation therapy toxicity staging system for oropharyngeal carcinoma. Laryngoscope. 2015;125(4):869–76.
13. Clark JM, Holmes EM, O'Connell DA, Harris J, Seikaly H, Biron VL. Long-term survival and swallowing outcomes in advanced stage oropharyngeal squamous cell carcinomas. Papillomavirus Res. 2019;7:1–10.
14. Bosetti C, Carioli G, Santucci C, et al. Global trends in oral and pharyngeal cancer incidence and mortality. Int J Cancer. 2020;147(4):1040–9.
15. Chaturvedi AK, Anderson WF, Lortet-Tieulent J, et al. Worldwide trends in incidence rates for oral cavity and oropharyngeal cancers. J Clin Oncol. 2013;31(36):4550–9.
16. Stein AP, Saha S, Kraninger JL, et al. Prevalence of human papillomavirus in oropharyngeal cancer: a systematic review. Cancer J. 2015;21(3):138–46.
17. Chaturvedi AK, Engels EA, Pfeiffer RM, et al. Human papillomavirus and rising oropharyngeal cancer incidence in the United States. J Clin Oncol. 2011;29(32):4294–301.
18. Gupta B, Johnson NW, Kumar N. Global epidemiology of head and neck cancers: a continuing challenge. Oncology. 2016;91(1):13–23.
19. Rehm J, Manthey J, Shield KD, Ferreira-Borges C. Trends in substance use and in the attributable burden of disease and mortality in the WHO European Region, 2010-16. Eur J Pub Health. 2019;29(4):723–8.
20. (CDC) CfDCaP. Cigarette smoking among adults—United States, 2006. MMWR Morb Mortal Wkly Rep. 2007;56(44):1157–61.
21. Chaturvedi AK, Engels EA, Anderson WF, Gillison ML. Incidence trends for human papillomavirus-related and -unrelated oral squamous cell carcinomas in the United States. J Clin Oncol. 2008;26(4):612–9.
22. Mehanna H, Franklin N, Compton N, et al. Geographic variation in human papillomavirus-related oropharyngeal cancer: data from 4 multinational randomized trials. Head Neck. 2016;38(Suppl 1):E1863–9.
23. Fakhry C, Westra WH, Li S, et al. Improved survival of patients with human papillomavirus-positive head and neck squamous cell carcinoma in a prospective clinical trial. J Natl Cancer Inst. 2008;100(4):261–9.
24. Rischin D, Young RJ, Fisher R, et al. Prognostic significance of p16INK4A and human papillomavirus in patients with oropharyngeal cancer treated on TROG 02.02 phase III trial. J Clin Oncol. 2010;28(27):4142–8.
25. Dayyani F, Etzel CJ, Liu M, Ho CH, Lippman SM, Tsao AS. Meta-analysis of the impact of human papillomavirus (HPV) on cancer risk and overall survival in head and neck squamous cell carcinomas (HNSCC). Head Neck Oncol. 2010;2:15.
26. Nygård M, Aagnes B, Bray F, Møller B, Mork J. Population-based evidence of increased survival in human papillomavirus-related head and neck cancer. Eur J Cancer. 2012;48(9):1341–6.
27. Pfister DG, Spencer S, Adelstein D, et al. Head and neck cancers, version 2.2020, NCCN clinical practice guidelines in oncology. J Natl Compr Cancer Netw. 2020;18(7):873–98.
28. Jäckel MC, Martin A, Steiner W. Twenty-five years experience with laser surgery for head and neck tumors: report of an international symposium, Göttingen, Germany, 2005. Eur Arch Otorhinolaryngol. 2007;264(6):577–85.
29. Sapundzhiev NR, Dünne AA, Ramaswamy A, Sitter H, Davis RK, Werner JA. Lymph node metastasis in an animal model: effect of piecemeal laser surgical resection. Lasers Surg Med. 2005;36(5):371–6.
30. Grant DG, Salassa JR, Hinni ML, Pearson BW, Perry WC. Carcinoma of the tongue base treated by transoral laser microsurgery, part one: untreated tumors, a prospective analysis of oncologic and functional outcomes. Laryngoscope. 2006;116(12):2150–5.
31. Henstrom DK, Moore EJ, Olsen KD, Kasperbauer JL, McGree ME. Transoral resection for squamous cell carcinoma of the base of the tongue. Arch Otolaryngol Head Neck Surg. 2009;135(12):1231–8.
32. Haughey BH, Hinni ML, Salassa JR, et al. Transoral laser microsurgery as primary treatment for advanced-

stage oropharyngeal cancer: a United States multicenter study. Head Neck. 2011;33(12):1683–94.
33. Weinstein GS, O'Malley BW, Rinaldo A, Silver CE, Werner JA, Ferlito A. Understanding contraindications for transoral robotic surgery (TORS) for oropharyngeal cancer. Eur Arch Otorhinolaryngol. 2015;272(7):1551–2.
34. de Almeida JR, Li R, Magnuson JS, et al. Oncologic outcomes after transoral robotic surgery: a multi-institutional study. JAMA Otolaryngol Head Neck Surg. 2015;141(12):1043–51.
35. Hanna J, Morse E, Brauer PR, Judson B, Mehra S. Positive margin rates and predictors in transoral robotic surgery after federal approval: a national quality study. Head Neck. 2019;41(9):3064–72.
36. Poon H, Li C, Gao W, Ren H, Lim CM. Evolution of robotic systems for transoral head and neck surgery. Oral Oncol. 2018;87:82–8.
37. Van Abel KM, Yin LX, Price DL, Janus JR, Kasperbauer JL, Moore EJ. One-year outcomes for da Vinci single port robot for transoral robotic surgery. Head Neck. 2020;42(8):2077–87.
38. Chan JYK, Tsang RK, Holsinger FC, et al. Prospective clinical trial to evaluate safety and feasibility of using a single port flexible robotic system for transoral head and neck surgery. Oral Oncol. 2019;94:101–5.
39. Holsinger FC, Magnuson JS, Weinstein GS, et al. A next-generation single-port robotic surgical system for transoral robotic surgery: results from prospective nonrandomized clinical trials. JAMA Otolaryngol Head Neck Surg. 2019;145(11):1027–34.
40. Park YM, Kim DH, Kang MS, et al. The first human trial of transoral robotic surgery using a single-port robotic system in the treatment of laryngo-pharyngeal cancer. Ann Surg Oncol. 2019;26(13):4472–80.
41. Moore EJ, Olsen KD, Kasperbauer JL. Transoral robotic surgery for oropharyngeal squamous cell carcinoma: a prospective study of feasibility and functional outcomes. Laryngoscope. 2009;119(11):2156–64.
42. Weinstein GS, O'Malley BW, Snyder W, Sherman E, Quon H. Transoral robotic surgery: radical tonsillectomy. Arch Otolaryngol Head Neck Surg. 2007;133(12):1220–6.
43. El Asmar M, Tsai HL, Fakhry C, et al. The prognostic impact of pathologic lymph nodes in HPV-positive oropharyngeal cancers. Oral Oncol. 2019;89:23–9.
44. Chen MM, Roman SA, Kraus DH, Sosa JA, Judson BL. Transoral robotic surgery: a population-level analysis. Otolaryngol Head Neck Surg. 2014;150(6):968–75.
45. Chung EJ, Oh JI, Choi KY, et al. Pattern of cervical lymph node metastasis in tonsil cancer: predictive factor analysis of contralateral and retropharyngeal lymph node metastasis. Oral Oncol. 2011;47(8):758–62.
46. Olzowy B, Tsalemchuk Y, Schotten KJ, Reichel O, Harréus U. Frequency of bilateral cervical metastases in oropharyngeal squamous cell carcinoma: a retrospective analysis of 352 cases after bilateral neck dissection. Head Neck. 2011;33(2):239–43.
47. Lim YC, Lee SY, Lim JY, et al. Management of contralateral N0 neck in tonsillar squamous cell carcinoma. Laryngoscope. 2005;115(9):1672–5.
48. Kato MG, Ellis MA, Nguyen SA, Day TA. Predictors of contralateral-bilateral nodal disease in oropharyngeal cancer: a National Cancer Data Base Study. Head Neck. 2018;40(2):338–48.
49. Miccio JA, Verma V, Kelly J, et al. Impact of contralateral lymph nodal involvement and extranodal extension on survival of surgically managed HPV-positive oropharyngeal cancer staged with the AJCC eighth edition. Oral Oncol. 2019;99:104447.
50. Möckelmann N, Busch CJ, Münscher A, Knecht R, Lörincz BB. Timing of neck dissection in patients undergoing transoral robotic surgery for head and neck cancer. Eur J Surg Oncol. 2015;41(6):773–8.
51. Cannon RB, Houlton JJ, Patel S, et al. Patterns of cervical node positivity, regional failure rates, and fistula rates for HPV+ oropharyngeal squamous cell carcinoma treated with transoral robotic surgery (TORS). Oral Oncol. 2018;86:296–300.
52. Frenkel CH, Yang J, Zhang M, Altieri MS, Telem DA, Samara GJ. Compared outcomes of concurrent versus staged transoral robotic surgery with neck dissection. Otolaryngol Head Neck Surg. 2017;157(5):791–7.
53. Kucur C, Durmus K, Gun R, et al. Safety and efficacy of concurrent neck dissection and transoral robotic surgery. Head Neck. 2016;38(Suppl 1):E519–23.
54. Moore EJ, Olsen KD, Martin EJ. Concurrent neck dissection and transoral robotic surgery. Laryngoscope. 2011;121(3):541–4.
55. Repanos C, Mirza AH, George M, Karkos PD. Timing of neck dissection in association with transoral surgery: a systematic review. Head Neck. 2017;39(5):1020–32.
56. Mandal R, Duvvuri U, Ferris RL, Kaffenberger TM, Choby GW, Kim S. Analysis of post-transoral robotic-assisted surgery hemorrhage: frequency, outcomes, and prevention. Head Neck. 2016;38(Suppl 1):E776–82.
57. Bollig CA, Gilley DR, Ahmad J, Jorgensen JB. Prophylactic arterial ligation following transoral robotic surgery: a systematic review and meta-analysis. Head Neck. 2020;42(4):739–46.
58. Sharbel DD, Abkemeier M, Sullivan J, et al. Transcervical arterial ligation for prevention of postoperative hemorrhage in transoral oropharyngectomy: systematic review and meta-analysis. Head Neck. 2021;43(1):334–44.
59. Stokes W, Ramadan J, Lawson G, Ferris FRL, Holsinger FC, Turner MT. Bleeding complications after transoral robotic surgery: a meta-analysis and systematic review. Laryngoscope. 2021;131(1):95–105.
60. Pignon JP, Bourhis J, Domenge C, Designé L. Chemotherapy added to locoregional treatment for head and neck squamous-cell carcinoma: three meta-analyses of updated individual data. MACH-NC Collaborative Group. Meta-Analysis of Chemotherapy on Head and Neck Cancer. Lancet. 2000;355(9208):949–55.

61. Eisbruch A, Lyden T, Bradford CR, et al. Objective assessment of swallowing dysfunction and aspiration after radiation concurrent with chemotherapy for head-and-neck cancer. Int J Radiat Oncol Biol Phys. 2002;53(1):23–8.
62. Nguyen NP, Moltz CC, Frank C, et al. Dysphagia following chemoradiation for locally advanced head and neck cancer. Ann Oncol. 2004;15(3):383–8.
63. Hunter KU, Schipper M, Feng FY, et al. Toxicities affecting quality of life after chemo-IMRT of oropharyngeal cancer: prospective study of patient-reported, observer-rated, and objective outcomes. Int J Radiat Oncol Biol Phys. 2013;85(4):935–40.
64. Ang KK, Harris J, Wheeler R, et al. Human papillomavirus and survival of patients with oropharyngeal cancer. N Engl J Med. 2010;363(1):24–35.
65. Bossi P, Orlandi E, Miceli R, et al. Treatment-related outcome of oropharyngeal cancer patients differentiated by HPV dictated risk profile: a tertiary cancer centre series analysis. Ann Oncol. 2014;25(3):694–9.
66. Lörincz BB, Jowett N, Knecht R. Decision management in transoral robotic surgery: indications, individual patient selection, and role in the multidisciplinary treatment for head and neck cancer from a European perspective. Head Neck. 2016;38(Suppl 1):E2190–6.
67. O'Hara J, Goff D, Cocks H, et al. One-year swallowing outcomes following transoral laser microsurgery +/− adjuvant therapy versus primary chemoradiotherapy for advanced stage oropharyngeal squamous cell carcinoma. Clin Otolaryngol. 2016;41(2):169–75.
68. Hutcheson KA, Holsinger FC, Kupferman ME, Lewin JS. Functional outcomes after TORS for oropharyngeal cancer: a systematic review. Eur Arch Otorhinolaryngol. 2015;272(2):463–71.
69. Yeh DH, Tam S, Fung K, et al. Transoral robotic surgery vs. radiotherapy for management of oropharyngeal squamous cell carcinoma—a systematic review of the literature. Eur J Surg Oncol. 2015;41(12):1603–14.
70. More YI, Tsue TT, Girod DA, et al. Functional swallowing outcomes following transoral robotic surgery vs primary chemoradiotherapy in patients with advanced-stage oropharynx and supraglottis cancers. JAMA Otolaryngol Head Neck Surg. 2013;139(1):43–8.
71. Dziegielewski PT, Teknos TN, Durmus K, et al. Transoral robotic surgery for oropharyngeal cancer: long-term quality of life and functional outcomes. JAMA Otolaryngol Head Neck Surg. 2013;139(11):1099–108.
72. Lörincz BB, Möckelmann N, Busch CJ, et al. Two-year survival analysis of 50 consecutive head and neck cancer patients treated with transoral robotic surgery in a single European centre. Ann Surg Oncol. 2015;22(Suppl 3):S1028–33.
73. Hararah MK, Stokes WA, Jones BL, et al. Nomogram for preoperative prediction of nodal extracapsular extension or positive surgical margins in oropharyngeal squamous cell carcinoma. Oral Oncol. 2018;83:73–80.
74. White HN, Moore EJ, Rosenthal EL, et al. Transoral robotic-assisted surgery for head and neck squamous cell carcinoma: one- and 2-year survival analysis. Arch Otolaryngol Head Neck Surg. 2010;136(12):1248–52.
75. Pignon JP, le Maître A, Maillard E, Bourhis J, Group M-NC. Meta-analysis of chemotherapy in head and neck cancer (MACH-NC): an update on 93 randomised trials and 17,346 patients. Radiother Oncol. 2009;92(1):4–14.
76. Moore EJ, Olsen SM, Laborde RR, et al. Long-term functional and oncologic results of transoral robotic surgery for oropharyngeal squamous cell carcinoma. Mayo Clin Proc. 2012;87(3):219–25.
77. Dalton CL, Milinis K, Houghton D, et al. Transoral laser microsurgery and radiotherapy for oropharyngeal squamous cell carcinoma: equitable survival and enhanced function compared with contemporary standards of care. Eur J Surg Oncol. 2020;46(11):2042–9.
78. Chen SY, Sinha P, Last A, et al. Outcomes of patients with single-node metastasis of human papillomavirus-related oropharyngeal cancer treated with transoral surgery. JAMA Otolaryngol Head Neck Surg. 2021;147(1):16–22.
79. Mahmoud O, Sung K, Civantos FJ, Thomas GR, Samuels MA. Transoral robotic surgery for oropharyngeal squamous cell carcinoma in the era of human papillomavirus. Head Neck. 2018;40(4):710–21.
80. Li H, Torabi SJ, Park HS, et al. Clinical value of transoral robotic surgery: nationwide results from the first 5 years of adoption. Laryngoscope. 2019;129(8):1844–55.
81. Nguyen AT, Luu M, Mallen-St Clair J, et al. Comparison of survival after transoral robotic surgery vs nonrobotic surgery in patients with early-stage oropharyngeal squamous cell carcinoma. JAMA Oncol. 2020;6(10):1555–62.
82. De Virgilio A, Costantino A, Mercante G, et al. Transoral robotic surgery and intensity-modulated radiotherapy in the treatment of the oropharyngeal carcinoma: a systematic review and meta-analysis. Eur Arch Otorhinolaryngol. 2020.
83. de Almeida JR, Byrd JK, Wu R, et al. A systematic review of transoral robotic surgery and radiotherapy for early oropharynx cancer: a systematic review. Laryngoscope. 2014;124(9):2096–102.
84. Castellano A, Sharma A. Systematic review of validated quality of life and swallow outcomes after transoral robotic surgery. Otolaryngol Head Neck Surg. 2019;161(4):561–7.
85. Hutcheson KA, Warneke CL, Yao CMKL, et al. Dysphagia after primary transoral robotic surgery with neck dissection vs nonsurgical therapy in patients with low- to intermediate-risk oropharyngeal cancer. JAMA Otolaryngol Head Neck Surg. 2019;145(11):1053–63.
86. Nichols AC, Theurer J, Prisman E, et al. Radiotherapy versus transoral robotic surgery and neck dissection for oropharyngeal squamous cell carcinoma

(ORATOR): an open-label, phase 2, randomised trial. Lancet Oncol. 2019;20(10):1349–59.
87. Bigelow EO, Seiwert TY, Fakhry C. Deintensification of treatment for human papillomavirus-related oropharyngeal cancer: current state and future directions. Oral Oncol. 2020;105:104652.
88. Seiwert TY, Foster CC, Blair EA, et al. OPTIMA: a phase II dose and volume de-escalation trial for human papillomavirus-positive oropharyngeal cancer. Ann Oncol. 2019;30(2):297–302.
89. Marur S, Li S, Cmelak AJ, et al. E1308: phase II trial of induction chemotherapy followed by reduced-dose radiation and weekly cetuximab in patients with HPV-associated resectable squamous cell carcinoma of the oropharynx- ECOG-ACRIN Cancer Research Group. J Clin Oncol. 2017;35(5):490–7.
90. Chera BS, Amdur RJ, Tepper JE, et al. Mature results of a prospective study of deintensified chemoradiotherapy for low-risk human papillomavirus-associated oropharyngeal squamous cell carcinoma. Cancer. 2018;124(11):2347–54.
91. Chen AM, Felix C, Wang PC, et al. Reduced-dose radiotherapy for human papillomavirus-associated squamous-cell carcinoma of the oropharynx: a single-arm, phase 2 study. Lancet Oncol. 2017;18(6):803–11.
92. Ma DJ, Price KA, Moore EJ, et al. Phase II evaluation of aggressive dose de-escalation for adjuvant chemoradiotherapy in human papillomavirus-associated oropharynx squamous cell carcinoma. J Clin Oncol. 2019;37(22):1909–18.
93. Swisher-McClure S, Lukens JN, Aggarwal C, et al. A phase 2 trial of alternative volumes of oropharyngeal irradiation for de-intensification (AVOID): omission of the resected primary tumor bed after transoral robotic surgery for human papilloma virus-related squamous cell carcinoma of the oropharynx. Int J Radiat Oncol Biol Phys. 2020;106(4):725–32.
94. Ferris RL, Flamand Y, Weinstein GS, et al. Transoral robotic surgical resection followed by randomization to low- or standard-dose IMRT in resectable p16+ locally advanced oropharynx cancer: a trial of the ECOG-ACRIN Cancer Research Group (E3311). J Clin Oncol. 2020;38(15_suppl):6500.

Advances in Radiation: Ion Therapy and Advanced Techniques

4

Bhanu P. Venkatesulu, Prashanth Girdhar, Henry S. Park, William M. Mendenhall, and Vivek Verma

Charged Particle Therapy

Proton Beam Therapy

Background

Protons are heavy particles with a unique dose deposition profile, in that the maximal dose deposition occurs at a fixed depth (Bragg peak), past which there is a steep dose fall off with minimal exit dose [1]. The initial work to implement PBT for therapeutic purposes arose from the pioneering efforts made at the Harvard cyclotron laboratory, Berkeley laboratory, and in Uppsala, Sweden in the 1940s and 1950s [2]. The world's first hospital-based PBT facility was opened at Clatterbridge Centre for Oncology in the UK in 1989 for treating ocular tumors [3]. As of the time of writing this chapter, there are 102 PBT facilities in the world, with 41 of them in the United States alone [4].

PBT has evolved from passive scattered beam delivery (analogous to forward-planned 3DCRT) to pencil beam scanning delivery (analogous to inverse-planned IMRT). The monoenergetic proton beam that exits the cyclotron has a sharp Bragg peak; the use of brass apertures, modulator wheels, and scattering foils help to create a spread-out Bragg peak that is planned to conformally treat the tumor target volume. This form of PBT is referred to as passive scattered proton therapy (PSPT).

PSPT is fraught with technical complexities when treating head and neck tumors: (1) lateral dose conformality in PSPT is not optimal and is largely achieved by brass apertures; (2) range uncertainties exist, which is only partially offset by compensator smearing; (3) achieving adequate conformality for complex tumor volumes necessitate beam patching or other measures; and (4) in order to achieve uniform tumor dose deposition at the Bragg peak, energy degradation is required to spread out the Bragg peak, which can result in higher subcutaneous dose deposition.

Many of these problems are addressed by pencil beam scanning (PBS), which can magnetically scan narrow pencil beams in various

B. P. Venkatesulu
Department of Radiation Oncology, Stritch School of Medicine, Loyola University, Chicago, Illinois and Edward Hines Veterans Affairs Hospital, Chicago, IL, USA

P. Girdhar
Department of Radiation Oncology, Tata Memorial Centre, Varanasi, India

H. S. Park
Department of Therapeutic Radiology, Yale School of Medicine, Smilow Cancer Hospital, New Haven, CT, USA

W. M. Mendenhall
Department of Radiation Oncology, University of Florida College of Medicine, Gainesville, FL, USA

V. Verma (✉)
Department of Radiation Oncology, University of Texas M.D. Anderson Cancer Center, Houston, TX, USA

R. A. Chandra, R. J. Li (eds.), *Multidisciplinary Management of Head and Neck Cancer*,
https://doi.org/10.1007/978-3-031-05973-5_4

layers of tumor tissue [5]. This facilitates intensity modulation and reduces the need for beam accessories such as those used for PSPT, which makes PBS the optimal tool for treating complex head and neck target volumes and multiple dose levels. PBS can be optimized to treat head and neck fields by single-field or multi-field optimization, which is referred to as intensity-modulated proton therapy (IMPT). IMPT allows the use the spread-out Bragg peaks of various beams to undergo simultaneous modulation of beam intensities, so that each beam treats the target volume heterogeneously while maintaining a uniform cumulative dose to a larger target volume [6].

Rationale

Photon IMRT has reduced the incidence of acute complications such as mucositis, dermatitis, xerostomia, as well as the incidence of late complications such as dysphagia, feeding tube dependency, and hypothyroidism compared to 3DCRT or 2DRT. In spite of significant reductions in toxicity, the rates of acute and chronic complications can still be as high as 40–50% with IMRT [7]. Additionally, the finding that patients with low- and intermediate-risk HPV-positive oropharyngeal cancers can have 3-year overall survival as high as 90–95% has shifted the spotlight on cancer survivorship and enhancing quality of life (QOL) with mitigation of long-term complications [8]. The complexity of head and neck target volumes and the close apposition of normal structures critical for swallowing, breathing, articulation of speech, and salivation has increased interest in PBT because the Bragg peak (point of maximal dose deposition) can be strategically placed based on the location of normal tissues that are desired to be spared of radiation.

Clinical Evidence

Oropharyngeal Cancer

The initial report on the utility of IMPT for oropharyngeal cancer was a prospective patient-reported QOL study comparing 50 patients treated with IMPT to 100 patients treated with IMRT who were matched for baseline characteristics. The majority of the patients were HPV-positive, nearly 40% received induction chemotherapy, and 80% of the patients received bilateral neck radiation. There were no differences in terms of overall survival (OS) or progression-free survival (PFS). The percutaneous gastrostomy tube (PEG) rate was 24% during RT, 12% at 3 months, and 2% at 1 year in the IMPT arm as compared to 38%, 23%, and 8% in the IMRT arm, respectively. The primary outcome was the composite endpoint of grade 3 weight loss or PEG dependency, which was reduced with IMPT at 3 months and at 1 year. There was also a reduced incidence of xerostomia (42% after IMPT vs 61% after IMRT) at 3 months [9]. In an updated analysis of 103 patients treated with PBS compared to 429 patients treated with IMRT, Cao et al. reported a reduced incidence of xerostomia (6% after IMPT vs 20% after IMRT) at 24 and 32 months of follow-up; the high-dose regions (V25-V70) in the oral cavity correlated with xerostomia outcomes [10].

MD Anderson Cancer Center reported on comparative outcomes between IMPT (35 patients) and IMRT (46 patients) in oropharyngeal cancer patients receiving concurrent chemotherapy using the MD Anderson Symptom Inventory for Head and Neck Cancer (MDASI-HN). IMPT resulted in fewer changes related to appetite and taste, both during subacute and long-term phases of posttreatment follow-up [11].

The University of Pennsylvania experience was a prospective QOL study that compared PBS vs. IMRT after trans-oral robotic surgery in 64 oropharyngeal cancer patients. PBS was associated with reduced dental problems at 3 and 6 months after completion of radiation, as well as reduced xerostomia at 6 and 12 months. The main reason attributed to the reduced incidence of side effects with PBS was the reduced mean dose to the composite oral cavity structure (21.2 Gy compared to 35.1 Gy with IMRT). Patients also reported reduced head and neck pain with PBS at 12 months as compared to IMRT [12].

Nasopharyngeal Cancer

In one of the earlier reports of PBT for nasopharyngeal cancer, MD Anderson Cancer Center reported a comparative analysis of ten patients treated with IMPT and 20 patients treated with IMRT. Similar to the above, the mean dose to the oral cavity (cutoff of 26 Gy) was associated with for the need for PEG placement. Twenty percent of patients in the IMPT group required a PEG, as compared to 65% in the IMRT arm [13].

Williams et al. reported a cohort of 26 patients with locally advanced nasopharyngeal cancer treated with IMPT. None of the patients experienced acute or late grade 4 or 5 toxicities; the outcomes were noteworthy, with 2-year locoregional control of 92% and overall survival of 85% [14]. McDonald et al. reported on 14 patients treated with PSPT and matched them with 26 patients treated with IMRT in patients with nasopharyngeal, nasal cavity, and paranasal sinus cancers. On multivariate analysis, PBT was associated with a reduced need for opioids as well as a lower rate of PEG dependence at completion of radiation as well as 3 months afterwards.

A phase 2 trial was conducted at the Massachusetts General Hospital evaluating cancer-specific outcomes and toxicity in 23 patients with stage III-IVB nasopharyngeal carcinoma treated with PBT and concurrent chemotherapy. The nasopharynx and upper neck lymphatic regions were treated with proton therapy and lower neck lymphatics with photon therapy. The 2-year OS was reported to be 100% and disease-free survival was 90%. Notably, there was no grade 3 xerostomia reported. A total of 48% of patients reported PEG placement during RT, but only one patient remained dependent on a PEG at 6 and 12 months [15]. In a similar study by Alterio et al., a mixed proton and photon approach resulted in a reduced incidence of mucositis and xerostomia compared to IMRT alone [16].

Sinonasal Malignancies

Yu et al. described 69 patients with sinonasal tumors as part of a proton collaborative group registry study. Forty-two patients received upfront RT and 27 were treated with re-RT. The median dose was 58.5 Gy (relative biological effectiveness, RBE) for upfront RT and 60 Gy (RBE) for the re-RT group. The freedom from disease progression was 84% in upfront RT and 47% in the re-RT groups, respectively. Eleven patients experienced acute grade 3 toxicities, with mucositis and pain being the most common. No late high-grade toxicities such as vision loss or brain necrosis were reported [17].

Memorial Sloan Kettering Cancer Center has published one of the largest series of nasal cavity and paranasal sinus patients treated with PSPT or PBS. They reported 68 RT-naïve and 18 re-RT patients, of whom 53% received IMPT. The 2-year local control rates were 83% and 77% for RT naïve and re-RT patients, respectively. IMPT significantly improved local control rates (91% vs. 72%) compared to PSPT, which could be related to more contemporary modalities of image guidance or improved target volume coverage with IMPT. Sixteen RT-naïve patients experienced grade 3 acute toxicities, with four of them experiencing late toxicities such as osteoradionecrosis, vision loss, soft tissue necrosis, and soft tissue fibrosis. Patients treated with re-RT experienced more late complications, with 11% having grade 3 or greater toxicities [18].

From a late toxicity standpoint, Zenda et al. reported on 90 patients with malignant tumors of the nasal cavity, paranasal sinuses, or skull base treated with postoperative or definitive intent PSPT. They reported grade 3 toxicities in 17 patients (19%) and grade 4 toxicities in six patients (7%), which included encephalomyelitis and optic nerve damage [19]. The Massachusetts General Hospital experience of 54 patients with stage III and IV squamous cell carcinomas of the nasal cavity and paranasal sinuses demonstrated no grade 4–5 toxicities. Seven patients developed grade 2 nasal stenosis, eight grade 2 neurologic toxicity, two grade 3 auditory toxicity, and one grade 3 bone toxicity [20].

Patel et al. reported a pooled analysis of 43 observational cohorts comparing charged particle therapy vs. photon therapy for malignancies of the nasal cavity and paranasal sinuses and showed that the pooled OS was significantly higher after charged particle therapy. Of the 43 cohorts, seven were proton-based treatments and three were mixed proton–photon therapy. Subgroup analysis also showed improved disease-free survival and locoregional control with charged particle therapy as compared to IMRT [21].

Salivary Gland Tumors

Some of the earliest reports of PBT pertain to salivary gland tumors. Memorial Sloan Kettering Cancer Center reported a single-institutional series of ipsilateral irradiation for salivary gland cancer or cutaneous squamous cell carcinoma. PBT was employed in 18 patients and IMRT in 23 patients. PBT reduced the maximum brainstem dose, maximum spinal cord dose, mean oral cavity, mean contralateral parotid, and mean contralateral submandibular gland dose, which translated into a significantly lower incidence of mucositis, nausea, and dysgeusia [22].

MD Anderson Cancer Center reported on 16 patients with adenoid cystic carcinoma treated with IMPT of whom 12 also received chemotherapy. The toxicity profile was favorable, with only three patients with acute grade 3 dermatitis and one with grade 3 mucositis. At last follow-up, 15 of 16 patients were without any evidence of disease [23].

One of the largest reports of PBT for salivary gland tumors comes from the Proton Therapy Collaborative Study Group. They reported on 105 patients, of whom 90 had parotid cancers and 15 had submandibular tumors. The median dose used was 66.5 Gy (RBE) in 33 fractions with concurrent chemotherapy used in 20% of the cases. The rates of acute complications were: dysgeusia, 4.8%; xerostomia, 7.6%; mucositis, 10.5%; and dysphagia, 10.5% [24].

Taken together, PBT in various head and neck subsites shows important potential in order to reduce acute and chronic complications and maintain similar disease control outcomes. However, the quality and quantity of evidence is low, as the majority of studies are single- or multi-institutional retrospective studies with inherent problems related to selection and/or publication biases. Nevertheless, the available data form the framework for prospective validation of PBT in enhancing QOL by reducing the probability of normal tissue complications.

Ongoing Clinical Trials

PBT was initially thought to be relatively cost-prohibitive. However, as with evolution of any modern technology, costs tend to decrease with time and more widespread availability. As PBT has expanded, so too have the number of phase 2/3 clinical trials examining the utility of PBT for head and neck malignancies (Table 4.1). A few notable trials will be briefly discussed.

There is an ongoing phase III noninferiority trial led by MD Anderson Cancer Center comparing IMPT vs. IMRT for oropharyngeal malignancies with a primary endpoint of PFS. The secondary objectives of the trial include both patient- and physician-reported outcomes for assessment of toxicity, quality-adjusted life-years, work productivity/impairment comparison, and cost–benefit economic analysis [25].

In Europe, there are two ongoing trials for oropharyngeal cancer—TORPEdO (TOxicity Reduction using Proton bEam therapy for Oropharyngeal cancer) and ARTSCAN V. The TORPEdO trial (UK) aims to randomize 180 patients in a 2:1 fashion to IMPT vs. IMRT. The primary endpoints of the study include a composite toxicity outcome measure of patient-reported University of Washington physical toxicity composite score, feeding tube dependency, or severe weight loss 12 months after treatment. The trial will also assess the cost-effectiveness of IMPT and will also prospectively validate an NTCP model as a biomarker for patient selection for IMPT [26]. The ARTSCAN V trial (Sweden) aims to enroll 100 patients and randomize to protons vs. photons for tonsillar cancers, clinical stage T1–2 (p16-positive or p16-negative) N0–1 (p16-positive)/N0-N2b (p16-negative) planned for ipsilateral neck radiotherapy. The primary aim of the study is to assess difference in acute and chronic complications of mucositis, pain, dysphagia, skin reactions, and other functional impairment between IMRT and proton therapy [27]. Lastly, another ongoing phase 2 trial is comparing photon vs. proton radiation for unilateral head and neck radiation for salivary gland, melanoma, and other skin cancers with a primary endpoint of acute grade 2 or greater toxicity [28].

Table 4.1 Summary of the ongoing randomized clinical trials evaluating proton therapy versus intensity-modulated radiation therapy for de novo nonmetastatic head and neck cancers

Reference	Phase	Sample size	Cancer type	Radiation dose	Primary endpoints
NCT04607694 (DAHANCA 35)	III	600	Squamous cell carcinoma of the pharynx or larynx	66–68 Gy in 33–34 fx over 6 weeks	Grade 2+ dysphagia and grade 4 xerostomia (6 months)
NCT02923570 (MSKCC)	II	132	Salivary gland cancer without the presence of extracapsular extension and/or positive surgical margin, skin cancer, melanoma	60–66 Gy in 30–33 fx over 6 weeks	Grade 2+ acute mucositis (1 year)
NCT01893307 (MDACC)	II/III	360	Oropharyngeal cancer	70 Gy in 33 fx over 6.5 weeks	Late grade 3+ toxicities (1.5 months to 2 years)
ISRCTN16424014 (TORPEdO, UK)	II	180	Oropharyngeal cancer	–	Patient-reported outcomes: (1) University of Washington physical toxicity composite score; (2) feeding tube dependence or severe weight loss (12 months)
NCT03829033 (ARTSCAN-V, Sweden)	II	100	Tonsillar cancers	–	Any acute (during therapy until 7 weeks from completion) and late (2–3 months through 5 years) effects

Gy Gray, *MSKCC* Memorial Sloan Kettering Cancer Center, *MDACC* MD Anderson Cancer Center

Lastly, two notable cooperative group trials are not specifically comparing photons versus PBT, but rather allow inclusion of PBT if desired. NRG HN001 is a randomized phase II/III trial for stage II-IVB nasopharyngeal cancer personalizing chemotherapy regimens based on EBV DNA status [29]. RTOG 1008 is a phase 2 study evaluating the role of postoperative radiation with or without weekly cisplatin for salivary gland tumors [30].

Carbon Ion Radiation Therapy

Background

As of June 2022, according to the Particle Therapy Co-Operative Group, there were only 13 active carbon ion radiation therapy (CIRT) facilities, all located in Europe (Germany, Austria, Italy) and Asia (Japan, China). Similar to PBT, CIRT has been studied for decades but has only relatively recently begun implementation in human subjects. Given its relative nascency in the oncologic arena, CIRT potentially represents the most advanced approach in the radiation oncology armamentarium, but remains without randomized evidence supporting its use in any disease site.

Nevertheless, CIRT for head and neck malignancies remains an active area of current investigation, with promising clinical data as well as ongoing trials. This section will examine CIRT from both the theoretical and experiential perspectives as a potential tool for the future, especially for technically challenging radiotherapeutic cases of the head and neck.

Rationale

Similar to PBT, CIRT is also marked by notable dosimetric advantages over photon therapy. This includes the aforementioned Bragg peak, which corresponds to the point of maximal dose deposition as the radiation beam traverses through tissue. Unlike protons, carbon ions experience defragmentation toward the end of the Bragg peak and as a result deposit a small (10–20%) proportion of dose distal to the Bragg peak. An additional dosimetric advantage of CIRT over PBT is the reduction of lateral scattering. This results in a sharper dose falloff in the lateral portions of the beam as compared to protons, which are lighter and have a relatively higher tendency to scatter throughout the periphery of the beam. This has implications for selecting beam angles; whereas direct beams are preferred for proton therapy, CIRT is often carried out using more orthogonal (sharper) beam angles in order to place the steep dose gradients in the lateral portion of the beam near radiation-sensitive organs.

The primary advantage of CIRT is that carbon ions are much more effective in creating DNA strand breaks than both protons and photons. Because radiation therapy acts on tumor cells by creating irreparable DNA double-strand breaks, CIRT is thought to have a higher "relative biological effectiveness" (RBE) than photon or proton therapy. RBE is a relative term, defined as 1.0 for photons. Whereas the RBE for protons is commonly thought to be only 1.1, the RBE of carbon ions is markedly higher and can range from 2.5 to 5 depending on the particular type of tissue and other factors [31].

Another theoretical advantage of CIRT is that it may better galvanize the immune system than conventional photon radiotherapy [32]. This notion is neither well understood nor well studied, but the role of the immune system in fighting cancer is becoming increasingly appreciated, along with the role of radiotherapy in enhancing immune activation and potentially improving outcomes in other disease sites [33]. Analogous studies are urgently needed in head and neck neoplasms in order to confirm the potential synergy between immune activation, radiotherapy, and CIRT.

Collectively, these advantages imply that CIRT may better allow for safer dose escalation and biological escalation of dose to tumor tissue, while being able to spare surrounding normal organs to a greater degree. In RT-specific terms, this is described as CIRT causing enhancement of the therapeutic ratio, which is long thought of as the "holy grail" in radiation oncology.

Patient Selection

With respect to head and neck malignancies, CIRT represents an attractive modality for several reasons. First, patients with head and neck cancers experience substantially more severe radiation-induced morbidities compared to other disease sites. The intricate anatomy of the head and neck results in significant amounts of scatter radiation from photons encroaching onto several normal organs (e.g., salivary glands, uninvolved oral cavity, pharyngeal constrictors), which can cause profound adverse events and alterations in quality-of-life. Taking advantage of the unique physical properties of heavy particles such as protons and carbon ions causes reduced dose exposure to normal organs. There are several dosimetric studies illustrating markedly reduced doses to several organs (e.g., parotid, optic apparatus, cochleae) with CIRT as compared to the standard-of-care photon techniques (e.g., intensity-modulated radiation therapy) and potentially even proton therapy [34, 35]. However, whether reduced dose to these organs results in a lower rate of clinician- or patient-reported toxicities remains undetermined.

Additionally, CIRT may better allow safe dose-escalation for patients with head and neck cancers. This is especially useful for bulky, invasive, and unresectable tumors in various sites of the head and neck, or for recurrent disease with or without previous irradiation. There are several factors that limit the maximum dose that can be delivered with photons for these patients, and the sharp dose fall-off (both along the Bragg peak and laterally along the beam) plays a key role in allowing for dose escalation. Moreover, the increased RBE of CIRT may be especially useful for large/unresectable tumors because it is well known that larger tumors are more difficult to control for a given radiotherapy dose; moreover, it is likely that large tumors have hypoxic areas, which are relatively radioresistant but less so to high RBE radiotherapy [36]. For recurrent cases, which presumably harbor more unfavorable and/or radioresistant tumor biology, the ability to deliver maximum safe doses is imperative; this is also the case in the previously irradiated head and neck, where patients are at high risk of developing high-grade toxicities.

Additional patient selection criteria in the treatment-naïve setting may include those most vulnerable to severe and/or long-term side effects from radiotherapy. These include both elderly/frail persons (who are often treated with suboptimal therapy as a result) and younger patients (who may be more likely affected by secondary malignancy risk) [37]. Postoperative patients who experience major surgical-related morbidities are also at high risk of developing further morbidities from adjuvant radiotherapy, and as a result, these patients may also benefit to a proportionally higher extent.

Clinical Evidence

The most widely utilized indications for CIRT based upon the existing literature is for re-irradiation and gross unresected disease, largely because meaningful therapeutic options for these patients is severely limited. There have been single- or multi-institutional retrospective reports of hundreds of patients reported for both of these indications. For instance, the German experience has described 229 patients with recurrent head and neck neoplasms status post CIRT reirradiation [38]. With a median follow-up of 29 months, the local control at 12 months (using death as a competing risk) was 60%, which is considerably higher than what one would expect from conventional photon re-irradiation. The median survival was 26 months, which is also substantially higher than expected in recurrent head and neck cancer patients. Late high-grade toxicities occurred in a nontrivial (15%) proportion of patients, but it is also no worse than expected with full-dose photon reirradiation. Additionally, the Japan Carbon-Ion Radiation Oncology Study Group has put forth several large-volume publications on definitive (with some adjuvant cases) therapy for sinonasal tumors, head and neck mucosal melanoma, adenoid cystic carcinoma (ACC) of the head and neck, and other neoplasms [39–41]. These high-volume, multicenter investigations have further supported the role of CIRT for advanced local disease with few other meaningful treatment options. Importantly, relatively larger-volume investigations of CIRT for rare head and neck tumors have also been reported, which is also valuable evidence for populations for whom prospective trials are unlikely to fully accrue [42, 43].

Table 4.2 illustrates pertinent details of prospective experiences of CIRT for head and neck malignancies. Most of these studies have evaluated recurrent disease and/or reirradiation [44, 45], salivary gland tumors [45, 46], or ACCs [46–48]. Importantly, every existing study has been nonrandomized and with heterogeneous patient populations, thereby making definitive interpretations difficult.

An early trial aiming to evaluate the maximum tolerated dose (MTD) now has long-term follow-up, showing that the MTD was 70.2 Gy (RBE) in 18 fractions over 6 weeks, or 64 Gy (RBE) in 16 fractions over 4 weeks. The latter has now been largely adopted as the standard dose and fractionation regimen in many centers in Asia [44]. The German experiences [45, 47, 49], however, have utilized a mixed photon–CIRT approach with a CIRT dose of 24 Gy (RBE) in eight fractions delivered initially, followed immediately by 50–54 Gy with IMRT. Collectively, the local control with this approach appears excellent (especially considering the high-risk nature of most enrolled patients), but out-of-field failures in these advanced patients remains a challenge and limits survival. Moreover, late toxicities seem relatively low, with an exception in the ACCEPT trial, which did not meet its primary endpoint likely because of concurrent cetuximab-related adverse events (the incidence of which was similar to historical data with cetuximab alone without CIRT)

Table 4.2 Published trials of carbon ion radiation therapy for head and neck neoplasms

Reference	Design	Sample size	Cancer type	Radiation dose	Systemic therapy	Median follow-up	Outcomes	High-grade RT-related toxicities
Mizoe et al. [44]	Phase I/II (dose escalation)	36	Mix of new and recurrent disease, mostly definitive	48.6–70.2 Gy (RBE) in 18 fractions (6 weeks); 52.8–64 Gy (RBE) in 16 fractions (4 weeks)	Not standardized and given in a minority of patients	90 months	5 years LC 75%, particularly high in salivary glands and melanoma, lower in pharyngeal and sinuses	Grade 3 skin reaction ($n = 7$) occurring in half of patients receiving 70.2 Gy (RBE) in 6 weeks and 67% of patients receiving 64 Gy (RBE) in 4 weeks
Jensen et al. [45]	Phase II	53	Advanced or metastatic malignant salivary gland tumors, incompletely resected or definitive RT	24 Gy (RBE) in 8 daily fractions followed by 50 Gy IMRT in 25 daily fractions	Not standardized and individually assessed	42 months	3 years LC 82%, 3y PFS 58%, 3 years OS 78%	Grade 3 mucositis in 26%, grade 3 dermatitis in 6%, one case of grade 4 carotid hemorrhage
Shirai et al. [48]	Observational	35	Mixed but 60% ACC; mix of adjuvant & definitive	All but 3 patients with 64 Gy (RBE) in 16 fractions	Not standardized and individually assessed	39 months	3y LC 93%, 3 years PFS 71%, 3 years OS 88%	Grade 3 acute & mucositis in 23% & 1%; grade 3 cataract in 6%, grade 3–4 visual impairment in 9%
Hauswald et al. [49]	Phase II	8	Mixed but $n = 5$ oropharynx; all definitive	24 Gy (RBE) in 8 daily fractions followed by 50 Gy IMRT in 25 daily fractions	Induction TPF and concurrent cetuximab	14 months	1 year LC, PFS, OS 100%	Grade 3 dysphagia ($n = 1$), grade 3 mucositis ($n = 4$)
Takayasu et al. [50]	Observational	21	Mucosal melanoma; majority unresected	All but 1 patient with 64 Gy (RBE) in 16 fractions	Concurrent and adjuvant DAV	16 months	3 years LC 93%, 3 years PFS 37%, 3 years OS 49%	Grade 3 mucositis in 10%
Vischioni et al. [46]	Phase II	51	Reirradiation; salivary gland tumors (75% ACC)	Various; median 60 Gy (RBE) in 12–20 fractions in 3–5 weeks	Not standardized and individually assessed	19 months	Tumor control maintained in 41% at last follow-up; 2 years PFS 52%, 2 years OS 64%	Grade 3 visual deficits ($n = 3$), grade 3 neuropathy ($n = 1$), grade 3 trismus (n = 4)
Adeberg et al. [47]	Phase I/II	23	ACC; both adjuvant and definitive	24 Gy (RBE) in 8 daily fractions followed by 54 Gy IMRT in 27 daily fractions	Concurrent cetuximab	39 months	Tumor control maintained in 78% at last follow-up; 3 years PFS 67%, 3 years OS 90%	Grade 3 dermatitis in 22%, grade 3 mucositis in 48%

RT radiation therapy, *Gy (RBE)* Gray relative biological effectiveness, *LC* local control, *IMRT* intensity-modulated radiation therapy, *PFS* progression-free survival, *OS* overall survival, *ACC* adenoid cystic carcinoma, *TPF* taxane/platin/5-fluorouracil, *DAV* dacarbazine, nimustine, vincristine

[47]. These principles have been reflected in trials from Japan [48, 50]. The only experience from Italy demonstrated relatively low rates of tumor control, but this was likely related to enrollment of only reirradiation cases [46]. Taken together, prospective data corroborate the largely positive retrospective experiences of CIRT for advanced and/or recurrent head and neck neoplasms and provide a strong foundation for the design of randomized trials (discussed below).

Unresolved Questions

The most pressing and urgent need regarding the future of CIRT in head and neck malignancies is randomized data, specifically comparing CIRT to photon therapy. There are a number of randomized trials currently accruing that are aiming to address this fundamental issue. One large randomized trial (ETOILE, NCT02838602) includes head and neck ACCs as part of several radioresistant tumors (e.g., sarcomas) in order to compare CIRT with IMRT [51]. The other large randomized trial (ACCO, NCT04214366) with an estimated accrual of 314 patients is being conducted in Germany with a primary endpoint of 5-year freedom from locoregional progression. Similar to the German paradigm of therapy as described above, one arm is a dose of 50 Gy in 25 fractions of IMRT and a 24 Gy (RBE) in eight fractions of CIRT boost; the experimental arm is a total dose of 66 Gy (RBE) in 22 fractions of CIRT. It is unclear why CIRT is included in both arms of the trial, but likely relates to the positive experience in Germany with CIRT as a boost after IMRT that has become standard for such cases [52]. The German group, however, is running a separate basket trial of reirradiation for recurrent and/or progressing head and neck cancers, which randomizes patients between photon or CIRT reirradiation (CARE, NCT04185974). The primary endpoint is toxicity, however, and it is unclear whether the trial will be powered for its secondary endpoints related to survival [53].

In addition to the above, there are also important randomized studies aiming to individualize CIRT dosing. This includes a trial in China (NCT04533620) aiming to evaluate the utility of an institutional predictive model, predicting the risk of developing mucosal necrosis and tailoring CIRT doses accordingly. The primary endpoint is PFS [54].

In addition to the above, there are several other important issues requiring continued attention. First, CIRT is an overall newer modality in the clinical realm, and thus many technical aspects require further refinement. In addition to range uncertainties (similar to proton therapy), estimations of RBE with CIRT are far from resolved. RBE for CIRT is estimated using modeling methods, but multiple models are utilized for these purposes, and the accuracy of these models is also not verified. Second, clinical data comparing CIRT to proton therapy are also greatly needed. There have been some retrospective studies which have not found differences in local control or other outcomes with CIRT as compared to proton therapy, which are somewhat concerning given that a major purported advantage of CIRT is its considerably higher RBE [55]. Third, CIRT facilities require immense costs to install and maintain, which has to date prohibited the establishment of a single operational facility in the United States thus far. Costs of PBT facilities have reduced over time [56, 57], but it is unclear how quickly this will occur in the future with CIRT centers.

Despite these and other limitations, CIRT is a potentially powerful and advanced tool for the management of challenging head and neck cancer cases. Although CIRT is currently in its infancy, further investigations on refinement of patient selection, as well as technical aspects related to CIRT will likely continue to be addressed going forward. Additionally, the eventual publication of randomized evidence could change the landscape of CIRT for head and neck cancer patients over the next few decades.

Stereotactic Body Radiation Therapy

Background and Rationale

SBRT delivers very high ablative doses of radiation in a few [1–5] fractions and has been extensively used in lung, prostate, pancreatic, and liver cancer. SBRT is an attractive treatment option for head and neck cancers given the protracted nature of conventionally fractionated RT for 6–7 weeks [58]. However, SBRT is largely utilized for smaller tumor volumes that could be challenging to apply in the context of complex and large irradiated volumes in close proximity to radiosensitive organs. The majority of data for SBRT are for recurrent head and neck cancers, with very few studies for primary head and neck cancer. SBRT for recurrent disease is discussed in another chapter, and hence, the focus herein will be on SBRT as primary treatment. This section will discuss only photon SBRT; although charged particle SBRT has been done in other disease sites, there are currently no data in head and neck neoplasms.

Clinical Evidence

A phase 1 study from the University of Texas Southwestern performed stepwise hypofractionated RT for early stage glottic cancer. The stepwise dose levels assessed were 50 Gy in 15 fractions, 45 Gy in ten fractions, and 42.5 Gy in five fractions. A total of 12 patients received SBRT using 42.5 Gy in five fractions. Two patients in the SBRT cohort developed significant toxicity, with one being a dose-limiting toxicity of vocal cord necrosis. The second patient had significant laryngeal mucositis requiring treatment with vitamin E and pentoxifylline. None of the patients in the SBRT group developed local recurrence [59].

Assaf et al. performed a retrospective review of 114 patients medically unfit for definitive therapy treated with SBRT, of whom 48 had previously untreated head and neck cancers. The dose ranged from 35 to 50 Gy in 4–6 fractions given twice weekly, whereas the elective nodal dose was 25 Gy in 4–6 fractions. The local control rate was 86%, with a median PFS of 24 months [60, 61].

Personalized Radiotherapy

Tumor Hypoxia

Tumor hypoxia has been widely studied in head and neck cancers and can be evaluated by invasive (e.g., biopsy) and noninvasive (e.g., PET scans, multi-parametric MRI) methods [62]. Hypoxic regions on MRI are characterized by lower apparent diffusion coefficient (ADC) and reduced perfusion k-trans at baseline. Multiple tracers have been found useful in diagnosing of hypoxia on PET imaging including but not limited to FMISO, FAZA, FHX4, and F-FETNIM [63–66].

Various questions arise with the use of PET scans, including the correct time to scan a patient after intravenous tracer injection, the time points (baseline versus during treatment versus after treatment) that best predict response or resistance to RT, and most importantly whether PET scans correlate with actual hypoxia in tissue. Small prospective studies seem to show that delayed imaging (1–2 h after injection) provide the best contrast between hypoxic tumor and tissue [67]. Further, recent studies have also shown that residual tumor hypoxia during chemoradiotherapy is a major driver of therapy resistance and that hypoxia after the second week of treatment as measured by FMISO-PET may serve as a biomarker for selection of patients at high risk for locoregional recurrence after radio chemotherapy [68]. However, studies also show that performing week FMISO PET shows only a low correlation with levels of hypoxic gene expression particularly after week 1 and 2 [69]. Therefore, on the one hand, FMISO PET seems to predict locoregional control with RT, but on the other hand, the biological rationale has not yet been unequivocally proven. More translational research is therefore necessary before these specialized PET scans may be routinely adopted in clinical practice.

The best means to evaluate hypoxia as of now appears to be biopsy to identify gene signatures [70]. A 15-gene hypoxic gene signature has been recently validated on patients in the DAHANCA 5 study for predicting a better response to hypoxia modification in therapy [71]. DAHANCA 30 is an ongoing randomized controlled trial regarding hypoxic modification with nimorazole in patients categorized as "more hypoxic" [72].

The publication of the 30 ROC trial has important implications for personalized RT approaches for oropharyngeal cancer. This study assessed whether functional imaging with pre- and intra-treatment dynamic fluorine-18-labeled FMISO PET can be used to de-escalate treatment in low-risk HPV-positive oropharyngeal cancer. Out of 19 patients, 15 had no hypoxia at baseline or during treatment, and their radiation dose was de-escalated to 30 Gy. Patients who had the de-escalated therapy had a planned neck dissection at 4 months. Altogether, 11/15 patients had a complete pathological response and the 2-year locoregional survival and overall survival were 94.4% and 94.7%, respectively. There were no grade 3 toxicities in the study group [73].

Immune Microenvironment

RT has been shown to induce "immunogenic" cell death in various subsites including HN. This immune sensitization appears to occur by mechanisms such as release of tumor-associated antigens, increased expression of DAMPs, and increased release of cytokines. it must be noted that head and neck neoplasms have innate mechanisms to escape immune attack. Resistance to T-cell recognition and killing can occur by altered antigen processing and presentation, absent or defective IFNγ receptor expression or downstream signaling, defective formation of the immunologic synapse, neutralization of granzyme released from T cells, resistance to death receptor-induced extrinsic apoptosis, caspase mutations, and intrinsic cell cycle alterations in response to cytotoxic insults. Genetic profiling to identify at least some of the most common resistance mechanisms to the above will help to triage patients that are expected or rather hypothesized to have better tumor control by immune sensitization by RT. It may also help in triaging patients that may respond to immune checkpoint inhibitors when patients progress.

In head and neck cancers, downregulated TAP1 expression (seen in 34% of cases) and loss of heterozygosity of chromosome 6p21 affect antigen presentation, and IFN-gamma receptor mutation and SOCS3 expression affect IFN-gamma expression and downstream signaling [74]. These mutations may have an important bearing in immune sensitization by RT. The optimal RT fractionation for immune sensitization is also not defined, but studies suggest hypofractionated RT to be the most effective [75]. This raises the question of feasibility of hypofractionation in head and neck cancers due to organs at risk. The role of partial-volume high-dose RT therefore also needs to be evaluated as well [76].

Radiosensitivity Index

The radiosensitivity index is a genomic expression assay that measures several genes associated with responsiveness to ionizing radiation, which is converted into a numerical score (lower scores correspond to greater radiosensitivity). Based on limited data, this metric appears to be an independent predictor of response to RT alone [77]. Multiple other gene signature panels have been developed and shown to predict improvement in locoregional control in "more radiosensitive" tumors (e.g., 2-year locoregional control of 86% vs. 61%; $p = 0.05$) [78–80]. Genomic adjusted radiation (GARD) dose is another tool being tested to personalize radiotherapy in these malignancies, but clinical data on GARD continues to evolve [81].

Conclusion

The landscape of RT for head and neck malignancies has undergone a myriad of changes, including rapid expansion of charged particle therapy with proton and carbon ions. SBRT for primary head and neck cancers is also under active investigation. A better understanding of tumor biology and genomics should enhance a shift toward more personalized radiotherapeutic approaches to these neoplasms. Further clinical data and other advancements in the coming years and decades could create major inflection points in the utility of these approaches toward oncologic management.

References

1. Patyal B. Dosimetry aspects of proton therapy. Technol Cancer Res Treat. 2007;6(4 Suppl):17–23.
2. A brief history of the Harvard University Cyclotrons—Richard Wilson [Internet]. [cited 2021 May 9]. https://www.hup.harvard.edu/catalog.php?isbn=9780674014602.
3. PTCOG—facilities in operation [Internet]. [cited 2022 Jun 2]. https://www.ptcog.ch/index.php/facilities-in-operation.
4. Current listing of Proton Therapy Centers in operation [Internet]. [cited 2022 Jun 2]. https://proton-therapy-centers.com/.
5. Beddok A, Vela A, Calugaru V, Tessonnier T, Kubes J, Dutheil P, et al. Proton therapy for head and neck squamous cell carcinomas: a review of the physical and clinical challenges. Radiother Oncol. 2020;147:30–9.
6. Moreno AC, Frank SJ, Garden AS, Rosenthal DI, Fuller CD, Gunn GB, et al. Intensity modulated proton therapy (IMPT)—the future of IMRT for head and neck cancer. Oral Oncol. 2019;88:66–74.
7. Machtay M, Moughan J, Trotti A, Garden AS, Weber RS, Cooper JS, et al. Factors associated with severe late toxicity after concurrent chemoradiation for locally advanced head and neck cancer: an RTOG analysis. J Clin Oncol. 2008;26(21):3582–9.
8. Nguyen-Tan PF, Zhang Q, Ang KK, Weber RS, Rosenthal DI, Soulieres D, et al. Randomized phase III trial to test accelerated versus standard fractionation in combination with concurrent cisplatin for head and neck carcinomas in the Radiation Therapy Oncology Group 0129 trial: long-term report of efficacy and toxicity. J Clin Oncol. 2014;32(34):3858–67.
9. Blanchard P, Garden AS, Gunn GB, Rosenthal DI, Morrison WH, Hernandez M, et al. Intensity-modulated proton beam therapy (IMPT) versus intensity-modulated photon therapy (IMRT) for patients with oropharynx cancer - a case matched analysis. Radiother Oncol. 2016;120(1):48–55.
10. Cao J, Zhang X, Jiang B, Chen J, Wang X, Wang L, et al. Intensity-modulated proton therapy for oropharyngeal cancer reduces rates of late xerostomia. Radiother Oncol. 2021;160:32–9.
11. Sio TT, Lin H-K, Shi Q, Gunn GB, Cleeland CS, Lee JJ, et al. Intensity modulated proton therapy versus intensity modulated photon radiation therapy for oropharyngeal cancer: first comparative results of patient-reported outcomes. Int J Radiat Oncol Biol Phys. 2016;95(4):1107–14.
12. Sharma S, Zhou O, Thompson R, Gabriel P, Chalian A, Rassekh C, et al. Quality of life of postoperative photon versus proton radiation therapy for oropharynx cancer. Int J Part Ther. 2018;5(2):11–7.
13. Holliday EB, Garden AS, Rosenthal DI, Fuller CD, Morrison WH, Gunn GB, et al. Proton therapy reduces treatment-related toxicities for patients with nasopharyngeal cancer: a case-match control study of intensity-modulated proton therapy and intensity-modulated photon therapy. Int J Part Ther. 2015;2(1):19–28.
14. Williams VM, Parvathaneni U, Laramore GE, Aljabab S, Wong TP, Liao JJ. Intensity-modulated proton therapy for nasopharynx cancer: 2-year outcomes from a single institution. Int J Part Ther. 2021;8(2):28–40. [cited 2021 Apr 25]. https://doi.org/10.14338/IJPT-20-00057.1.
15. Chan A, Adams JA, Weyman E, Parambi R, Goldsmith T, Holman A, et al. A phase II trial of proton radiation therapy with chemotherapy for nasopharyngeal carcinoma. Int J Radiat Oncol Biol Phys. 2012;84(3):S151–2.
16. Alterio D, D'Ippolito E, Vischioni B, Fossati P, Gandini S, Bonora M, et al. Mixed-beam approach in locally advanced nasopharyngeal carcinoma: IMRT followed by proton therapy boost versus IMRT-only. Evaluation of toxicity and efficacy. Acta Oncol. 2020;59(5):541–8.
17. Yu NY, Gamez ME, Hartsell WF, Tsai HK, Laramore GE, Larson GL, et al. A multi-institutional experience of proton beam therapy for sinonasal tumors. Adv Radiat Oncol. 2019;4(4):689–98.
18. Fan M, Kang JJ, Lee A, Fan D, Wang H, Kitpanit S, et al. Outcomes and toxicities of definitive radiotherapy and reirradiation using 3-dimensional conformal or intensity-modulated (pencil beam) proton therapy for patients with nasal cavity and paranasal sinus malignancies. Cancer. 2020;126(9):1905–16.
19. Zenda S, Kawashima M, Arahira S, Kohno R, Nishio T, Tahara M, et al. Late toxicity of proton beam therapy for patients with the nasal cavity, para-nasal sinuses, or involving the skull base malignancy: importance of long-term follow-up. Int J Clin Oncol. 2015;20(3):447–54.
20. Russo AL, Adams JA, Weyman EA, Busse PM, Goldberg SI, Varvares M, et al. Long-term outcomes after proton beam therapy for sinonasal squamous cell carcinoma. Int J Radiat Oncol Biol Phys. 2016;95(1):368–76.
21. Patel SH, Wang Z, Wong WW, Murad MH, Buckey CR, Mohammed K, et al. Charged particle therapy versus photon therapy for paranasal sinus and nasal cavity malignant diseases: a systematic review and meta-analysis. Lancet Oncol. 2014;15(9):1027–38.
22. Romesser PB, Cahlon O, Scher E, Zhou Y, Berry SL, Rybkin A, et al. Proton beam radiation therapy results in significantly reduced toxicity compared with intensity-modulated radiation therapy for head and neck tumors that require ipsilateral radiation. Radiother Oncol. 2016;118(2):286–92.
23. Holliday E, Bhattasali O, Kies MS, Hanna E, Garden AS, Rosenthal DI, et al. Postoperative intensity-modulated proton therapy for head and neck adenoid cystic carcinoma. Int J Part Ther. 2016;2(4):533–43.
24. Chuong M, Bryant J, Hartsell W, Larson G, Badiyan S, Laramore GE, et al. Minimal acute toxicity from proton beam therapy for major salivary gland cancer. Acta Oncol. 2020;59(2):196–200.

25. Frank SJ, Blanchard P, Lee JJ, Sturgis EM, Kies MS, Machtay M, et al. Comparing intensity-modulated proton therapy with intensity-modulated photon therapy for oropharyngeal cancer: the journey from clinical trial concept to activation. Semin Radiat Oncol. 2018;28(2):108–13.
26. Price J, Hall E, West C, Thomson D. TORPEdO—a phase III trial of intensity-modulated proton beam therapy versus intensity-modulated radiotherapy for multi-toxicity reduction in oropharyngeal cancer. Clin Oncol (R Coll Radiol). 2020;32(2):84–8.
27. Lund University Hospital. Photon therapy versus proton therapy in early tonsil cancer. [Internet]. clinicaltrials.gov; 2020 [cited 2021 May 9]. Report No.: NCT03829033. https://clinicaltrials.gov/ct2/show/NCT03829033.
28. Memorial Sloan Kettering Cancer Center. A phase II randomized study of proton versus photon beam radiotherapy in the treatment of unilateral head and neck cancer [Internet]. clinicaltrials.gov; 2020 [cited 2021 May 9]. Report No.: NCT02923570. https://clinicaltrials.gov/ct2/show/NCT02923570.
29. NRG Oncology. Randomized phase II and phase III studies of individualized treatment for nasopharyngeal carcinoma based on biomarker Epstein Barr Virus (EBV) deoxyribonucleic acid (DNA) [Internet]. clinicaltrials.gov; 2021 [cited 2021 Apr 22]. Report No.: NCT02135042. https://clinicaltrials.gov/ct2/show/NCT02135042.
30. Radiation Therapy Oncology Group. A randomized phase II/phase III study of adjuvant concurrent radiation and chemotherapy versus radiation alone in resected high-risk malignant salivary gland tumors [Internet]. clinicaltrials.gov; 2021 [cited 2021 May 9]. Report No.: NCT01220583. https://clinicaltrials.gov/ct2/show/NCT01220583.
31. Mohamad O, Sishc BJ, Saha J, Pompos A, Rahimi A, Story MD, et al. Carbon ion radiotherapy: a review of clinical experiences and preclinical research, with an emphasis on DNA damage/repair. Cancers (Basel). 2017;9(6):66.
32. Helm A, Ebner DK, Tinganelli W, Simoniello P, Bisio A, Marchesano V, et al. Combining heavy-ion therapy with immunotherapy: an update on recent developments. Int J Part Ther. 2018;5(1):84–93.
33. Theelen WSME, Chen D, Verma V, Hobbs BP, Peulen HMU, Aerts JGJV, et al. Pembrolizumab with or without radiotherapy for metastatic non-small-cell lung cancer: a pooled analysis of two randomised trials. Lancet Respir Med. 2021;9(5):467–75.
34. Suit H, DeLaney T, Goldberg S, Paganetti H, Clasie B, Gerweck L, et al. Proton vs carbon ion beams in the definitive radiation treatment of cancer patients. Radiother Oncol. 2010;95(1):3–22.
35. Wang L, Hu J, Liu X, Wang W, Kong L, Lu JJ. Intensity-modulated carbon-ion radiation therapy versus intensity-modulated photon-based radiation therapy in locally recurrent nasopharyngeal carcinoma: a dosimetric comparison. Cancer Manag Res. 2019;11:7767–77.
36. Valable S, Gérault AN, Lambert G, Leblond MM, Anfray C, Toutain J, et al. Impact of hypoxia on carbon ion therapy in glioblastoma cells: modulation by LET and hypoxia-dependent genes. Cancers (Basel). 2020;2(8):2019. [cited 2021 May 9]. https://www.ncbi.nlm.nih.gov/pmc/articles/PMC7464439/.
37. Verma V, Ganti AK. Concurrent chemoradiotherapy in older adults with squamous cell head & neck cancer: evidence and management. J Geriatr Oncol. 2016;7(3):145–53.
38. Held T, Windisch P, Akbaba S, Lang K, El Shafie R, Bernhardt D, et al. Carbon ion reirradiation for recurrent head and neck cancer: a single-institutional experience. Int J Radiat Oncol Biol Phys. 2019;105(4):803–11.
39. Koto M, Demizu Y, Saitoh J-I, Suefuji H, Tsuji H, Okimoto T, et al. Definitive carbon-ion radiation therapy for locally advanced sinonasal malignant tumors: subgroup analysis of a multicenter study by the Japan Carbon-Ion Radiation Oncology Study Group (J-CROS). Int J Radiat Oncol Biol Phys. 2018;102(2):353–61.
40. Koto M, Demizu Y, Saitoh J-I, Suefuji H, Tsuji H, Okimoto T, et al. Multicenter study of carbon-ion radiation therapy for mucosal melanoma of the head and neck: subanalysis of the Japan Carbon-Ion Radiation Oncology Study Group (J-CROS) Study (1402 HN). Int J Radiat Oncol Biol Phys. 2017;97(5):1054–60.
41. Sulaiman NS, Demizu Y, Koto M, Saitoh J-I, Suefuji H, Tsuji H, et al. Multicenter study of carbon-ion radiation therapy for adenoid cystic carcinoma of the head and neck: subanalysis of the Japan Carbon-Ion Radiation Oncology Study Group (J-CROS) Study (1402 HN). Int J Radiat Oncol Biol Phys. 2018;100(3):639–46.
42. Held T, Windisch P, Akbaba S, Lang K, Farnia B, Liermann J, et al. Rare entities in head-and-neck cancer: salvage re-irradiation with carbon ions. Radiat Oncol. 2019;14(1):202.
43. Yang J, Gao J, Qiu X, Hu J, Hu W, Wu X, et al. Intensity-modulated proton and carbon-ion radiation therapy in the management of head and neck sarcomas. Cancer Med. 2019;8(10):4574–86.
44. Mizoe J-E, Tsujii H, Kamada T, Matsuoka Y, Tsuji H, Osaka Y, et al. Dose escalation study of carbon ion radiotherapy for locally advanced head-and-neck cancer. Int J Radiat Oncol Biol Phys. 2004;60(2):358–64.
45. Jensen AD, Nikoghosyan AV, Lossner K, Haberer T, Jäkel O, Münter MW, et al. COSMIC: a regimen of intensity modulated radiation therapy plus dose-escalated, raster-scanned carbon ion boost for malignant salivary gland tumors: results of the prospective phase 2 trial. Int J Radiat Oncol Biol Phys. 2015;93(1):37–46.
46. Vischioni B, Dhanireddy B, Severo C, Bonora M, Ronchi S, Vitolo V, et al. Reirradiation of salivary gland tumors with carbon ion radiotherapy at CNAO. Radiother Oncol. 2020;145:172–7.
47. Adeberg S, Akbaba S, Lang K, Held T, Verma V, Nikoghosyan A, et al. The phase 1/2 ACCEPT trial:

concurrent cetuximab and intensity modulated radiation therapy with carbon ion boost for adenoid cystic carcinoma of the head and neck. Int J Radiat Oncol Biol Phys. 2020;106(1):167–73.
48. Shirai K, Saitoh J-I, Musha A, Abe T, Kobayashi D, Takahashi T, et al. Prospective observational study of carbon-ion radiotherapy for non-squamous cell carcinoma of the head and neck. Cancer Sci. 2017;108(10):2039–44.
49. Hauswald H, Jensen AD, Krauss J, Haselmann R, Lossner K, Hartmann S, et al. Phase II study of induction chemotherapy with docetaxel, cisplatin, 5-fluorouracil followed by radioimmunotherapy with cetuximab and intensity-modulated radiotherapy in combination with a carbon ion boost for locally advanced tumors of the oro-, hypopharynx and larynx. Clin Transl Radiat Oncol. 2018;13:64–73.
50. Takayasu Y, Kubo N, Shino M, Nikkuni O, Ida S, Musha A, et al. Carbon-ion radiotherapy combined with chemotherapy for head and neck mucosal melanoma: prospective observational study. Cancer Med. 2019;8(17):7227–35.
51. Randomized carbon ions vs standard radiotherapy for radioresistant tumors—no study results posted—ClinicalTrials.gov [Internet]. [cited 2021 Jun 9]. https://clinicaltrials.gov/ct2/show/results/NCT02838602.
52. MD KH. Adenoid cystic carcinoma and carbon ion only irradiation [Internet]. clinicaltrials.gov; 2019 [cited 2021 Jun 8]. Report No.: NCT04214366. https://clinicaltrials.gov/ct2/show/NCT04214366.
53. Debus J. Carbon ion re-radiotherapy in patients with recurrent or progressive locally advanced head-and-neck cancer: a phase-II study to evaluate toxicity and efficacy [Internet]. clinicaltrials.gov; 2020 [cited 2021 Jun 8]. Report No.: NCT04185974. https://clinicaltrials.gov/ct2/show/NCT04185974.
54. Lu JJ. Predictive-model based individualized carbon-ion radiotherapy for patients with locally recurrent nasopharyngeal carcinoma: a phase 2 randomized trial [Internet]. clinicaltrials.gov; 2020 [cited 2021 Jun 8]. Report No.: NCT04533620. https://clinicaltrials.gov/ct2/show/NCT04533620.
55. Takagi M, Demizu Y, Hashimoto N, Mima M, Terashima K, Fujii O, et al. Treatment outcomes of particle radiotherapy using protons or carbon ions as a single-modality therapy for adenoid cystic carcinoma of the head and neck. Radiother Oncol. 2014;113(3):364–70.
56. Verma V, Mishra MV, Mehta MP. A systematic review of the cost and cost-effectiveness studies of proton radiotherapy. Cancer. 2016;122(10):1483–501.
57. Verma V, Shah C, Rwigema J-CM, Solberg T, Zhu X, Simone CB. Cost-comparativeness of proton versus photon therapy. Chin Clin Oncol. 2016;5(4):56.
58. Lo SS, Teh BS, Lu JJ, Schefter TE. Stereotactic body radiation therapy. Berlin: Springer; 2012. 434 p
59. Sher DJ, Timmerman RD, Nedzi L, Ding C, Pham N-L, Zhao B, et al. Phase 1 fractional dose-escalation study of equipotent stereotactic radiation therapy regimens for early-stage glottic larynx cancer. Int J Radiat Oncol Biol Phys. 2019;105(1):110–8.
60. AL-Assaf H, Poon I, Lee JW, Karam I, Higgins K, Enepekides D. Stereotactic body radiotherapy (SBRT) for medically unfit head and neck cancer. Int J Radiat Oncol Biol Phys. 2017;99(2):E319.
61. Al-Assaf H, Erler D, Karam I, Lee JW, Higgins K, Enepekides D, et al. Stereotactic body radiotherapy for medically unfit patients with cancers to the head and neck. Head Neck. 2020;42(8):2050–7.
62. Wiedenmann N, Bunea H, Rischke HC, Bunea A, Majerus L, Bielak L, et al. Effect of radiochemotherapy on T2* MRI in HNSCC and its relation to FMISO PET derived hypoxia and FDG PET. Radiat Oncol. 2018;13(1):159. [cited 2021 May 9]. https://www.ncbi.nlm.nih.gov/pmc/articles/PMC6114038/.
63. Betts HM, O'Connor RA, Christian JA, Vinayakamoorthy V, Foweraker K, Pascoe AC, et al. Hypoxia imaging with [18F]HX4 PET in squamous cell head and neck cancers: a pilot study for integration into treatment planning. Nucl Med Commun. 2019;40(1):73–8.
64. Hu M, Xie P, Lee NY, Li M, Ho F, Lian M, et al. Hypoxia with 18F-fluoroerythronitroimidazole integrated positron emission tomography and computed tomography (18F-FETNIM PET/CT) in locoregionally advanced head and neck cancer: hypoxia changes during chemoradiotherapy and impact on clinical outcome. Medicine (Baltimore). 2019;98(40):e17067.
65. Servagi-Vernat S, Differding S, Hanin F-X, Labar D, Bol A, Lee JA, et al. A prospective clinical study of ^{18}F-FAZA PET-CT hypoxia imaging in head and neck squamous cell carcinoma before and during radiation therapy. Eur J Nucl Med Mol Imaging. 2014;41(8):1544–52.
66. Rajendran JG, Schwartz DL, O'Sullivan J, Peterson LM, Ng P, Scharnhorst J, et al. Tumor hypoxia imaging with [F-18] fluoromisonidazole positron emission tomography in head and neck cancer. Clin Cancer Res. 2006;12(18):5435–41.
67. Chatterjee A, Gupta T, Rangarajan V, Purandare N, Kunder S, Murthy V, et al. Optimal timing of fluorine-18-fluoromisonidazole positron emission tomography/computed tomography for assessment of tumor hypoxia in patients with head and neck squamous cell carcinoma. Nucl Med Commun. 2018;39(9):859–64.
68. Löck S, Linge A, Seidlitz A, Bandurska-Luque A, Nowak A, Gudziol V, et al. Repeat FMISO-PET imaging weakly correlates with hypoxia-associated gene expressions for locally advanced HNSCC treated by primary radiochemotherapy. Radiother Oncol. 2019;135:43–50.
69. Löck S, Perrin R, Seidlitz A, Bandurska-Luque A, Zschaeck S, Zöphel K, et al. Residual tumour hypoxia in head-and-neck cancer patients undergoing primary radiochemotherapy, final results of a prospective trial on repeat FMISO-PET imaging. Radiother Oncol. 2017;124(3):533–40.

70. Eustace A, Mani N, Span PN, Irlam JJ, Taylor J, Betts GNJ, et al. A 26-gene hypoxia signature predicts benefit from hypoxia-modifying therapy in laryngeal cancer but not bladder cancer. Clin Cancer Res. 2013;19(17):4879–88.
71. Toustrup K, Sørensen BS, Metwally MAH, Tramm T, Mortensen LS, Overgaard J, et al. Validation of a 15-gene hypoxia classifier in head and neck cancer for prospective use in clinical trials. Acta Oncol. 2016;55(9–10):1091–8.
72. Danish Head and Neck Cancer Group. DAHANCA 30: a randomized non-inferiority trial of hypoxia-profile guided hypoxic modification with nimorazole during radiotherapy/chemoradiotherapy of squamous cell carcinoma of the head and neck. [Internet]. clinicaltrials.gov; 2021 [cited 2021 May 6]. Report No.: NCT02661152. https://clinicaltrials.gov/ct2/show/NCT02661152.
73. Riaz N, Sherman E, Pei X, Schöder H, Grkovski M, Paudyal R, et al. Precision radiotherapy: reduction in radiation for oropharyngeal cancer in the 30 ROC trial. J Natl Cancer Inst. 2021;113(6):742–51.
74. Greene S, Patel P, Allen CT. How patients with an intact immune system develop head and neck cancer. Oral Oncol. 2019;92:26–32.
75. Lee Y, Auh SL, Wang Y, Burnette B, Wang Y, Meng Y, et al. Therapeutic effects of ablative radiation on local tumor require CD8+ T cells: changing strategies for cancer treatment. Blood. 2009;114(3):589–95.
76. Markovsky E, Budhu S, Samstein RM, Li H, Russell J, Zhang Z, et al. An antitumor immune response is evoked by partial-volume single-dose radiation in 2 murine models. Int J Radiat Oncol Biol Phys. 2019;103(3):697–708.
77. Torres-Roca JF, Fulp WJ, Caudell JJ, Servant N, Bollet MA, van de Vijver M, et al. Integration of a radiosensitivity molecular signature into the assessment of local recurrence risk in breast cancer. Int J Radiat Oncol Biol Phys. 2015;93(3):631–8.
78. Kim SI, Kang JW, Noh JK, Jung HR, Lee YC, Lee JW, et al. Gene signature for prediction of radiosensitivity in human papillomavirus-negative head and neck squamous cell carcinoma. Radiat Oncol J. 2020;38(2):99–108.
79. Schmidt S, Linge A, Zwanenburg A, Leger S, Lohaus F, Krenn C, et al. Development and validation of a gene signature for patients with head and neck carcinomas treated by postoperative radio(chemo)therapy. Clin Cancer Res. 2018;24(6):1364–74.
80. Eschrich SA, Pramana J, Zhang H, Zhao H, Boulware D, Lee J-H, et al. A gene expression model of intrinsic tumor radiosensitivity: prediction of response and prognosis after chemoradiation. Int J Radiat Oncol Biol Phys. 2009;75(2):489–96.
81. Scott JG, Berglund A, Schell MJ, Mihaylov I, Fulp WJ, Yue B, et al. A genome-based model for adjusting radiotherapy dose (GARD): a retrospective, cohort-based study. Lancet Oncol. 2017;18(2):202–11.

5 Systemic Therapy Advances in Head and Neck Cancer

Perrin E. Romine and Cristina P. Rodriguez

Key Points

- Pembrolizumab is now the first-line therapy for recurrent/metastatic SCCHN alone (in CPS ≥ 1) or in combination with platinum/5-fluorouracil doublet (independent of CPS).
- Pembrolizumab and nivolumab are approved as the second-line therapy in advanced/metastatic SCCHN.
- Trials of dual-checkpoint inhibition are ongoing, but there is not data to support this approach at this time.
- Organ preservation remains the standard of care for laryngeal carcinomas based on long-term follow-up data of RTOG 91-11.
- De-escalation of therapy with cetuximab/RT in HPV-positive OPC is inferior to the standard of care cisplatin/RT.
- Trials of adjuvant immunotherapy or combination immunotherapy/radiation therapy are ongoing in locally advanced SCCHN; however, early results have been negative.

Introduction

Head and neck cancers represent a varied group of malignancies. While histologically the vast majority are squamous cell carcinomas (SCCHN), individual management is not uniform and is driven largely by the anatomic location of the tumor, extent of disease, and patient fitness. Tumor behavior is also dependent on antecedent drivers of malignancy, specifically cancers arising from prior tobacco and alcohol use versus those that are virally mediated. Despite the diversity in location and pathogenesis, head and neck cancers share an inherent chemosensitivity [1].

Multiple chemotherapy agents, both used as a single agent or in combination, have been proven to be effective in the treatment of head and neck cancer, including the platinum agents cisplatin and carboplatin, methotrexate, 5-fluorouracil, and the taxanes paclitaxel and docetaxel [2–5]. Systemic therapy as a single modality, however, is not curative, and the treatment of medically fit locally advanced SCCHN requires a multidisciplinary approach. Clarifying the goal of treatment, specifically curative versus palliative, is of utmost importance as this dictates the degree of acceptable toxicity associated with therapy. This chapter will focus on the role of systemic therapy within the multidisciplinary care of head and neck cancer, with an emphasis on recent advances to the field. Specific focus will be placed on both

P. E. Romine (✉) · C. P. Rodriguez
Division of Medical Oncology, University of Washington, Seattle, WA, USA
e-mail: perrinr@uw.edu; rodrigcr@uw.edu

R. A. Chandra, R. J. Li (eds.), *Multidisciplinary Management of Head and Neck Cancer*,
https://doi.org/10.1007/978-3-031-05973-5_5

the recurrent/metastatic setting and the role of systemic therapy within the locally advanced setting.

Systemic Therapy for Recurrent/Metastatic Head and Neck Cancer

Locally recurrent and metastatic squamous cell carcinoma of the head and neck carries an extremely poor prognosis of 6–9 months with associated significant morbidity [6]. Numerous trials examining combination chemotherapy versus single-agent regimens have been done attempting to improve outcomes. Historically these have consistently showed improved response rates and toxicity with no significant change in survival [7, 8]. The introduction of the anti-EGFR inhibitor cetuximab in combination with a platinum and 5-fluorouracil regimen was the first phase III trial to demonstrate significant survival improvement, but median overall survival (OS) remained poor at 10.1 months with substantial associated treatment toxicity [9].

Immunotherapy for Recurrent/Metastatic Head and Neck Cancer

One of the major advances in the field of head and neck cancer over the past decade is the introduction of immunotherapy. Checkpoint inhibitors are monoclonal antibodies designed to block the interaction between host immune cells and tumor cells and are widely used in many solid tumors. Pembrolizumab and nivolumab, both anti-PD-1 antibodies, target the interaction between PD-1 and PD-1 ligand (PD-L1). PD-1 is a transmembrane protein found on T cells, B cells, and NK cells, which binds to PD-L1 expression on host cells as well as cancer cells inhibiting apoptosis. By effectively blocking the downregulatory signal from PD-1/PD-L1, anti-PD-1 checkpoint inhibitors reconstitute the cytotoxic T-cell function, enabling host immune-mediated tumor destruction.

Checkpoint inhibitors were first evaluated as a second-line therapy following prior platinum based therapy in SCCHN. The phase III Checkmate 141 trial randomized patients with refractory head and neck cancer progressing less than 6 months following platinum therapy to single agent nivolumab versus single-agent chemotherapy [10]. With a median follow-up of 5.1 months, median OS was statistically longer in the nivolumab arm at 7.5 months than in the chemotherapy arm at 5.1 months. Response rates were higher in the nivolumab arm (13.3% versus 5.8%), quality of life indices were improved, and toxicity was notably less. In a planned subset analysis of PD-L1 status and p16/HPV-positive patients, patients with a PD-L1 score of 1% or more had a marked reduction in hazard ratio (HR) for death, while those with a PD-L1 expression of <1% had no difference in HR. No association was found between p16 status and survival.

Similar results were seen in the phase III Keynote-040 trial evaluating pembrolizumab versus single-agent chemotherapy for recurrent or metastatic squamous cell carcinoma of the head and neck following platinum therapy [11]. While including patients with either platinum refractory or recurrent disease, this study showed improved OS (8.4 versus 6.9 months) in the pembrolizumab arm versus standard of care. Response rates were comparable between the two arms (14.6% and 10.1%, respectively). Patients with a PD-L1 score of <1% again showed worse OS (median OS 6.3 months versus 7.0 months). Toxicity was significantly less in the pembrolizumab arm. Results from both Checkmate 141 and Keynote-040 led to the approval of nivolumab and pembrolizumab in the second-line setting (after platinum-based therapy) independent of PDL1 status.

Subsequent work has been done to move immunotherapy into the first-line setting. Keynote-048 reports result from a phase III trial evaluating patients with newly diagnosed recurrent or metastatic squamous cell carcinoma of the head and neck randomized to first-line pembrolizumab alone, pembrolizumab with a platinum and 5-fluorouracil doublet, or standard of care cetuximab plus platinum and 5-fluorouracil dou-

Table 5.1 Ongoing immunotherapy trials in recurrent/metastatic SCCHN

Trial	Intervention	Phase	*N*	Primary endpoint	Status
Checkmate 714 (NCT02823574)	Nivolumab + ipilimumab vs. nivolumab + placebo	II	675	ORR in platinum refractory subgroup, duration of response	Accrual complete, preliminary results published
Checkmate 651 (NCT02741570)	Nivolumab + ipilimumab vs. EXTREME regimen	III	930	OS/PFS	Accrual complete
KESTREL (NCT02551159)	Durvalumab ± tremelimumab vs. EXTREME regimen	III	823	OS	Accrual complete, preliminary results published
NCT03669718	Cemiplumab vs. cemiplumab + ISA101b	II	194	ORR/treatment related AE rate in HPV+ OPC with CPS ≥ 1	Accrual ongoing
INDUCE-3 (NCT04128696)	Pembrolizumab vs. pembrolizumab + GSK3359609	III	600	OS in PD-L1 positive patients	Accrual ongoing
LEAP-10 (NCT04199104)	Pembrolizumab ± lenvatinib	III	500	ORR/OS/PFS	Accrual ongoing

ORR overall response rate, *EXTREME regimen* cetuximab, 5-fluorouracil, and cisplatin, *OS* overall survival, *PFS* performance-free survival, *HPV* human papilloma virus, *OPC* oropharyngeal carcinoma

blet (known as the EXTREME regimen). Results were stratified based on PD-L1 status and reported as the combined proportion score (CPS). In the CPS ≥1 group, median OS was significantly improved to 12.3 months in patients treated with single-agent pembrolizumab compared with 10.3 months in the cetuximab plus chemotherapy arm. Similar trends were seen in the pembrolizumab plus chemotherapy versus the EXTREME regimen regardless of CPS status. By demonstrating improved OS compared with prior standard of care multidrug chemotherapy regimens, results from Keynote-048 have moved pembrolizumab into the first-line setting, either as monotherapy for patients with a CPS score of ≥1 or in combination with a platinum/5-fluorouracil doublet. It is worthwhile noting that although pembrolizumab monotherapy had a more favorable toxicity profile compared to EXTREME, pembrolizumab/platinum, and 5-fluorouracil combinations resulted similar high-grade toxicities in approximately 85% of patients.

Combination immunotherapy, specifically the addition of anti-CTLA-4 agents to anti-PD-1 agents, has demonstrated improved response rates and survival in other solid tumor malignancies, specifically melanoma and renal cell carcinoma. CTLA-4 is a transmembrane protein expressed on T lymphocytes where it acts as a downregulator of activation of T cells in the presence of antigen-presenting cells. By blocking this interaction, anti-CTLA-4 monoclonal antibodies effectively remove the physiologic brake on T-cell activation. While this improves response rate in certain malignancies, dual blockade is associated with a substantial increase in toxicity. Trials evaluating dual checkpoint inhibition in recurrent/metastatic squamous cell carcinoma of the head and neck have thus far proved negative, although full results have not yet been published from CheckMate 714 or KESTREL [12–14]. Ultimately it remains to be seen if there is a role for dual checkpoint inhibition in head and neck malignancies, but data does not support its use at this time. A full list of ongoing immunotherapy trials in recurrent/metastatic SCCHN can be found in Table 5.1.

Targeted Agents for Recurrent/Metastatic Head and Neck Cancer

Squamous cell carcinoma of the head and neck is known to overexpress the epidermal growth factor receptor (EGFR) [15]. Targeted therapy with the anti-EGFR monoclonal antibody cetuximab has previously been demonstrated to have a role in recurrent/metastatic head and neck cancer both with and without platinum agents [9, 16–18]. More recently, the results of the GORTEC 2014-01 trial evaluated the use of docetaxel with a platinum/cetuximab combination compared with the standard of care cisplatin/5-fluorouracil/cetuximab EXTREME regimen. This trial demonstrated comparable survival outcomes; however, improved toxicity was seen with the docetaxel regimen, suggesting an alternative regimen for patients with a borderline performance status [19].

Subsequent work has been done to establish a role for panitumumab, also an anti-EGFR monoclonal antibody. Phase III trial results evaluating the use of panitumumab with cisplatin and 5-fluorouracil versus cisplatin and 5-fluorouracil alone failed to demonstrate an overall survival advantage despite increased response rates and toxicities [20]. Similar results were seen with the addition of bevacizumab, an anti-VEGF monoclonal antibody, to platinum doublet therapy in the first-line setting [21]. While further biomarker-driven work is necessary to say whether there is a role for anti-VEGF inhibitors such as bevacizumab in squamous cell carcinoma of the head and neck, data does not support its use at this time.

Extrapolating data from the results of cetuximab in SCCHN, multiple studies have evaluated the use of oral tyrosine kinase inhibitors (TKIs) targeting the EGFR pathway, either alone or in combination with single-agent chemotherapy. Early results demonstrated that gefitinib alone was not superior to single-agent methotrexate in recurrent SCCHN [22]. Subsequent phase III trial results evaluating the combination of gefitinib with docetaxel in the recurrent or metastatic setting failed to demonstrate a survival benefit with the addition of gefitinib [23]. Similar work done with single-agent afatinib versus methotrexate in both the first- and second-line setting demonstrated progression-free survival (PFS) improvement with afatinib, but failed to demonstrate an OS benefit [24, 25]. In the absence of clear OS benefit, there is no current role for oral TKIs in recurrent or metastatic SCCHN outside of clinical trials.

Palliative Metronomic Chemotherapy

Despite the advent of immunotherapy, outcomes for platinum-refractory recurrent or metastatic SCCHN remain dismal. Metronomic chemotherapy, specifically the use of low dose chemotherapy with minimal to no interruption in dosing, was initially designed as a means of overcoming drug resistance by refocusing the therapeutic effect on tumor angiogenesis [26, 27]. Initial phase II studies of metronomic methotrexate and celecoxib versus single-agent platinum therapy in the first-line, resource-limited setting demonstrated activity of the metronomic regimen for SCCHN, but subsequent work demonstrated poor activity in the platinum refractory setting [28, 29]. Hypothesizing that resistance was driven by upregulation of ABCG2, a multidrug transporter involved in methotrexate excretion from cells, subsequent work has evaluated the use of erlotinib, an ABCG2 inhibitor, in combination with metronomic methotrexate and celecoxib. Phase II trial results in platinum refractory SCCHN treated with this triple-drug metronomic regimen demonstrated improved PFS and OS compared with historic controls, but fail to account for improved outcomes with the first- and second-line immunotherapy options [30]. Further randomized studies are necessary prior to integration into routine practice.

Locally Advanced Squamous Cell Carcinoma of the Head and Neck

Representing the majority of SCCHN diagnosis within the US, patients with locally advanced SCCHN have potentially curable disease when treated with a thoughtful multidisciplinary approach. Historically this has included surgical resection for all patients deemed to be surgical candidates, followed by adjuvant radiation therapy with or without concurrent chemotherapy. We will review recent changes in the use of systemic therapy within this multidisciplinary approach below, starting with definitive chemoradiation therapy approaches and followed by postoperative chemoradiation therapy.

Organ Preservation in Locally Advanced Squamous Cell Carcinoma of the Larynx

For certain tumor locations, particularly laryngeal carcinomas, surgical resection carries substantial long-term morbidity. The publication of the VA Cooperative Study in 1991 was a paradigm shift in management of these patients, demonstrating comparable survival outcomes in patients with stage III-IV laryngeal carcinoma treated with upfront induction chemotherapy with cisplatin/5-fluorouracil followed by radiation therapy with the option of salvage total laryngectomy in non-responders [31]. Despite comparable OS outcomes, this approach was associated with higher locoregional failure rates.

In an effort to address the higher locoregional failure rates seen with induction chemotherapy followed by radiation therapy, the phase III RTOG 91-11 compared neoadjuvant chemotherapy followed by radiation, concurrent chemoradiation therapy with cisplatin, and single modality radiation therapy alone in patients with stage III or IV squamous cell carcinoma of the larynx [32]. While OS did not differ between the arms, rates of salvage laryngectomy were significantly lower in the arm treated with concurrent chemoradiation therapy, effectively making this approach standard of care in cisplatin eligible patients. Ten-year follow-up data from RTOG 91-11 is now available for evaluation [33]. This supports comparable OS rates in the induction chemotherapy versus concomitant cisplatin/radiation therapy (RT) arm, improved locoregional control and larynx preservation in the concomitant cisplatin/RT arm, with no difference in late treatment toxicities between the arms. It is important to highlight patient selection when considering an organ sparing approach. Specifically, patients with high volume T4 tumors (with >1 cm invasion into the base of tongue and/or penetration through cartilage) were excluded from RTOG 91-11, and results from this study should not be extrapolated to this subset of patients.

Subsequent phase II work aimed to improve results for organ preservation by comparing induction chemotherapy using docetaxel/cisplatin/5-fluorouracil followed by concurrent chemoradiation therapy with either cetuximab or cisplatin in locally advanced laryngeal carcinoma [34]. No clear survival advantage was seen between the two arms, although fewer local failures were seen in the cisplatin/RT arm. As this study did not compare induction chemotherapy with concurrent cisplatin/RT with concurrent chemoradiation therapy alone, we are unable to draw direct comparisons between these two approaches. Baring a clear survival advantage, however, we would not advocate for an induction regimen at this time.

Definitive Chemoradiation Therapy in Unresectable Locally Advanced SCCHN

While surgical resection remains the mainstay of therapy for locally advanced SCCHN, a substantial percentage of patients will be deemed to be unresectable, due to either tumor location or comorbidities. Drawing from prior organ preservation studies, work by the Intergroup Study established the role of definitive cisplatin/RT in unresectable SCCHN patients in the early 2000s [35]. In platinum ineligible patients, cetuximab with radiation therapy was similarly demonstrated to improve survival outcomes in the IMCL-9815 trial compared with radiation therapy alone [36]. While previously never compared directly to each other in a randomized fashion, results from the MACH-NC meta-analysis suggest an improved survival benefit with cisplatin-based regimens, although this comes with increased associated toxicity [37].

De-escalation of Therapy in HPV-Positive Oropharyngeal Carcinoma

HPV status has repeatedly been shown to be a strong and independent prognostic survival factor for patients with oropharyngeal carcinoma (OPC) [38–40]. Due to the increased short- and long-term toxicities associated with cisplatin/RT treatment, the younger age at diagnosis of HPV-positive OPC patients, and their improved survival outcomes, there has been substantial interest over the past decade in de-escalating therapy for the HPV-positive OPC subgroup. Two large randomized phase III studies have been published evaluating the use of definitive cetuximab/RT compared with cisplatin/RT in locally advanced OPC. Results from both demonstrated inferior overall survival in the cetuximab/RT arm compared with the cisplatin/RT arm. Locoregional control was similarly inferior, and interestingly toxicity was comparable [41, 42]. As such, cisplatin/RT regimens remain the standard of care for definitive chemoradiation therapy for SCCHN, including the HPV-positive subpopulation.

Neoadjuvant Chemotherapy Followed by Definitive Chemoradiation Therapy

Despite the curative intent of chemoradiation in locally advanced SCCHN, locoregional, and distant failure rates remain elevated. In order to improve on this in higher risk patients, several studies have evaluated the addition of neoadjuvant chemotherapy to definitive chemoradiation therapy. The PARADIGM study evaluated the use of neoadjuvant docetaxel, cisplatin, and 5-fluorouracil (TPF) followed by chemoradiation therapy using docetaxel or carboplatin versus standard chemoradiation therapy with cisplatin in patients with N2–3 or T3–4 disease [43]. The DeCIDE trial randomized patients with N2–3 disease to concurrent chemoradiation therapy using docetaxel, 5-fluorouracil, and hydroxyurea versus neoadjuvant TPF followed by concurrent chemoradiation therapy with docetaxel, 5-fluorouracil, and hydroxyurea [44]. Lastly, the GORTEC 2007-02 trial randomized patients with N2b-N3 SCCHN to neoadjuvant TPF followed by definitive cetuximab/RT versus concurrent carboplatin/fluorouracil RT [45]. All three studies failed to show an improvement in OS or locoregional control. Unfortunately, all three studies were designed prior to publication of de-escalation trial results in HPV-positive OPC demonstrating clear survival advantages to cisplatin/RT regimens, and as such none of these studies utilized cisplatin/RT in both arms. This makes extrapolation of their data difficult with current standard of care practices. Regardless, based on these results, neoadjuvant chemotherapy is not recommended routinely in high risk SCCHN.

Alternative Anti-EGFR Agents in Concurrent Chemoradiation Therapy

Drawing on the results of the IMCL-9815 trial, numerous studies evaluating the use of alternative anti-EGFR agents concurrent with radiation therapy have been performed over the past decade. Studies evaluating the use of panitumumab in combination with cisplatin as well as single-agent panitumumab with radiation failed to demonstrate either an overall survival or a locoregional control benefit compared with standard cisplatin-based chemoradiation therapy regimens [46, 47]. Similar work done comparing cisplatin/RT alone versus erlotinib, a small molecular EGRF inhibitor, combined with cisplatin/RT failed to demonstrate improved response rates or PFS [48].

Afatinib, an oral irreversible ERBB inhibitor that binds to both EGFR and HER2 and HER4, has activity against EGFR-mutated NSCLC. As previously discussed, afatinib has been shown to improve PFS when used following cisplatin in recurrent/metastatic SCCHN [24, 25]. A subsequent phase III trial evaluated patients with intermediate- to high-risk locally advanced SCCHN previously treated with definitive platinum-based chemoradiation therapy randomized between adjuvant afatinib and placebo. There was no significant change in disease-free survival or OS, but a substantial increase in toxicity was seen in the afatinib arm [49]. With the preponderance of evidence arguing against the use of anti-EGFR agents aside from cetuximab, we would not advocate for their use in SCCHN at this time.

Immunotherapy in Locally Advanced SCCHN

Given the significant results seen using immunotherapy in recurrent/metastatic head and neck cancer, there has been understandable excitement surrounding moving IO into early stages. Thus far, however, trials evaluating the use of concurrent immunotherapy with chemoradiation therapy in locally advanced SCCHN have proven negative [50, 51]. The JAVELIN Head and Neck 100 trial evaluated the use of avelumab, an anti-PDL1 monoclonal antibody, with concurrent cisplatin-based radiation therapy for locally advanced SCCHN followed by up to 1 year of adjuvant avelumab. The GORTEC 2015-01 trial, in contrast, randomized patients who were cisplatin ineligible to either concurrent cetuximab/radiation therapy or pembrolizumab/radiation therapy. While final results of GORTEC 2015-01 have yet to be published, preliminary results from this trial and finalized results from JAVELIN Head and Neck 100 have failed to demonstrate improvements in PFS or OS. A list of ongoing larger randomized trials evaluating combined immunotherapy/radiation therapy can be found in Table 5.2.

Additional studies are evaluating the use of immunotherapy as neo-adjuvant/adjuvant treatment following chemoradiation therapy, including in the good risk, HPV-positive OPC subpopulation (Table 5.3). While awaiting results from these studies, however, there is no current indication for the use of immunotherapy in locally advanced SCCHN.

Table 5.2 Ongoing immunotherapy/radiation therapy trials in locally advanced SCCHN

Trial	Intervention	Phase	*N*	Primary endpoint	Status
KEYNOTE-412 (NCT03040999)	Pembrolizumab + cisplatin/RT vs. placebo + cisplatin/RT	III	780	EFS	Accrual complete
REACH (NCT02999087)	Avelumab + cisplatin/RT vs. cisplatin/RT and avelumab + cetuximab/RT vs. cetuximab/RT	III	688	PFS	Accrual complete
GORTEC 2015-01 (PembroRad)	Pembrolizumab/RT vs. cetuximab/RT in cisplatin ineligible patients	II	133	LRC	Accrual complete, preliminary results published
KEYCHAIN (NCT03383094)	Pembrolizumab/RT vs. cisplatin/RT in intermediate/advanced risk HPV+ patients	II	114	PFS	Accrual ongoing
HN005 (NCT03952585)	Cisplatin/RT vs. nivolumab/de-esclRT vs. cisplatin/de-esclRT in good risk HPV+ OPC	II/III	711	PFS/QOL	Accrual ongoing
CompARE (NCT04116047)	Cisplatin/RT vs. cisplatin/dose-esclRT vs. durvalumab/cisplatin/RT	III	695	EFS/OS	Accrual ongoing
HN004 (NCT03258554)	Cetuximab/RT vs. durvalumab/RT in cisplatin ineligible patients	II/III	474	DLT/PFS/OS	Accrual ongoing

RT radiation therapy, *EFS* event-free survival, *PFS* progression-free survival, *LRC* locoregional control, *HPV* human papilloma virus, *de-esclRT* de-escalating radiation therapy, *QOL* quality of life, *dose-escl* dose-escalating radiation therapy, *OS* overall survival, *DLT* dose-limiting toxicity

Table 5.3 Ongoing adjuvant/neoadjuvant immunotherapy trials in locally advanced SCCHN

Trial	Intervention	Phase	*N*	Primary endpoint	Status
IMSTAR-HN (NCT03700905)	Neoadjuvant nivolumab, surgery, cisplatin/RT, and adjuvant nivolumab + ipilimumab vs. surgery and cisplatin/RT	III	276	DFS	Accrual complete
KEYNOTE-689 (NCT03765918)	Neoadjuvant pembrolizumab, surgery, and pembrolizumab/ cisplatin/RT vs. surgery and cisplatin/RT	III	704	mPR, EFS	Accrual ongoing
IMvoke010 (NCT03452137)	Adjuvant atezolizumab vs. placebo following chemo/RT	III	400	EFS/OS	Accrual ongoing
ECOG-ACRIN 3161 (NCT03811015)	Adjuvant nivolumab vs. observation following chemo/RT	II/III	744	PFS/OS	Accrual ongoing
PATHWay (NCT02841748)	Adjuvant pembrolizumab vs. placebo following chemo/RT	II	100	PFS	Accrual ongoing

RT radiation therapy, *DFS* disease-free survival, *mPR* major pathologic response, *EFS* event-free survival, *OS* overall survival, *PFS* progression-free survival

Advances in Postoperative Chemoradiation Therapy

For patients deemed to be surgical resection candidates, postoperative radiation therapy is standard of care for locally advanced disease. Prior work has demonstrated a clear survival and locoregional control advantage with the addition of cisplatin to postoperative radiation therapy in patients with SCCHN and high-risk features, specifically positive margins and/or extranodal extension [52, 53]. More recent work has been done to see if by adjusting cisplatin dosing regimens, we can improve toxicity while maintaining outcomes.

Standard cisplatin dosing involves given 100 mg/m^2 every 3 weeks for 2–3 cycles during radiation therapy. Alternative regimens have evaluated the use of weekly lower doses of cisplatin. Results from this approach have been conflicting. In one trial evaluating the use of cisplatin 30 mg/m^2 weekly versus cisplatin 100 mg/m^2 every 3 weeks in locally advanced SCCHN patients with extracapsular extension, close or positive margins, more than two positive nodes, or T4 disease, inferior locoregional control rates were seen in the weekly cisplatin arm [54]. However, subsequent work done using a more stringent criteria for high risk (only positive margins and/or extranodal extension) and a higher weekly cisplatin dose of 40 mg/m^2 demonstrated comparable survival (both overall and relapse free) with substantially improved toxicity profile in the weekly arm [55]. While awaiting final publication of this data, it is reasonable to consider weekly dosing of cisplatin 40 mg/m^2 concurrent with radiation therapy in the appropriately selected patient population.

Management of "intermediate"-risk patients who undergo definitive resection and postoperative radiation therapy is an ongoing area of research. While prior work has not shown a benefit of cisplatin-based radiation therapy in this context, the ongoing RTOG 0920 evaluates the use of concurrent cetuximab radiation therapy in patients with particular intermediate-risk factors, including perineural invasion, lymphovascular invasion, nodal involvement (a single node >3 cm or more than one node involved), close margins, T3–T4a tumors, or >5 mm depth of invasion in the oral cavity [56]. Results from this trial are eagerly anticipated and have the potential to further change management for postoperative therapy.

Future Directions

Despite the advances seen over the past decade, much work remains to be done in improving outcomes in both locally advanced and recurrent/

metastatic SCCHN. Within the context of recurrent/metastatic SCCHN, two major ongoing research themes emerge including investigations exploring various checkpoint inhibitors combinations in the first-line [12–14] and therapeutic strategies in the management of immunotherapy refractory patients. Monalizumab, a novel immune checkpoint inhibitor that targets NKG2A on cytotoxic CD8-positive T cells, is currently in phase III trials combined with cetuximab for patients who have progressed through two lines or more of treatment including immunotherapy [57].

Targeted therapies for specific molecular aberrations are not represented in the therapeutic armamentarium for recurrent/metastatic SCCHN, however promising phase II work done demonstrating survival benefit and a 71% RR in heavily pretreated recurrent/metastatic SCCHN patients harboring HRAS mutations when treated with Tipifarnib, a farnesyltransferase inhibitor that downregulates HRAS' downstream signaling [58]. As of the writing of this manuscript, Tipifarnib has received FDA breakthrough designation but is unapproved for this indication.

Finally, early phase studies of cellular therapeutics are gaining traction with active investigation in adoptive cellular therapy, and particularly tumor-infiltrating lymphocytes (TILs) and engineered T-cell receptor (TCR) therapy in recurrent/metastatic SCCHN, but these approaches have yet to be widely validated [59].

Within the realm of locally advanced disease, there are ongoing trials evaluating ways to improve chemoradiation therapy, particularly within the cisplatin ineligible population. The REACH study, an ongoing phase III trial evaluating avelumab combined with cetuximab/radiation therapy versus standard of care cisplatin/radiation therapy or cetuximab/radiation therapy (in platinum ineligible patients), has completed recruitment although we continue to await preliminary results [60]. Similarly, the NRG HN004 evaluates the use of durvalumab/radiation therapy versus standard of care cetuximab/radiation therapy in cisplatin ineligible patients with locally advanced SCCHN [61].

Additional studies are aimed at the role of escalating therapy for patients with locally advanced SCCHN and known high-risk features. Phase II work evaluating the addition of Debio 1143, an oral antagonist of inhibition of apoptosis that carries additional radiosensitization through activation of inflammatory pathways, to definitive cisplatin/RT. This study demonstrated improved locoregional control and PFS, but not OS, in the patients treated with combined Debio 1143 and cisplatin/RT versus those treated with cisplatin/RT alone [62]. There is an ongoing Phase III trial evaluating Debio 1143 in combination with standard of care cisplatin/RT, but preliminary results are not yet available [63].

Cases of locally recurrent SCCHN remain therapeutically challenging. While salvage surgery remains the treatment of choice, many patients are not candidates for this. For these patients, re-irradiation can be considered, although it too carries substantial morbidity and is not feasible for many patients. Ongoing trials are evaluating the use of immunotherapy concurrent with radiation therapy in the locally recurrent setting [64, 65] as well as adjuvant immunotherapy post-salvage surgery [66–69].

References

1. Adelstein DJ, Tan EH, Lavertu P. Treatment of head and neck cancer: the role of chemotherapy. Crit Rev Oncol Hematol. 1996;24:97–116.
2. Rozencweig M, von Hoff DD, Slavik M, Muggia FM. Cis-diamminedichloroplatinum (II). A new anticancer drug. Ann Intern Med. 1977;86:803–12.
3. Bertino JR, Mosher MB, DeConti RC. Chemotherapy of cancer of the head and neck. Cancer. 1973;31:1141–9.
4. Al-Sarraf M. Head and neck cancer: chemotherapy concepts. Semin Oncol. 1988;15:70–85.
5. Guardiola E, Peyrade F, Chaigneau L, Cupissol D, Tchiknavorian X, Bompas E, et al. Results of a randomised phase II study comparing docetaxel with methotrexate in patients with recurrent head and neck cancer. Eur J Cancer. 2004;40:2071–6.
6. Argiris A, Li Y, Forastiere A. Prognostic factors and long-term survivorship in patients with recurrent or metastatic carcinoma of the head and neck. Cancer. 2004;101:2222–9.
7. Forastiere AA, Metch B, Schuller DE, Ensley JF, Hutchins LF, Triozzi P, et al. Randomized compari-

son of cisplatin plus fluorouracil and carboplatin plus fluorouracil versus methotrexate in advanced squamous-cell carcinoma of the head and neck: a Southwest Oncology Group study. J Clin Oncol. 1992;10:1245–51.
8. Jacobs C, Lyman G, Velez-García E, Sridhar KS, Knight W, Hochster H, et al. A phase III randomized study comparing cisplatin and fluorouracil as single agents and in combination for advanced squamous cell carcinoma of the head and neck. J Clin Oncol. 1992;10:257–63.
9. Vermorken JB, Mesia R, Rivera F, Remenar E, Kawecki A, Rottey S, et al. Platinum-based chemotherapy plus cetuximab in head and neck cancer. N Engl J Med. 2008;359:1116–27.
10. Ferris RL, Blumenschein G, Fayette J, Guigay J, Colevas AD, Licitra L, et al. Nivolumab for recurrent squamous-cell carcinoma of the head and neck. N Engl J Med. 2016;375:1856–67.
11. Cohen EEW, Soulières D, Le Tourneau C, Dinis J, Licitra L, Ahn M-J, et al. Pembrolizumab versus methotrexate, docetaxel, or cetuximab for recurrent or metastatic head-and-neck squamous cell carcinoma (KEYNOTE-040): a randomised, open-label, phase 3 study. Lancet. 2019;393:156–67.
12. Ferris RL, Haddad R, Even C, Tahara M, Dvorkin M, Ciuleanu TE, et al. Durvalumab with or without tremelimumab in patients with recurrent or metastatic head and neck squamous cell carcinoma: EAGLE, a randomized, open-label phase III study. Ann Oncol. 2020;31:942–50.
13. National Library of Medicine. Study of nivolumab in combination with ipilimumab versus nivolumab in combination with ipilimumab placebo in patients with recurrent or metastatic squamous cell carcinoma of the head and neck (CheckMate 714) [Internet]. Report No.: NCT02823574. https://clinicaltrials.gov/ct2/show/results/NCT02823574.
14. National Library of Medicine. Phase III open label study of MEDI 4736 with/without tremelimumab versus standard of care (SOC) in recurrent/metastatic head and neck cancer (KESTREL) [Internet]. Report No.: NCT02551159. https://clinicaltrials.gov/ct2/show/NCT02551159.
15. Chung CH, Ely K, McGavran L, Varella-Garcia M, Parker J, Parker N, et al. Increased epidermal growth factor receptor gene copy number is associated with poor prognosis in head and neck squamous cell carcinomas. J Clin Oncol. 2006;24:4170–6.
16. Vermorken JB, Trigo J, Hitt R, Koralewski P, Diaz-Rubio E, Rolland F, et al. Open-label, uncontrolled, multicenter phase II study to evaluate the efficacy and toxicity of cetuximab as a single agent in patients with recurrent and/or metastatic squamous cell carcinoma of the head and neck who failed to respond to platinum-based therapy. J Clin Oncol. 2007;25:2171–7.
17. Herbst RS, Arquette M, Shin DM, Dicke K, Vokes EE, Azarnia N, et al. Phase II multicenter study of the epidermal growth factor receptor antibody cetuximab and cisplatin for recurrent and refractory squamous cell carcinoma of the head and neck. J Clin Oncol. 2005;23:5578–87.
18. Baselga J, Trigo JM, Bourhis J, Tortochaux J, Cortés-Funes H, Hitt R, et al. Phase II multicenter study of the antiepidermal growth factor receptor monoclonal antibody cetuximab in combination with platinum-based chemotherapy in patients with platinum-refractory metastatic and/or recurrent squamous cell carcinoma of the head and neck. J Clin Oncol. 2005;23:5568–77.
19. Guigay J, Aupérin A, Fayette J, Saada-Bouzid E, Lafond C, Taberna M, et al. Cetuximab, docetaxel, and cisplatin versus platinum, fluorouracil, and cetuximab as first-line treatment in patients with recurrent or metastatic head and neck squamous-cell carcinoma (GORTEC 2014-01 TPExtreme): a multicentre, open-label, randomised, phase 2 trial. Lancet Oncol. 2021;22:463–75.
20. Vermorken JB, Stöhlmacher-Williams J, Davidenko I, Licitra L, Winquist E, Villanueva C, et al. Cisplatin and fluorouracil with or without panitumumab in patients with recurrent or metastatic squamous-cell carcinoma of the head and neck (SPECTRUM): an open-label phase 3 randomised trial. Lancet Oncol. 2013;14:697–710.
21. Argiris A, Li S, Savvides P, Ohr JP, Gilbert J, Levine MA, et al. Phase III randomized trial of chemotherapy with or without bevacizumab in patients with recurrent or metastatic head and neck cancer. J Clin Oncol. 2019;37:3266–74.
22. Stewart JSW, Cohen EEW, Licitra L, Van Herpen CML, Khorprasert C, Soulieres D, et al. Phase III study of gefitinib compared with intravenous methotrexate for recurrent squamous cell carcinoma of the head and neck [corrected]. J Clin Oncol. 2009;27:1864–71.
23. Argiris A, Ghebremichael M, Gilbert J, Lee J-W, Sachidanandam K, Kolesar JM, et al. Phase III randomized, placebo-controlled trial of docetaxel with or without gefitinib in recurrent or metastatic head and neck cancer: an eastern cooperative oncology group trial. J Clin Oncol. 2013;31:1405–14.
24. Machiels J-PH, Haddad RI, Fayette J, Licitra LF, Tahara M, Vermorken JB, et al. Afatinib versus methotrexate as second-line treatment in patients with recurrent or metastatic squamous-cell carcinoma of the head and neck progressing on or after platinum-based therapy (LUX-Head & Neck 1): an open-label, randomised phase 3 trial. Lancet Oncol. 2015;16:583–94.
25. Guo Y, Ahn M-J, Chan A, Wang C-H, Kang J-H, Kim S-B, et al. Afatinib versus methotrexate as second-line treatment in Asian patients with recurrent or metastatic squamous cell carcinoma of the head and neck progressing on or after platinum-based therapy (LUX-Head & Neck 3): an open-label, randomised phase III trial. Ann Oncol. 2019;30:1831–9.
26. Browder T, Butterfield CE, Kräling BM, Shi B, Marshall B, O'Reilly MS, et al. Antiangiogenic scheduling of chemotherapy improves efficacy against experimental drug-resistant cancer. Cancer Res. 2000;60:1878–86.

27. Bertolini F, Paul S, Mancuso P, Monestiroli S, Gobbi A, Shaked Y, et al. Maximum tolerable dose and low-dose metronomic chemotherapy have opposite effects on the mobilization and viability of circulating endothelial progenitor cells. Cancer Res. 2003;63:4342–6.
28. Patil VM, Noronha V, Joshi A, Muddu VK, Dhumal S, Bhosale B, et al. A prospective randomized phase II study comparing metronomic chemotherapy with chemotherapy (single agent cisplatin), in patients with metastatic, relapsed or inoperable squamous cell carcinoma of head and neck. Oral Oncol. 2015;51:279–86.
29. Patil VM, Noronha V, Joshi A, Pinninti R, Dhumal S, Bhattacharjee A, et al. Metronomic chemotherapy in platinum-insensitive failures and/or early failures postmultimodality management in oral cancers. Indian J Med Paediatr Oncol. 2015;36:161–5.
30. Patil VM, Noronha V, Joshi A, Dhumal S, Mahimkar M, Bhattacharjee A, et al. Phase I/II study of palliative triple metronomic chemotherapy in platinum-refractory/early-failure oral cancer. J Clin Oncol. 2019;37:3032–41.
31. Department of Veterans Affairs Laryngeal Cancer Study Group, Wolf GT, Fisher SG, Hong WK, Hillman R, Spaulding M, et al. Induction chemotherapy plus radiation compared with surgery plus radiation in patients with advanced laryngeal cancer. N Engl J Med. 1991;324:1685–90.
32. Forastiere AA, Goepfert H, Maor M, Pajak TF, Weber R, Morrison W, et al. Concurrent chemotherapy and radiotherapy for organ preservation in advanced laryngeal cancer. N Engl J Med. 2003;349:2091–8.
33. Forastiere AA, Zhang Q, Weber RS, Maor MH, Goepfert H, Pajak TF, et al. Long-term results of RTOG 91-11: a comparison of three nonsurgical treatment strategies to preserve the larynx in patients with locally advanced larynx cancer. J Clin Oncol. 2013;31:845–52.
34. Lefebvre JL, Pointreau Y, Rolland F, Alfonsi M, Baudoux A, Sire C, et al. Induction chemotherapy followed by either chemoradiotherapy or bioradiotherapy for larynx preservation: the TREMPLIN randomized phase II study. J Clin Oncol. 2013;31:853–9.
35. Adelstein DJ, Li Y, Adams GL, Wagner H, Kish JA, Ensley JF, et al. An intergroup phase III comparison of standard radiation therapy and two schedules of concurrent chemoradiotherapy in patients with unresectable squamous cell head and neck cancer. J Clin Oncol. 2003;21:92–8.
36. Bonner JA, Harari PM, Giralt J, Azarnia N, Shin DM, Cohen RB, et al. Radiotherapy plus cetuximab for squamous-cell carcinoma of the head and neck. N Engl J Med. 2006;354:567–78.
37. Pignon J-P, le Maître A, Maillard E, Bourhis J, MACH-NC Collaborative Group. Meta-analysis of chemotherapy in head and neck cancer (MACH-NC): an update on 93 randomised trials and 17,346 patients. Radiother Oncol. 2009;92:4–14.
38. Ang KK, Harris J, Wheeler R, Weber R, Rosenthal DI, Nguyen-Tân PF, et al. Human papillomavirus and survival of patients with oropharyngeal cancer. N Engl J Med. 2010;363:24–35.
39. Fakhry C, Zhang Q, Nguyen-Tan PF, Rosenthal D, El-Naggar A, Garden AS, et al. Human papillomavirus and overall survival after progression of oropharyngeal squamous cell carcinoma. J Clin Oncol. 2014;32:3365–73.
40. Rosenthal DI, Harari PM, Giralt J, Bell D, Raben D, Liu J, et al. Association of human papillomavirus and p16 status with outcomes in the IMCL-9815 phase III registration trial for patients with locoregionally advanced oropharyngeal squamous cell carcinoma of the head and neck treated with radiotherapy with or without cetuximab. J Clin Oncol. 2016;34:1300–8.
41. Gillison ML, Trotti AM, Harris J, Eisbruch A, Harari PM, Adelstein DJ, et al. Radiotherapy plus cetuximab or cisplatin in human papillomavirus-positive oropharyngeal cancer (NRG Oncology RTOG 1016): a randomised, multicentre, non-inferiority trial. Lancet. 2019;393:40–50.
42. Mehanna H, Robinson M, Hartley A, Kong A, Foran B, Fulton-Lieuw T, et al. Radiotherapy plus cisplatin or cetuximab in low-risk human papillomavirus-positive oropharyngeal cancer (De-ESCALaTE HPV): an open-label randomised controlled phase 3 trial. Lancet. 2019;393:51–60.
43. Haddad R, O'Neill A, Rabinowits G, Tishler R, Khuri F, Adkins D, et al. Induction chemotherapy followed by concurrent chemoradiotherapy (sequential chemoradiotherapy) versus concurrent chemoradiotherapy alone in locally advanced head and neck cancer (PARADIGM): a randomised phase 3 trial. Lancet Oncol. 2013;14:257–64.
44. Cohen EEW, Karrison TG, Kocherginsky M, Mueller J, Egan R, Huang CH, et al. Phase III randomized trial of induction chemotherapy in patients with N2 or N3 locally advanced head and neck cancer. J Clin Oncol. 2014;32:2735–43.
45. Geoffrois L, Martin L, De Raucourt D, Sun XS, Tao Y, Maingon P, et al. Induction chemotherapy followed by cetuximab radiotherapy is not superior to concurrent chemoradiotherapy for head and neck carcinomas: results of the GORTEC 2007-02 phase III randomized trial. J Clin Oncol. 2018;36:3077–83.
46. Mesía R, Henke M, Fortin A, Minn H, Yunes Ancona AC, Cmelak A, et al. Chemoradiotherapy with or without panitumumab in patients with unresected, locally advanced squamous-cell carcinoma of the head and neck (CONCERT-1): a randomised, controlled, open-label phase 2 trial. Lancet Oncol. 2015;16:208–20.
47. Giralt J, Trigo J, Nuyts S, Ozsahin M, Skladowski K, Hatoum G, et al. Panitumumab plus radiotherapy versus chemoradiotherapy in patients with unresected, locally advanced squamous-cell carcinoma of the head and neck (CONCERT-2): a randomised, controlled, open-label phase 2 trial. Lancet Oncol. 2015;16:221–32.
48. Martins RG, Parvathaneni U, Bauman JE, Sharma AK, Raez LE, Papagikos MA, et al. Cisplatin and radiotherapy with or without erlotinib in locally advanced squamous cell carcinoma of the head and neck: a randomized phase II trial. J Clin Oncol. 2013;31:1415–21.

49. Burtness B, Haddad R, Dinis J, Trigo J, Yokota T, de Souza Viana L, et al. Afatinib vs placebo as adjuvant therapy after chemoradiotherapy in squamous cell carcinoma of the head and neck: a randomized clinical trial. JAMA Oncol. 2019;5(8):1170–80.
50. Cohen EEW, Ferris RL, Psyrri A, Haddad R, Tahara M, Bourhis J, et al. Primary results of the phase III JAVELIN head & neck 100 trial: avelumab plus chemoradiotherapy (CRT) followed by avelumab maintenance vs CRT in patients with locally advanced squamous cell carcinoma of the head and neck (LA SCCHN). Ann Oncol. 2020;31:S599–628.
51. Tao Y, Sun X, Sire C, Martin L, Alfonsi M, Prevost JB, et al. LBA38—pembrolizumab versus cetuximab, concomitant with radiotherapy (RT) in locally advanced head and neck squamous cell carcinoma (LA-HNSCC): results of the GORTEC 2015-01 "PembroRad" randomized trial. Ann Oncol. 2020;31:S1142–215.
52. Bernier J, Domenge C, Ozsahin M, Matuszewska K, Lefèbvre J-L, Greiner RH, et al. Postoperative irradiation with or without concomitant chemotherapy for locally advanced head and neck cancer. N Engl J Med. 2004;350:1945–52.
53. Cooper JS, Pajak TF, Forastiere AA, Jacobs J, Campbell BH, Saxman SB, et al. Postoperative concurrent radiotherapy and chemotherapy for high-risk squamous-cell carcinoma of the head and neck. N Engl J Med. 2004;350:1937–44.
54. Noronha V, Joshi A, Patil VM, Agarwal J, Ghosh-Laskar S, Budrukkar A, et al. Once-a-week versus once-every-3-weeks cisplatin chemoradiation for locally advanced head and neck cancer: a phase III randomized noninferiority trial. J Clin Oncol. 2018;36:1064–72.
55. Kiyota N, Tahara M, Fujii H, Yamazaki T, Mitani H, Iwae S, et al. Phase II/III trial of post-operative chemoradiotherapy comparing 3-weekly cisplatin with weekly cisplatin in high-risk patients with squamous cell carcinoma of head and neck (JCOG1008). JCO. 2020;38:6502.
56. National Library of Medicine. A phase III study of postoperative radiation therapy (IMRT) +/− cetuximab for locally-advanced resected head and neck cancer [Internet]. Report No.: NCT00956007. https://clinicaltrials.gov/ct2/show/NCT00956007.
57. National Library of Medicine. A phase 3 randomized, double-blind, multicenter, global study of monalizumab or placebo with cetuximab in patients with recurrent or metastatic squamous cell carcinoma of the head and neck previously treated with an immune checkpoint inhibitor. Report No.: NCT04590963.
58. Ho AL, Chau N, Bauman JE, Bible K, Chintakuntlawar A, Cabanillas ME, et al. Preliminary results from a phase 2 trial of tipifarnib in squamous cell carcinomas (SCCs) with HRAS mutations. Ann Oncol. 2018;29:viii372–99.
59. Qureshi HA, Lee SM. Immunotherapy approaches beyond PD-1 inhibition: the future of cellular therapy for head and neck squamous cell carcinoma. Curr Treat Options in Oncol. 2019;20:31.
60. National Library of Medicine. A phase III randomized trial of avelumab-cetuximab-radiotherapy versus standards of care in locally advanced squamous cell carcinoma of the head and neck. Report No.: NCT02999087.
61. National Library of Medicine. Randomized phase II/III trial of radiotherapy with concurrent MEDI4736 (durvalumab) vs. radiotherapy with concurrent cetuximab in patients with locoregionally advanced head and neck cancer with a contraindication to cisplatin. Report No.: NCT03258554.
62. Sun X-S, Tao Y, Le Tourneau C, Pointreau Y, Sire C, Kaminsky M-C, et al. Debio 1143 and high-dose cisplatin chemoradiotherapy in high-risk locoregionally advanced squamous cell carcinoma of the head and neck: a double-blind, multicentre, randomised, phase 2 study. Lancet Oncol. 2020;21:1173–87.
63. National Library of Medicine. A randomized, double-blind placebo-controlled, phase 3 study of Debio 1143 in combination with platinum-based chemotherapy and standard fractionation intensity-modulated radiotherapy in patients with locally advanced squamous cell carcinoma of the head and neck, suitable for definitive chemoradiotherapy. Report No.: NCT04459715.
64. National Library of Medicine. Winship4221-17: Phase II study of IMRT re-irradiation with concurrent/adjuvant nivolumab in patients with locoregionally recurrent or second primary squamous cell cancer of the head and neck. Report No.: NCT03521570.
65. National Library of Medicine. A phase II trial of reirradiation combined with open label pembrolizumab in patients with locoregional inoperable recurrence or second primary squamous cell carcinoma of the head and neck. Report No.: NCT02289209.
66. National Library of Medicine. An open-label, phase 2 efficacy study with window of opportunity immune assessment of pembrolizumab in relapsed, locally recurrent squamous cell carcinoma of the head and neck. Report No.: NCT02769520.
67. National Library of Medicine. A phase II randomized trial of adjuvant therapy with pembrolizumab after resection of recurrent/second primary head and neck squamous cell carcinoma with high risk features. Report No.: NCT04671667.
68. National Library of Medicine. A single arm phase 2 study of adjuvant nivolumab after salvage resection in head and neck squamous cell carcinoma patients previously treated with definitive therapy. Report No.: NCT03355560.
69. National Library of Medicine. Adjuvant immunotherapy after salvage surgery in head and neck squamous cell carcinoma: phase 2 trial evaluating the efficacy and the toxicity of nivolumab alone, and the combination nivolumab and ipilimumab. Report No.: NCT03406247.

Advances in Molecular, Functional, and Anatomical Head and Neck Imaging

6

Akash Deelip Shah, Ramesh Paudyal, and Amita Shukla-Dave

Key Points

- US is useful for the evaluation of thyroid, salivary gland, and nodal disease. US-guided FNA is the most accurate method for nodal staging. Newer US elastography technology is promising for characterization of malignant thyroid nodules and lymph nodes.
- Contrast-enhanced CT is the best imaging modality for anatomic definition of tumors in the oropharynx, hypopharynx, and larynx. Multidetector CT technology minimizes artifact and allows for easy multiplanar visualization of tumors. Newer dual energy CT technology provides better contrast and differentiation of tissues.
- Multisequence contrast-enhanced MR provides superior soft tissue detail for tumors in the nasopharynx, oral cavity, sinonasal cavity, and salivary glands. Higher soft tissue contrast in MR provides a window into bone marrow, vascular, and perineural infiltration.
- Functional MRI can provide noninvasive biomarkers for assessing tumor histology, aggressiveness, and prediction of clinical outcomes.
- Metabolic imaging with FDG PET/CT is vital for staging, treatment monitoring, and surveillance in advanced head and neck squamous cell carcinoma (HNSCC), is excellent at identifying unknown primary tumor sites, and identifies tumor recurrence earlier than conventional imaging.

Introduction

The anatomic and functional complexity of the head and neck region makes accurate diagnosis challenging. Head and neck cancer management is largely dependent on the locoregional anatomic extension and distant spread of tumors. Imaging provides critical detail that allows for cancer staging in accordance with the tumor-node-metastasis (TNM) system, treatment selection, and prognosis. While mucosal lesions are readily detected by clinical examination, submucosal extension and regional and distant spread cannot be assessed solely by clinical examination and require imaging.

Akash Deelip Shah and Ramesh Paudyal contributed equally with all other contributors.

A. D. Shah (✉) · R. Paudyal · A. Shukla-Dave
Memorial Sloan Kettering Cancer Center, New York, NY, USA
e-mail: shaha@mskcc.org; paudyalr@mskcc.org; davea@mskcc.org

R. A. Chandra, R. J. Li (eds.), *Multidisciplinary Management of Head and Neck Cancer*,
https://doi.org/10.1007/978-3-031-05973-5_6

Imaging for head and neck cancers follows a multimodal approach incorporating ultrasonography (US), computed tomography (CT), magnetic resonance imaging (MRI), and combined positron emission tomography and CT (PET/CT) to assess the anatomic and functional status of disease [1]. Modality selection and utilization is tailored depending on the organ of interest and whether imaging is acquired for diagnosis, staging, treatment planning, and/or surveillance.

While a thorough review of imaging in head and neck cancers is outside of the scope of this text, this chapter will explore state-of-the-art imaging techniques, the latest evidence, and recent advances in the application of anatomic, functional, and metabolic imaging to head and neck cancers.

Ultrasound

In head and neck cancers, ultrasonography is most often used in the characterization of superficial primary and nodal disease, including evaluation of thyroid nodules, local spread of thyroid cancer, salivary gland neoplasms, and neck lymphadenopathy [2]. The primary advantages of US in comparison to other modalities are low cost and easy accessibility, rapid scan times, high spatial resolution, and absence of ionizing radiation. Disadvantages include the inability to assess deeper neck structures and operator dependence.

Ultrasound is the first-line imaging tool for the evaluation of thyroid nodules. Because of the high prevalence of thyroid nodules and the typically indolent nature of papillary type thyroid cancers, non-evidence-based management of thyroid nodules has the potential to place significant burden on healthcare costs and increase patient anxiety without improving outcomes. Therefore, in 2015, a standardized Thyroid Imaging, Reporting, and Data System (TIRADS) was formalized to provide guidance on management of thyroid nodules based on sonographic features, allowing risk stratification of recommendations for follow-up and biopsy [3, 4]. The focus of TIRADS is identifying six categories of imaging features for risk stratification: nodule composition (solid or predominantly solid, cystic or predominantly cystic, and spongiform), echogenicity (hyperechoic, isoechoic, hypoechoic, very hypoechoic), shape (wider-than-tall or taller-than-wide), margins (smooth, ill-defined, lobulated or irregular, extrathyroidal extension), and presence of echogenic foci (microcalcifications, rim calcifications, or punctate echogenic foci). Points are assigned based on the presence of these features, and recommendations for continued follow-up or tissue diagnosis are made based on the TIRADS level and maximum nodule diameter. Cystic and spongiform composition, hyperechoic echogenicity, wider-than-tall shape, smooth margins, and absence of microcalcifications or punctate echogenicities favor benign cytology. A recent meta-analysis of the accuracy of the TIRADS level 5 (highly suspicious) categorization demonstrated 70% sensitivity and 89% specificity.

Recent advances in the piezoelectric crystal technology have enabled higher frequency and bandwidth sonographic imaging which allows for better spatial and contrast resolution. For example, newer ultrasound probes can better assess thyroid and salivary gland masses for spiculated and infiltrative margins and thyroid, salivary, and nodal masses for extracapsular extension. With improvements in near-field resolution, US can play a role in detecting extralaryngeal disease in laryngeal neoplasms and visualize vocal cord movement [5]. Improved depth penetration has allowed transcervical evaluation of the oropharynx, enabling detection of small tumors [6]. These applications may be useful in patients for whom conventional CT and MR are not feasible.

Ultrasound elastography (USE) aims to characterize tissue based on its elasticity, operating on a principle similar to clinical palpation. Strain USE (sUSE) maps identify tissues that undergo mild deformation from the transducer during ultrasound interrogation, allowing qualitative and semi-quantitative assessment of tissue stiffness. Shear-wave elastography (SWE) allows for quantitative assessment of tissue elasticity in response to subtle motion in response to acoustic impulses. Tissue stiffness is measured utilizing the parameters of elasticity score (ES), strain ratio (SR), and SWE indices. USE can increase the accuracy of conventional ultrasound for thyroid nodules [7]. Papillary thyroid carcinomas demonstrate greater stiffness than benign nodules (higher ES, SR, and SWE indices) [8, 9]. A recent meta-analysis of thyroid nodules evaluated using USE demonstrated sensitivities of 83% and 78% and specificities of 83% and 82% for sUSE and SWE, respectively [8]. At present, operator, patient, and tissue variability contributes to frequent false positives and false negatives, precluding widespread adoption of USE; however, USE can play an important role in further characterization of indeterminate thyroid nodules on conventional US. USE also shows promise in identifying malignant lymph nodes as characterized by increased stiffness. Numerous studies have shown variable accuracies, for example, ranging from 62% to 94% for one SWE technique [10].

US also provides real-time imaging guidance for fine needle aspiration cytology (FNAC) and core-needle biopsies (CNB) in the head and neck. US-guided FNAC is the most accurate nodal staging method in most head and neck cancers [11, 12]. However, US-guided CNB is the optimal biopsy technique when lymphoma is clinically suspected; the tissue volume allows for accurate histological and immunohistochemical analysis [13, 14]. It is preferred over excisional biopsies in high-risk patients, including those for whom anesthesia poses a risk or those who have a history of radiation therapy or surgery in the neck due to increased wound healing complications [15]. Additionally, there are growing indications for US-CNB, including identifying the human papillomavirus (HPV) status of head and neck squamous cell carcinoma (HNSCC) tumors [10].

Lastly, there are several emerging applications of ultrasound for head and neck cancer. Transoral ultrasound can assist with tumor staging and anatomical delineation for oral cavity and oropharyngeal masses; tumor thickness in oral cancer is a predictor of nodal metastasis [16]. Contrast-enhanced ultrasound has yet to find a useful role in the management of head and neck cancer but has shown some utility in differentiating benign from malignant lymphadenopathy [17]. Studies are ongoing to assess the potential value of CEUS in diagnosis and staging of head and neck cancers.

Computed Tomography and Magnetic Resonance Imaging

Computed tomography (CT) and magnetic resonance (MR) cross-sectional imaging are invaluable in pretherapeutic staging and treatment planning for head and neck cancers. Both the modalities can provide complementary information about the location and extent of tumors, relationship of tumors to surrounding structures, and nodal involvement, allowing for accurate staging, treatment planning, and determining prognosis. Cross-sectional imaging is also routinely used in treatment monitoring and surveillance.

CT

CT is the most widely used imaging technique in head and neck neoplasms, although it is most commonly used in the evaluation of palpable neck masses and lesions within the oropharynx, hypopharynx, and larynx. CT can determine tumor extent and size, identify nodal disease, assess treatment response, and identify recurrence. Compared to MRI, CT has several advantages. Operationally, it is more widely available, costs less, is faster, and is easily reproducible. Reduced scan time is particularly important in the head and neck region, which is heavily susceptible to respiratory and swallowing motion artifact, improving accuracy of interpretation in addition to allowing for a better patient experience. Diagnostically, CT provides superior bone detail and intratumoral calcium detection. Easy multiplanar reformatting allows for easier interpretation. However, disadvantages of CT include inferior soft tissue contrast resolution to MR, the need for iodinated contrast (which should be avoided in treatment-naïve thyroid cancer), ionizing radiation exposure, and dental artifacts.

The major development in CT over the past two decades has been the introduction and refinement of multidetector spiral CT (MDCT), which enables rapid acquisition of volumetric data that can be reconstructed into multiple planes to optimize the signal to noise ratio. MDCT results in reduced scan time and patient radiation dose.

Patients are scanned with the neck in slight extension during quiet respiration. Slice thicknesses of 0.6–1.25 mm are generally used. Three-dimensional display of volumetric data is also possible with this technique and is primarily used in the setting of surgical planning and virtual endoscopic visualization of tumors. Dynamic maneuvers can be used to improve visualization of certain anatomic structures. For example, a modified Valsalva maneuver dilates the hypopharynx and accentuates the pyriform sinuses and postcricoid region to assess lesions that are otherwise obscured by apposition of mucosal surfaces. Phonation can improve visualization of small lesions in the vocal cords. Open mouth instruction can allow visualization of lesions otherwise obscured by dental artifact [18].

Intravenous iodinated contrast is critical to provide contrast between soft tissue, vasculature, and pathology. A single bolus of 80–100 cc contrast injected at 1–2 cc/s suffices. The main contraindication to intravenous iodinated contrast is severe renal failure (eGFR <30, mL/min/1.73 m^2); however, measures to mitigate the risk of contrast in these patients can be undertaken in cases where imaging is absolutely necessary.

The median radiation dose for multidetector CT scans of the neck is 3.9 mSv [19]. For reference, the average radiation dose that a person living in the United States receives annually is 6.2 mSv [20].

Dual Energy CT

Dual energy CT (DECT) is a newer technique that offers further differentiation of tissue based on composition. In conventional CT, the attenuation of tissues with differing elemental compositions can be similar; for example, iodine and calcium have overlapping CT densities. It can be difficult to differentiate vascular calcification from vascular contrast. The linear attenuation coefficient of a given CT voxel is related not only to material composition but also to the photon beam energy and mass density of the material [21]. With DECT, different energy spectra can be utilized to differentiate and quantify material composition. For example, iodine and bone have similar linear attenuation coefficients at 100 KeV; however, simultaneously obtaining additional images at 50 KeV can differentiate the two. Newer DECT protocols operate without increasing radiation dose to the patient. The standard display of DECT images uses virtual monochromatic maps that simulate a CT obtained at one energy spectrum. There are several promising applications of DECT, including the utilization of lower energy spectra to accentuate iodine contrast enhancement, basis material decomposition maps to label the concentration of iodine and other material within tissues, generation of virtual noncontrast images from contrast-enhanced images, and generation of virtual noncalcium maps to assess bone marrow edema. DECT has been used in the diagnosis and staging of head and neck cancers. HNSCC demonstrate improved visibility on virtual monochromatic maps reconstructed at energies lower than 65–70 KeV [22]. Similarly, low energy virtual monochromatic maps and iodine maps have been used to differentiate nonossified thyroid cartilage from thyroid cartilage tumor invasion, recurrent tumor from posttreatment change, and malignant from benign lymph nodes [23–25].

MRI

MRI is often the imaging modality of choice in regions that benefit from high tissue contrast such as the nasopharynx, oropharynx and oral cavity, sinonasal cavity, and salivary glands. It is particularly important in treatment planning prior to radiation therapy, allowing for precise delineation of radiation fields to spare surrounding structures. MRI is excellent at assessing soft tissue tumor infiltration, bone marrow infiltration, perineural spread, vascular invasion, and nodal disease [26]. Although MRI provides superior soft tissue contrast resolution and avoids ionizing radiation, longer scan times can lead to excessive patient motion and discomfort, and costs are higher. Use is also limited in patients with ferromagnetic implants, claustrophobia, and renal insufficiency, the latter due to the risk of nephrogenic systemic fibrosis.

Anatomical MR images are acquired using standard imaging protocols including fat-saturated T_2-weighted, precontrast T_1-weighted, and postcontrast T_1-weighted sequences in multiple planes. High T_1 signal can derive from fat, methemoglobin, melanin, proteinaceous fluid, and some paramagnetic substances. Low T_1 signal derives from air, fluid collections, calcifications, scar tissue/fibrosis, and vascular flow voids. T_1-weighted images benefit from the increased conspicuity of fat, thereby readily visualizing low signal tumoral tissue that infiltrates or effaces fat planes. Some head and neck lesions demonstrate characteristic signal on T_2; for example, fibrous tissue demonstrates low T_2 signal, and fluid collections and edema demonstrate high T_2 signal. High T_2 signal additionally derives from deoxyhemoglobin and sometimes fat. Low T_2 signal additionally derives from calcification/mineralization, hemosiderin, paramagnetic substances, and vascular flow voids. Dynamic contrast-enhanced (DCE)-MRI utilizes gadolinium-based contrast agent (CA) administered intravenously that enhances the tissue water protons relaxation rate constant ($R_1 = 1/T_1$) [27]. Diffusion-weighted imaging (DWI-MRI) is used to identify areas of restricted diffusion as can be seen in hypercellular tumors and pathologic lymph nodes [28] on the basis of diffusion of water molecules in tissue.

Contraindications to MRI include patients with cardiac implantable electronic devices (although newer MR safe devices are available), metallic intraocular foreign bodies, implantable neurostimulators, cochlear implants, drug infusion pumps, and cerebral aneurysm clips. There are several additional relative contraindications including coronary stents, programmable shunts, intrauterine devices, and IVC filters. The reader is referred to MRIsafety.com for details on individual devices and scenarios. Contraindications to gadolinium contrast administration include severe renal failure (eGFR <30 mL/min/1.73 m^2) and pregnancy [29].

Functional Imaging

CT and MRI are optimal for delineating tumor anatomy; however, histopathological identification, detection of small nodal metastases, distinguishing posttreatment change from residual or recurrent tumor, and assessment of treatment response remain major challenges for cross-sectional imaging. Over the past decade, functional MRI techniques including DW- and DCE-MRI have come to the forefront in head and neck cancers. These techniques enable both qualitative and quantitative evaluation of the functional status of tumors and posttreatment response. These emerging technologies will play a growing role in histopathological identification, predicting treatment response to chemotherapy and radiation, identifying recurrent disease, treatment monitoring, and surveillance.

DW-MRI

The diffusion-weighted magnetic resonance imaging (DW-MRI) technique generates the signal contrast by capturing random motion (i.e., Brownian motion) of water molecules in tissue [30]. The development of DW-MRI techniques has resulted in better diagnostic performance in detecting primary and recurrent head and neck (HN) cancer. The cell membranes, intracellular organelles, and macromolecules hinder and restrict water molecules' movement in tissue. Tissue microstructural restrictions and microcirculation contribute to DW signal attenuation, reflecting abnormalities in tissue organization at the cellular level (e.g., tissue microstructure and cellularity) [31]. Therefore, quantitative DW images are used for lesion characterization, prognosis, and evaluation of treatment response.

DW-MRI techniques are rapidly evolving, with improvements in quality and speed of data acquisition [32]. Single-shot diffusion-weighted echo-planar imaging is commonly used for DWI data acquisition because of its short duration and good signal–noise ratio but is limited by lower resolution, especially at the neck region due to air–tissue interfaces. However, susceptibility variations across the neck regions cause local magnetic field inhomogeneities and can lead to image distortions [33].

The degree of water mobility in tissue can be quantified by calculating the apparent diffusion coefficient (ADC, [mm^2/s]) from the DW signal with at least two or more b-values. The DW signal as a function of the b-values can be fitted for each voxel in the image using the following monoexponential model (Eq. 6.1):

$$S_b = S_0 \mathrm{e}^{-b\times \mathrm{ADC}} \quad (6.1)$$

where S_b and S_0 denote signal intensities with and without diffusion weighting, respectively; b is the diffusion-sensitizing factor; ADC is a surrogate marker of tumor cellularity.

At higher b-values, the signal decay fits the multiexponential model in vivo. In contrast, at lower b-values ($b < 100$ s/mm^2), the ADC is a composite metric influenced by tissue microperfusion and can lead to overestimation of the ADC. Thus, the choice of the optimal b-values will affect the signal in the DWI images and modify the ADC value. Le Bihan introduced the intravoxel incoherent motion (IVIM) model to account for thermally driven tissue diffusion and blood flow microcirculation of the randomly oriented capillaries [34]. Le Bihan hypothesized that signal attenuation due to perfusion is involved at low b-values ($b \leq 100$ s/mm^2), relating to the signal from the fraction of the perfusion space. In contrast, true diffusion signal attenuation dominates at higher b-values ($b > 100$ s/mm^2) [34].

A biexponential model that describes the signal arising from the two-compartment tissue (i.e., intravascular and extravascular space) model as a function of b-values for the IVIM model without injection of CA is given by (Eq. 6.2) [34]:

$$S_b = S_0 \left[f\,\mathrm{e}^{-bD^*} + (1-f)\,\mathrm{e}^{-bD} \right] \quad (6.2)$$

where D is the true diffusion coefficient, D^* is the pseudo diffusion coefficient, and f is the perfusion fraction.

The complex cellular structures of tissue membranes alter the displacement of a water molecule that substantially deviates diffusion from a Gaussian nature (non-Gaussian [NG]) and readily observable at high b-values ($b > 100$ s/mm^2). Jansen et al. apply the concept of diffusion kurtosis imaging (DKI) to an oncological setting, in particular HNSCC [35].

The DW signal data as a function of b-value to the DKI model are fitted as follows (Eq. 6.3) [36, 37]:

$$S(b) = S_0 \left[\mathrm{e}^{-b\times D_{\mathrm{app}} + \frac{1}{6} K_{app} \left(b\times D_{\mathrm{app}}\right)^2} \right] \quad (6.3)$$

where D_{app} and K_{app} are the apparent diffusion (mm^2/s) and kurtosis (unitless) coefficients, respectively.

The hindered and restricted diffusion can be described simultaneously by incorporating the diffusion kurtosis into the IVIM model called NG-IVIM. The NG-IVIM DW model provides estimates of the quantitative metrics f, D, D^*, and K. The DW signal vs. b-value is fitted to NG-IVIM as follows [34, 38]:

$$S(b) = S_0 \left[f\,\mathrm{e}^{-bD^*} + (1-f)\,\mathrm{e}^{-b\times D + \frac{1}{6} K \left(b\times D\right)^2} \right] \quad (6.4)$$

where K is the kurtosis coefficient.

DCE-MRI

DCE-MRI acquires sequential images before, during, and after injection of a CA [39]. The DCE signal enhancement is associated with the spin-lattice or longitudinal relaxation time (T_1) of water protons by a short-range interaction (nm). The DCE signal can be modeled either semi-quantitatively or quantitatively. The semi-quantitative analysis uses the signal enhancement curve to calculate the upslope or washout phase, an initial area under the curve (AUC), and time to peak [39]. In contrast, the quantitative analysis utilizes the commonly used Tofts pharmacokinetic model to estimate the quantitative metrics that characterize underlying tumor physiology such as perfusion and permeability [40]. Accurate quantification of the multiple kinetic parameters requires selection of an appropriate pharmacokinetic model [41–43]. The extended Tofts model (ETM) estimates the volume transfer constant of a CA, K^{trans} (min^{-1}), the volume fraction of the extravascular extracellular space (EES), v_e, and volume fraction of blood plasma space, v_p. K^{trans}

represents plasma flow (F_p), when $F_p \ll$ PS and PS permeability surface area product (PS) when $F_p \gg$ PS [41]. The flow and permeability limited conditions can be seen in leaky vascular organs such as liver and largely intact blood–brain barrier, respectively.

A linear relationship is assumed between the change of longitudinal relaxation rate, R_1 ($\Delta R_1 = 1/\Delta T_1$), and the total amount of CA in the tissue, such that water exchange is in the fast exchange limit (FXL) [40]:

$$R_1(t) = R_{10} + r_1 C_t(t) \rightarrow \Delta R_1(t) = r_1 C_t(t) \quad (6.5)$$

where t is time, $R_1(t)$ is the time course of tissue R_1, R_{10} (s) is the precontrast R_1, r_1 ($\mathrm{mM^{-1}\,s^{-1}}$) is the longitudinal relaxivity of the CA, and $C_t(t)$ is the tissue concentration of CA (mM). The longitudinal relaxivity is assumed to be a constant and is independent of its location in the tissue. It has been reported that relaxivity is dependent on CA macromolecular content [44, 45].

The relaxation rate constant of the EES for FXL is given by:

$$R_{1e}(t) = R_{10e} + r_1 C_e(t) \quad (6.6)$$

where $R_{1e}(t)$ is the time course of EES R_1, R_{10e} (s) is the precontrast R_{10}, and $C_e(t)$ is the EES concentration of CA (mM).

The ETM assumes that the CA exchanges between the vascular space and EES. The total tissue CA concentration is given by [40] (Eq. 6.6):

$$C_t(t) = K^{\mathrm{trans}} \int_0^t C_p(\tau) e^{-k_{ep}(t-\tau)} d\tau + v_p C_p \quad (6.7)$$

where $k_{ep} = K^{\mathrm{trans}}/v_e$ is the rate constant of CA transport from the vascular space, v_e is the volume fraction of EES, and C_p is the concentration–time course of contrast agent in the blood plasma, known as the arterial input function.

The TM model tissue CA concentration can be readily obtained from Eq. 6.6 for a weakly vascularized tissue (i.e., $v_p \sim 0$) [40]:

$$C_t(t) = K^{\mathrm{trans}} \int_0^t C_p(\tau) e^{-k_{ep}(t-\tau)} d\tau \quad (6.8)$$

T_1w DCE-MRI accounts for equilibrium water exchange across the vascular wall (between intravascular space and extravascular space) and the cellular wall (between intracellular space [ICS] and EES). Notably, CAs do not enter the cell. Therefore, the relation between the relaxation rate constant and CA concentration is not so straightforward. The intercompartmental equilibrium water exchange kinetics can be described by a linear three-site two-exchange [3S2X] model for modeling longitudinal relaxation rate of tissue water protons which is adapted from the Bloch–McConnell equations [46, 47]. The full three-compartment model has five parameters, including K^{trans}, v_e, v_p, and the two rate constants of water exchange across the vascular endothelium and the cellular wall [48]. The solution of Bloch–McConnell's equations for two-site water exchange model (i.e., intracellular space and EES) yields two Eigenvalues of biexponential signal, representing the two longitudinal relaxation rate constants that provide estimates of three parameters, K^{trans}, v_e, and the mean lifetime of intracellular water protons, τ_i [49]. One of the Eigenvalues represents the longitudinal relaxation rate constant for the fast exchange regime model (FXR), also called the shutter speed model (SSM). The observable $R_{1t}(t)$ for the FXR regime is given by [49]:

$$R_{1t}(t) = \frac{1}{2}\left[\left(R_{1i} + k_{ie} + R_{1e} + k_{ei}\right) - \sqrt{\left(R_{1i} + k_{ie} - R_{1e} - k_{ei}\right)^2 + 4 k_{ie} k_{ei}}\right] \quad (6.9)$$

where R_{1i} and R_{1e} are the relaxation rates of ICS and ESS, k_{ie} ($k_{ie} = 1/\tau_i$) and k_{ei} are the rates of water exchange from the ICS to EES, and vice versa.

Molecular Spectroscopy

Magnetic resonance spectroscopy (MRS), including both phosphorous-31 (^{31}P) and proton (^{1}H), characterizes the tumor tissue metabolism at a cellular level [50]. MRS technique has the unique ability to assess tumor physiology at the molecular level by evaluating the presence of specific metabolites [51]. Phosphorous MRS (^{31}P MRS) is used to assess tissue bioenergetics and metabolism of membrane phospholipids [52]. In contrast, proton MRS provides information about cellular metabolism, describing the underlying biologic and pathophysiologic events associated with tumors [53]. The biochemical pathways involved in ^{1}H MRS of choline may be different from the phospholipid metabolites seen on ^{31}P MRS, and thus the two MRS techniques may provide complementary information on the tumor metabolism. MRS has shown promise to differentiate nonmalignant from malignant tumors and lymph nodes and to differentiate between residual malignancies from postradiation changes in head and neck cancers [53].

Tumor Characterization with Quantitative DWI and DCE-MRI

Monoexponential and NG-IVIM-derived ADC/D maps correlate with tumor cellularity due to the restricted free diffusion of water in tumors with increased cellular density; these tumors tend to have lower ADC values. Less dense tumors with necrotic and cystic elements have higher ADC values [54].

Several studies have demonstrated that DWI can distinguish HN tumor types, including differentiating nasopharyngeal squamous cell carcinoma from nasopharyngeal lymphoma, head and neck cysts from tumors, and benign from malignant head and neck tumors [55–57]. DWI has also been applied in the salivary glands to distinguish pleomorphic adenoma from carcinomas [58]. Warthin's tumors show overlap with carcinomas, possibly due to the presence of lymphoid tissue. There is evidence for the utility of DWI in distinguishing benign and malignant thyroid nodules; this limited utility may be related to heterogeneity in the histologic composition of abnormal thyroid nodules. In general, head and neck malignancies demonstrate low ADC values due to hypercellularity, enlarged nuclei, and hyperchromatism. Head and neck lymphomas have the lowest ADC values, and benign lesions outside of the thyroid gland tend to have higher ADC values than malignant lesions. DWI has also been used to identify HPV-positive oropharyngeal squamous cell carcinomas, which carry a better prognosis [59, 60]. Nonkeratinizing and basaloid differentiated histology may contribute to lower ADC values in these tumors [61]. Additional studies have demonstrated potential corelates between ADC values and tumoral expression of Ki-67, EGFR, VEGF, p53, p16, and HER2 [62, 63]. Such results may facilitate the use of DWI in creating individualized tailored treatment plans with targeted agents. Representative monoexponential and NG-IVIM model-derived parametric maps are shown in Fig. 6.1.

Similar efforts have been undertaken to evaluate the role of DCE-MRI in differentiating head and neck tumor types. DCE-MRI has proved useful in distinguishing SCC from lymphoma. SCC demonstrates increased tumoral perfusion and capillary permeability, possibly due to its lower cellularity in comparison to lymphoma [64, 65]. DCE-MRI parameters have been studied in differentiating paragangliomas from schwannomas in the carotid space, with paragangliomas demonstrating decreased perfusion parameters; this may reflect the poor perfusion environment in paragangliomas due to the presence of pathologic vasculature and extensive arteriovenous shunting [66, 67].

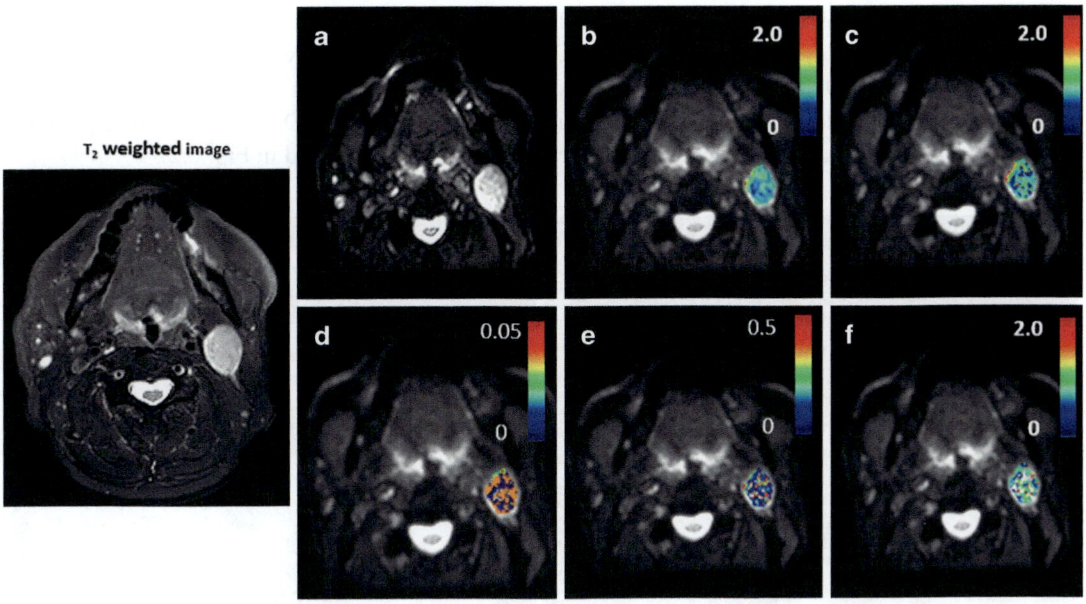

Fig. 6.1 Left: Representative T_2-weighted image from a patient with head and neck cancer (65 years, male). Right: (**a**) diffusion weighted image ($b = 0$ s/mm^2), (**b**). Apparent diffusion coefficient (ADC × 10^{-3} [mm^2/s]), (**c**) true diffusion coefficient (D × 10^{-3} [mm^2/s], (**d**) pseudo-diffusion coefficient (D^* × 10^{-3} [mm^2/s]), (**e**) perfusion fraction (f), and (**f**) kurtosis coefficient (K). ADC was derived from a monoexponential model, and D, D^*, f, and K were derived from non-Gaussian intravoxel incoherent motion model

Tumor Staging

Nodal disease is a key component of TNM staging and treatment planning. Conventional cross-sectional CT and MR have limited sensitivity and specificity in the detection of malignant lymph nodes, as an evaluation solely based on size and morphology can miss active disease. DWI has been applied to differentiating benign from malignant lymph nodes; nodal staging is more accurate with the addition of DWI to conventional MRI. ADC values tend to be lower in malignant nodes [68–71]. DCE-MRI has also shown utility here, with malignant nodal tissue demonstrating decreased transit of contrast and reduced volume of the extravascular space; in one study, these nodes demonstrated longer time to perfusion, lower peak enhancement, and slower washout [72].

Therapy Response Assessment

Tumors with lower cellularity, increased necrotic and cystic components with poor oxygenation, higher stromal content, and HPV-negative status demonstrate greater resistance to treatment and poorer outcomes; these tumors tend to demonstrate higher ADC and D values [59, 73]. Higher baseline ADC and D values in HNSCC can predict poor local control and poor treatment response, correlating to increased risk of recurrence [74]. Evaluating changes between baseline ADC and D values during treatment may be more clinically relevant, as it minimizes variability due to differences in individual scanners and site protocols. Greater rise in ADC and D values following treatment is predictive of tumor response [59, 75].

Disordered tumoral angiogenesis with the development of leaky and tortuous vessels mediated by the release of VEGF results in a hypoxic tumoral environment [76]. Tumor hypoxia is associated with treatment resistance, aggressive disease, and poor clinical outcomes [77]. These hypoxic environments can develop in areas of high cellular density in addition to poor blood perfusion [78]. DCE-MRI parameters can be used to characterize tumor hypoxia. Several studies have demonstrated better treatment response in head and neck tumors with high baseline and posttreatment perfusion, likely through improved chemotherapeutic agent and oxygen delivery.

These studies have shown higher rates of local control and complete response [38, 79, 80]. Higher perfusion on DCE-MRI also correlates with better lymph node treatment response in metastatic HNSCC [81, 82]. The mean lifetime of intracellular water protons derived from the FXR model have shown promise as a prognostic marker for patients with HN cancer [83]. Representative FXR DCE-MRI-derived parametric maps are displayed in Fig. 6.2.

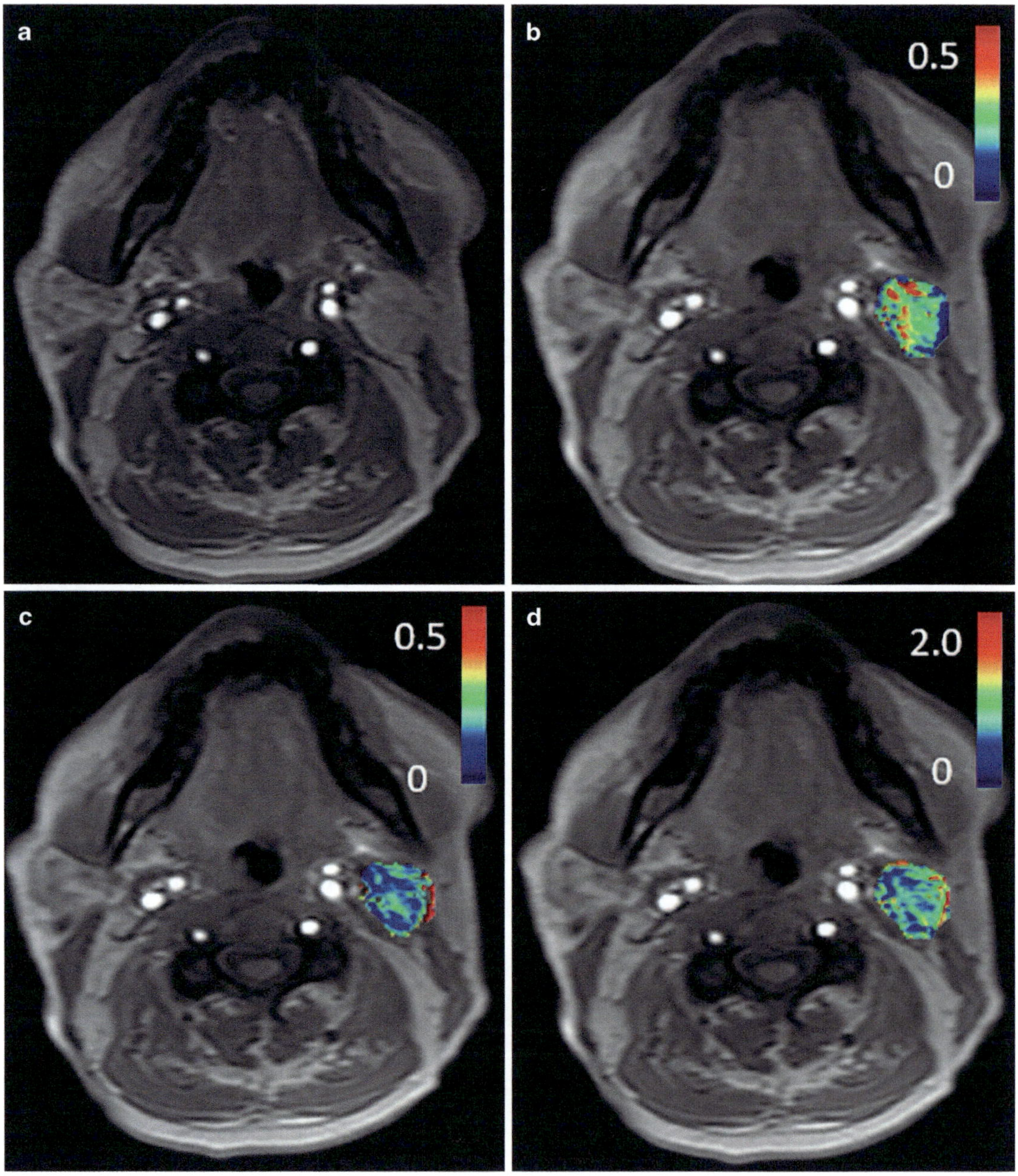

Fig. 6.2 (**a**) Representative T_1-weighted (T_1w) image of an early dynamic phase after injection of contrast agent, (**b**) volume transfer constant (K^{trans} [min^{-1}]), (**c**) volume fraction of the extravascular extracellular space (v_e), and (**d**) mean lifetime of intracellular water molecules (τ_i [s]) from a patient with head and neck squamous cell carcinoma (65 years, male). Parametric maps generated from the fast exchange model were overlaid on precontrast T_1w images

Posttreatment Change

Anatomic distortion and inflammation from surgical and radiation therapy limit the utility of conventional CT and MRI in detecting underlying residual or recurrent tumor. Residual HNSCC after treatment demonstrates lower ADC than posttreatment fibrosis, likely secondary to increased cellularity within residual disease [84, 85]. DCE-MRI has also been used to identify residual disease posttreatment. Posttreatment fibrosis has been found to have higher permeability surface area, longer time to peak, lower relative washout ratio, and greater contrast uptake and enhancement ratio [86–88]. Enhancement in residual tumor is earlier and more intense due to differing perfusion microenvironments. Dose de-escalation approaches that have been proposed for HPV-related and p16+ SCC, wherein subclinical and nodal targets receive doses of 30 Gy instead of the standard 70 Gy, may minimize treatment-related toxicity and facilitate improved identification of residual or recurrent tumor [89].

Metabolic Imaging

Imaging of in vivo metabolic pathways and receptor-ligand interactions provides important information in the evaluation of neoplasms; PET achieves this by imaging the biodistribution of positron-emitting radioisotopes. PET/CT is the optimal imaging modality for staging, treatment monitoring, and surveillance of advanced HNSCC [26]. It is also useful for identifying synchronous or metachronous lesions in the head and neck as well as identifying unknown primary sites. Various radioisotopes can be used; however, ^{18}F-fluorodeoxyglucose (FDG) is the most common. Similar to blood glucose, FDG is transported into cells with high glucose metabolism, thus identifying neoplastic, infectious, and inflammatory tissue. Because glycolytic activity in inflammatory tissue can result in false-positive FDG uptake, PET/CT is typically delayed until 8–12 weeks following radiotherapy. The most accurate imaging technique is to combine PET with contrast-enhanced CT in the head and neck, which allows for better anatomic localization. Simultaneous acquisition with CT-based attenuation correction reduces imaging time. Current standard indications for PET/CT for head and neck cancer are the evaluation of T_3 and T_4 tumors, clinically suspected nodal or distant metastatic disease, treatment monitoring, and surveillance to assess for recurrent tumor [90]. Patients must fast for at least 4–6 h prior to imaging to minimize blood glucose competing with FDG, which decreases image quality; 14–18 mCi FDG is injected and images are obtained 1 h after injection.

Unknown Primary

Five to ten percent of patients with HNSCC present with metastatic neck lymphadenopathy of unknown primary site [91]. PET/CT can identify primary tumor sites in approximately 25% of these patients [92]. A recent meta-analysis demonstrated 97% sensitivity and 68% specificity for PET/CT in the detection of primary tumor sites [93]. It is important to note that PET/CT frequently results in false negatives for small lesions less than 1 cm in size owing to its limited resolution and volume averaging effects.

Staging and Response Assessment

Accurate T staging requires precise anatomic definition from cross-sectional CT and MR imaging; the limited resolution of FDG PET precludes its application to T staging. Background physiologic FDG activity within the pharyngeal tissue can further obscure identification of the primary lesion. However, PET/CT has demonstrated value in T staging of oral cavity cancers because of its ability to readily detect mandibular involvement, which is an important determinant of surgical planning; one study demonstrated a sensitivity of 100% and specificity of 85% in the detection of mandibular invasion from intraoral squamous cell carcinoma [94].

Several studies have established the utility of PET/CT in nodal staging for HNSCC, with sensitives ranging from 86% to 98% and specificities ranging from 88% to 99%, with higher accuracy for nodal staging than CT or MR [95, 96]. CT and MR can fail to detect nonenlarged and morphologically normal lymph nodes with active disease. The recently completed ACRIN 6685 trial demonstrated a high negative predictive value of 94% for FDG PET/CT for T_2 to T_4 and N0 disease. Findings from PET/CT changed surgical management in 22% of these patients. Thus, FDG PET may be able to spare a subgroup of patients from undergoing elective neck dissections for accurate staging or empiric radiation therapy.

The incidence of distant metastases in HNSCC is approximately 2–18% [97]. Patients with three or more lymph node metastases, lymph nodes larger than 6 cm, bilateral nodal disease, and regional recurrence are at higher risk for distant metastasis [98]. PET/CT has a reported negative predictive value of 99% in the identification of distant metastases [99] and specificity and sensitivity up to 92% and 93%, respectively [100], and therefore is an important pretreatment step to avoid unnecessary treatment.

Postoperative and postradiation edema, fibrosis, and inflammation can mimic residual or recurrent tumor using conventional CT and MR in head and neck cancers, thereby limiting its utility in therapy response assessment. Furthermore, many newer treatments including immunotherapeutic agents are cytostatic, and therefore, tumor size may not serve as an adequate marker of response. FDG PET/CT provides utility for response assessment as it can evaluate the presence of viable metabolically active tumor. Both quantitative and qualitative assessments have shown high accuracy and reliability in detection of treatment response. A large meta-analysis of the performance of FDG PET/CT in posttreatment response assessment demonstrated 94% sensitivity and 82% specificity in the detection of residual tumor, with 95% negative predictive value [101]. Positive predictive value was lower at 75%, possibly reflecting a higher rate of false positives from posttreatment inflammatory tissue, which merits the need for both delayed scans 8–12 weeks after radiation treatment and careful attention to clinical and anatomic imaging findings. Representative CT, PET, and T_1-weighted MR images are shown in Fig. 6.3.

PET/CT is superior to CT for the detection of viable tumor within residual lymphadenopathy. One study reported a negative predictive value of 97% for PET/CT in the detection of residual nodal disease [102]. Several studies support the utility of negative FDG PET findings despite persistent enlarged or morphologically abnormal lymph nodes on CT after definitive CRT [103].

A recent randomized controlled trial of patients with HNSCC and advanced nodal dis-

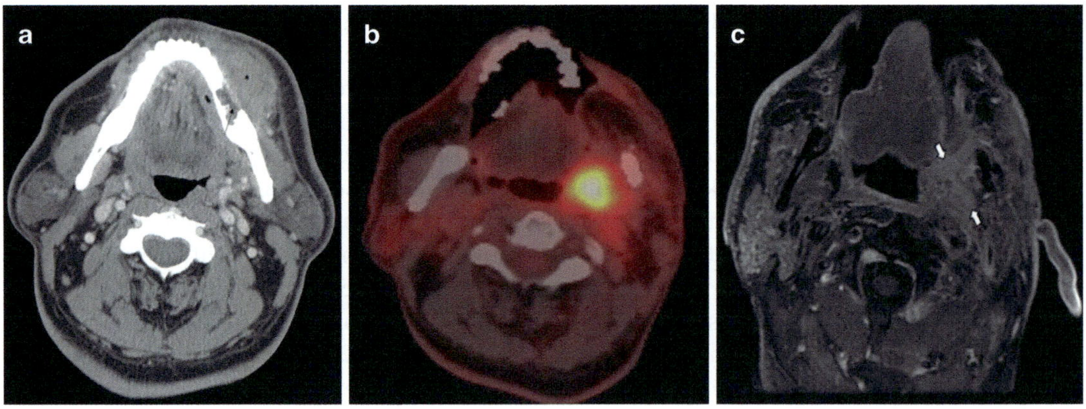

Fig. 6.3 A 56-year-old male with left mandibular gingival tumor (* on CT) with osseous erosion of the mandible evident (black arrow on CT), status postsurgery and radiation therapy, with new ill-defined enhancement in the left oropharynx at the flap margin (open white arrows on MRI), indeterminate for posttreatment changes vs. recurrent tumor. Fused images from PET/CT demonstrate hypermetabolic activity in the region of enhancement in the left oropharynx, suspicious for recurrence. Biopsy demonstrated recurrent squamous cell carcinoma

ease demonstrated similar 2-year overall survival rates for patients who underwent PET/CT surveillance vs. planned neck dissection after completion of chemoradiotherapy. This approach can significantly reduce patient morbidity from additional surgeries and is more cost-effective [104]. PET/CT identifies local, regional, and distant recurrence earlier than conventional CT and MR. In a recent prospective trial, FDG PET/CT detected 99% of recurrences in HNSCC treated with curative surgery of definitive chemoradiotherapy [105]. A more recent meta-analysis demonstrated a sensitivity of 92% and specificity of 87% for PET/CT in the detection of recurrences. The accuracy was higher in patients in whom recurrence was not clinically suspected.

PET/MR

Simultaneous acquisition of PET and MR images has been established as an increasingly viable alternative to PET/CT in recent years, performing equally in the staging of head and neck cancer and radiation therapy planning [106]. Its role in the clinical setting has been limited by high costs and logistical requirements.

Conclusion

Multimodality imaging of the head and neck is a necessary tool for optimal management of head and neck cancers. Recent advances in imaging technology have enabled clinicians to provide accurate tumor diagnosis and staging, effective treatment planning, and improved surveillance. US is valuable in the assessment of superficial tumors including within the thyroid gland, salivary glands, and lymph nodes. US-guided biopsies are a cost-effective and safer alternative to excisional biopsies. CT and MR are excellent for precise anatomic delineation of tumors and soft tissue detail for staging and treatment purposes. Advanced MRI and PET/CT provide complementary information on functional and metabolic pathways within tumors that can help guide tumor detection and differentiation, treatment planning and monitoring, and surveillance. As technology improves in the future, these tools will continue to develop and are likely to see increasingly widespread clinical utility.

References

1. Daisne JF, Duprez T, Weynand B, Lonneux M, Hamoir M, Reychler H, Gregoire V. Tumor volume in pharyngolaryngeal squamous cell carcinoma: comparison at CT, MR imaging, and FDG PET and validation with surgical specimen. Radiology. 2004;233:93–100. https://doi.org/10.1148/radiol.2331030660.
2. Senchenkov A, Staren ED. Ultrasound in head and neck surgery: thyroid, parathyroid, and cervical lymph nodes. Surg Clin North Am. 2004;84:973–1000, v. https://doi.org/10.1016/j.suc.2004.04.007.
3. Tessler FN, Middleton WD, Grant EG, Hoang JK, Berland LL, Teefey SA, Cronan JJ, Beland MD, Desser TS, Frates MC, et al. ACR thyroid imaging, reporting and data system (TI-RADS): white paper of the ACR TI-RADS Committee. J Am Coll Radiol. 2017;14:587–95. https://doi.org/10.1016/j.jacr.2017.01.046.
4. Grant EG, Tessler FN, Hoang JK, Langer JE, Beland MD, Berland LL, Cronan JJ, Desser TS, Frates MC, Hamper UM, et al. Thyroid ultrasound reporting lexicon: white paper of the ACR thyroid imaging, reporting and data system (TIRADS) committee. J Am Coll Radiol. 2015;12:1272–9. https://doi.org/10.1016/j.jacr.2015.07.011.
5. Carneiro-Pla D, Miller BS, Wilhelm SM, Milas M, Gauger PG, Cohen MS, Hughes DT, Solorzano CC. Feasibility of surgeon-performed transcutaneous vocal cord ultrasonography in identifying vocal cord mobility: a multi-institutional experience. Surgery. 2014;156:1597–604. https://doi.org/10.1016/j.surg.2014.08.071.
6. Coquia SF, Hamper UM, Holman ME, DeJong MR, Subramaniam RM, Aygun N, Fakhry C. Visualization of the oropharynx with transcervical ultrasound. Am J Roentgenol. 2015;205:1288–94. https://doi.org/10.2214/Ajr.15.14299.
7. Trimboli P, Guglielmi R, Monti S, Misischi I, Graziano F, Nasrollah N, Amendola S, Morgante SN, Deiana MG, Valabrega S, et al. Ultrasound sensitivity for thyroid malignancy is increased by real-time elastography: a prospective multicenter study. J Clin Endocr Metab. 2012;97:4524–30. https://doi.org/10.1210/jc.2012-2951.
8. Tian WG, Hao S, Gao B, Jiang Y, Zhang XH, Zhang S, Guo LJ, Yan J, Luo DL. Comparing the diagnostic accuracy of RTE and SWE in differentiating malignant thyroid nodules from benign ones: a meta-analysis. Cell Physiol Biochem. 2016;39:2451–63. https://doi.org/10.1159/000452513.

9. Ghajarzadeh M, Sodagari F, Shakiba M. Diagnostic accuracy of sonoelastography in detecting malignant thyroid nodules: a systematic review and meta-analysis. Am J Roentgenol. 2014;202:W379–89. https://doi.org/10.2214/Ajr.12.9785.
10. McQueen AS, Bhatia KS. Head and neck ultrasound: technical advances, novel applications and the role of elastography. Clin Radiol. 2018;73:81–93. https://doi.org/10.1016/j.crad.2017.08.003.
11. van den Brekel MW, Castelijns JA, Stel HV, Luth WJ, Valk J, van der Waal I, Snow GB. Occult metastatic neck disease: detection with US and US-guided fine-needle aspiration cytology. Radiology. 1991;180:457–61. https://doi.org/10.1148/radiology.180.2.2068312.
12. de Bondt RB, Nelemans PJ, Hofman PA, Casselman JW, Kremer B, van Engelshoven JM, Beets-Tan RG. Detection of lymph node metastases in head and neck cancer: a meta-analysis comparing US, USgFNAC, CT and MR imaging. Eur J Radiol. 2007;64:266–72. https://doi.org/10.1016/j.ejrad.2007.02.037.
13. Skelton E, Jewison A, Okpaluba C, Sallomi J, Lowe J, Ramesar K, Grace R, Howlett DC. Image-guided core needle biopsy in the diagnosis of malignant lymphoma. Eur J Surg Oncol. 2015;41:852–8. https://doi.org/10.1016/j.ejso.2015.04.015.
14. Burke C, Thomas R, Inglis C, Baldwin A, Ramesar K, Grace R, Howlett DC. Ultrasound-guided core biopsy in the diagnosis of lymphoma of the head and neck. A 9 year experience. Br J Radiol. 2011;84:727–32. https://doi.org/10.1259/bjr/60580076.
15. Huang PC, Liu CY, Chuang WY, Shih LY, Wan YL. Ultrasound-guided core needle biopsy of cervical lymphadenopathy in patients with lymphoma: the clinical efficacy and factors associated with unsuccessful diagnosis. Ultrasound Med Biol. 2010;36:1431–6. https://doi.org/10.1016/j.ultrasmedbio.2010.05.018.
16. Lodder WL, Teertstra HJ, Tan IB, Pameijer FA, Smeele LE, van Velthuysen MLF, van den Brekel MWM. Tumour thickness in oral cancer using an intra-oral ultrasound probe. Eur Radiol. 2011;21:98–106. https://doi.org/10.1007/s00330-010-1891-7.
17. Slaisova R, Benda K, Jarkovsky J, Petrasova H, Szturz P, Valek V. Contrast-enhanced ultrasonography compared to gray-scale and power doppler in the diagnosis of peripheral lymphadenopathy. Eur J Radiol. 2013;82:693–8. https://doi.org/10.1016/j.ejrad.2012.12.008.
18. Henrot P, Blum A, Toussaint B, Troufleau P, Stines J, Roland J. Dynamic maneuvers in local staging of head and neck malignancies with current imaging techniques: principles and clinical applications. Radiographics. 2003;23:1201–13. https://doi.org/10.1148/rg.235025045.
19. Smith-Bindman R, Lipson J, Marcus R, Kim KP, Mahesh M, Gould R, Berrington de Gonzalez A, Miglioretti DL. Radiation dose associated with common computed tomography examinations and the associated lifetime attributable risk of cancer. Arch Intern Med. 2009;169:2078–86. https://doi.org/10.1001/archinternmed.2009.427.
20. Agency, U.S.E.P. Radiation protection. 2021.
21. McCollough CH, Leng S, Yu L, Fletcher JG. Dual- and multi-energy CT: principles, technical approaches, and clinical applications. Radiology. 2015;276:637–53. https://doi.org/10.1148/radiol.2015142631.
22. Tynan C, Cardea JM. Home health hazard assessment. J Gerontol Nurs. 1987;13:25–8. https://doi.org/10.3928/0098-9134-19871001-07.
23. Liang H, Li A, Li Y, Cheng H, Zhao Q, Li J, Wang Q. A retrospective study of dual-energy CT for clinical detecting of metastatic cervical lymph nodes in laryngeal and hypopharyngeal squamous cell carcinoma. Acta Otolaryngol. 2015;135:722–8. https://doi.org/10.3109/00016489.2015.1015164.
24. Forghani R, Levental M, Gupta R, Lam S, Dadfar N, Curtin HD. Different spectral hounsfield unit curve and high-energy virtual monochromatic image characteristics of squamous cell carcinoma compared with nonossified thyroid cartilage. AJNR Am J Neuroradiol. 2015;36:1194–200. https://doi.org/10.3174/ajnr.A4253.
25. Yamauchi H, Buehler M, Goodsitt MM, Keshavarzi N, Srinivasan A. Dual-energy CT-based differentiation of benign posttreatment changes from primary or recurrent malignancy of the head and neck: comparison of spectral Hounsfield units at 40 and 70 keV and iodine concentration. Am J Roentgenol. 2016;206:580–7. https://doi.org/10.2214/Ajr.15.14896.
26. Vishwanath V, Jafarieh S, Rembielak A. The role of imaging in head and neck cancer: an overview of different imaging modalities in primary diagnosis and staging of the disease. J Contemp Brachyther. 2020;12:512–8. https://doi.org/10.5114/jcb.2020.100386.
27. Tofts PS, Berkowitz B, Schnall MD. Quantitative-analysis of dynamic Gd-Dtpa enhancement in breast-tumors using a permeability model. Magn Reson Med. 1995;33:564–8. https://doi.org/10.1002/mrm.1910330416.
28. Chawla S, Kim S, Wang S, Poptani H. Diffusion-weighted imaging in head and neck cancers. Future Oncol. 2009;5:959–75. https://doi.org/10.2217/fon.09.77.
29. Ghadimi M, Sapra A. Magnetic resonance imaging contraindications. Treasure Island, FL: StatPearls; 2021.
30. Le Bihan D. Diffusion/perfusion MR imaging of the brain: from structure to function. Radiology. 1990;177:328–9. https://doi.org/10.1148/radiology.177.2.2217762.
31. Chenevert TL, McKeever PE, Ross BD. Monitoring early response of experimental brain tumors to therapy using diffusion magnetic resonance imaging. Clin Cancer Res. 1997;3:1457–66.

32. Chilla GS, Tan CH, Xu C, Poh CL. Diffusion weighted magnetic resonance imaging and its recent trend-a survey. Quant Imaging Med Surg. 2015;5:407–22. https://doi.org/10.3978/j.issn.2223-4292.2015.03.01.
33. Bammer R. Basic principles of diffusion-weighted imaging. Eur J Radiol. 2003;45:169–84. https://doi.org/10.1016/S0720-048x(02)00303-0.
34. Le Bihan D. Intravoxel incoherent motion imaging using steady-state free precession. Magn Reson Med. 1988;7:346–51.
35. Jansen JF, Stambuk HE, Koutcher JA, Shukla-Dave A. Non-Gaussian analysis of diffusion-weighted MR imaging in head and neck squamous cell carcinoma: a feasibility study. AJNR Am J Neuroradiol. 2010;31:741–8. https://doi.org/10.3174/ajnr.A1919.
36. Jensen JH, Helpern JA, Ramani A, Lu H, Kaczynski K. Diffusional kurtosis imaging: the quantification of non-Gaussian water diffusion by means of magnetic resonance imaging. Magn Reson Med. 2005;53:1432–40. https://doi.org/10.1002/mrm.20508.
37. Jansen JF, Koutcher JA, Shukla-Dave A. Non-invasive imaging of angiogenesis in head and neck squamous cell carcinoma. Angiogenesis. 2010;13:149–60.
38. Agrawal S, Awasthi R, Singh A, Haris M, Gupta RK, Rathore RKS. An exploratory study into the role of dynamic contrast-enhanced (DCE) MRI metrics as predictors of response in head and neck cancers. Clin Radiol. 2012;67:E1–5. https://doi.org/10.1016/j.crad.2012.03.005.
39. Parker GJM, Padhani AR. T-1-w DCE-MRI: T-1-weighted dynamic contrast-enhanced MRI. In: Quantitative MRI of the brain: measuring changes caused by disease; 2003. p. 341–64. https://doi.org/10.1002/0470869526.ch10.
40. Tofts PS, Brix G, Buckley DL, Evelhoch JL, Henderson E, Knopp MV, Larsson HB, Lee TY, Mayr NA, Parker GJ, et al. Estimating kinetic parameters from dynamic contrast-enhanced T(1)-weighted MRI of a diffusable tracer: standardized quantities and symbols. J Magn Reson Imaging. 1999;10:223–32.
41. Sourbron SP, Buckley DL. On the scope and interpretation of the Tofts models for DCE-MRI. Magn Reson Med. 2011;66:735–45. https://doi.org/10.1002/mrm.22861.
42. Ewing JR, Bagher-Ebadian H. Model selection in measures of vascular parameters using dynamic contrast-enhanced MRI: experimental and clinical applications. NMR Biomed. 2013;26:1028–41. https://doi.org/10.1002/nbm.2996.
43. Paudyal R, Lu Y, Hatzoglou V, Moreira A, Stambuk HE, Oh JH, Cunanan KM, Aramburu Nunez D, Mazaheri Y, Gonen M, et al. Dynamic contrast-enhanced MRI model selection for predicting tumor aggressiveness in papillary thyroid cancers. NMR Biomed. 2020;33:e4166. https://doi.org/10.1002/nbm.4166.
44. Bagher-Ebadian H, Paudyal R, Nagaraja TN, Croxen RL, Fenstermacher JD, Ewing JR. MRI estimation of gadolinium and albumin effects on water proton. NeuroImage. 2011;54:S176–9. https://doi.org/10.1016/j.neuroimage.2010.05.032.
45. Bagher-Ebadian H, Jafari-Khouzani K, Mitsias PD, Lu M, Soltanian-Zadeh H, Chopp M, Ewing JR. Predicting final extent of ischemic infarction using artificial neural network analysis of multi-parametric MRI in patients with stroke. PLoS One. 2011;6:e22626. https://doi.org/10.1371/journal.pone.0022626.
46. Mcconnell HM. Reaction rates by nuclear magnetic resonance. J Chem Phys. 1958;28:430–1. https://doi.org/10.1063/1.1744152.
47. Paudyal R, Bagher-Ebadian H, Nagaraja TN, Fenstermacher JD, Ewing JR. Modeling of look-locker estimates of the magnetic resonance imaging estimate of longitudinal relaxation rate in tissue after contrast administration. Magn Reson Med. 2011;66:1432–44. https://doi.org/10.1002/mrm.22852.
48. Paudyal R, Poptani H, Cai K, Zhou R, Glickson JD. Impact of transvascular and cellular-interstitial water exchange on dynamic contrast-enhanced magnetic resonance imaging estimates of blood to tissue transfer constant and blood plasma volume. J Magn Reson Imaging. 2013;37:435–44. https://doi.org/10.1002/jmri.23837.
49. Do RK, Reyngold M, Paudyal R, Oh JH, Konar AS, LoCastro E, Goodman KA, Shukla-Dave A. Diffusion-weighted and dynamic contrast-enhanced MRI derived imaging metrics for stereotactic body radiotherapy of pancreatic ductal adenocarcinoma: preliminary findings. Tomography. 2020;6:261–71. https://doi.org/10.18383/j.tom.2020.00015.
50. Abdel Razek AA, Poptani H. MR spectroscopy of head and neck cancer. Eur J Radiol. 2013;82:982–9. https://doi.org/10.1016/j.ejrad.2013.01.025.
51. Shah GV, Gandhi D, Mukherji SK. Magnetic resonance spectroscopy of head and neck neoplasms. Top Magn Reson Imaging. 2004;15:87–94. https://doi.org/10.1097/01.rmr.0000130599.19502.ba.
52. Chawla S, Kim S, Loevner LA, Quon H, Wang S, Mutale F, Weinstein G, Delikatny EJ, Poptani H. Proton and phosphorous MR spectroscopy in squamous cell carcinomas of the head and neck. Acad Radiol. 2009;16:1366–72. https://doi.org/10.1016/j.acra.2009.06.001.
53. Mukherji SK, Schiro S, Castillo M, Kwock L, Muller KE, Blackstock W. Proton MR spectroscopy of squamous cell carcinoma of the extracranial head and neck: in vitro and in vivo studies. AJNR Am J Neuroradiol. 1997;18:1057–72.
54. Surov A, Meyer HJ, Wienke A. Apparent diffusion coefficient for distinguishing between malignant and benign lesions in the head and neck region: a systematic review and meta-analysis.

Front Oncol. 2019;9:1362. https://doi.org/10.3389/fonc.2019.01362.
55. Fong D, Bhatia KS, Yeung D, King AD. Diagnostic accuracy of diffusion-weighted MR imaging for nasopharyngeal carcinoma, head and neck lymphoma and squamous cell carcinoma at the primary site. Oral Oncol. 2010;46:603–6. https://doi.org/10.1016/j.oraloncology.2010.05.004.
56. Sakamoto J, Imaizumi A, Sasaki Y, Kamio T, Wakoh M, Otonari-Yamamoto M, Sano T. Comparison of accuracy of intravoxel incoherent motion and apparent diffusion coefficient techniques for predicting malignancy of head and neck tumors using half-Fourier single-shot turbo spin-echo diffusion-weighted imaging. Magn Reson Imaging. 2014;32:860–6. https://doi.org/10.1016/j.mri.2014.05.002.
57. Sakamoto J, Yoshino N, Okochi K, Imaizumi A, Tetsumura A, Kurohara K, Kurabayashi T. Tissue characterization of head and neck lesions using diffusion-weighted MR imaging with SPLICE. Eur J Radiol. 2009;69:260–8. https://doi.org/10.1016/j.ejrad.2007.10.008.
58. Yabuuchi H, Matsuo Y, Kamitani T, Setoguchi T, Okafuji T, Soeda H, Sakai S, Hatakenaka M, Nakashima T, Oda Y, et al. Parotid gland tumors: can addition of diffusion-weighted MR imaging to dynamic contrast-enhanced MR imaging improve diagnostic accuracy in characterization? Radiology. 2008;249:909–16. https://doi.org/10.1148/radiol.2493072045.
59. Paudyal R, Oh JH, Riaz N, Venigalla P, Li J, Hatzoglou V, Leeman J, Nunez DA, Lu Y, Deasy JO, et al. Intravoxel incoherent motion diffusion-weighted MRI during chemoradiation therapy to characterize and monitor treatment response in human papillomavirus head and neck squamous cell carcinoma. J Magn Reson Imaging. 2017;45:1013–23. https://doi.org/10.1002/jmri.25523.
60. Riaz N, Sherman E, Pei X, Schoder H, Paudyal R, Katabi N, Ma D, Tsai C, McBride S, Morris L, et al. Genetic and micro-environmental factors influencing response to definitive 30Gy chemo-radiotherapy (chemoRT) in HPV positive oropharyngeal cancer (OPC). Int J Radiat Oncol Biol Phys. 2020;106:1138.
61. Payabvash S. Quantitative diffusion magnetic resonance imaging in head and neck tumors. Quant Imaging Med Surg. 2018;8:1052–65. https://doi.org/10.21037/qims.2018.10.14.
62. Meyer HJ, Leifels L, Hamerla G, Hohn AK, Surov A. ADC-histogram analysis in head and neck squamous cell carcinoma. Associations with different histopathological features including expression of EGFR, VEGF, HIF-1 alpha, Her 2 and p53. A preliminary study. Magn Reson Imaging. 2018;54:214–7. https://doi.org/10.1016/j.mri.2018.07.013.
63. Surov A, Meyer HJ, Wienke A. Can imaging parameters provide information regarding histopathology in head and neck squamous cell carcinoma? A meta-analysis. Transl Oncol. 2018;11:498–503. https://doi.org/10.1016/j.tranon.2018.02.004.
64. Kitamoto E, Chikui T, Kawano S, Ohga M, Kobayashi K, Matsuo Y, Yoshiura T, Obara M, Honda H, Yoshiura K. The application of dynamic contrast-enhanced MRI and diffusion-weighted MRI in patients with maxillofacial tumors. Acad Radiol. 2015;22:210–6. https://doi.org/10.1016/j.acra.2014.08.016.
65. Lee FKH, King AD, Ma BBY, Yeung DKW. Dynamic contrast enhancement magnetic resonance imaging (DCE-MRI) for differential diagnosis in head and neck cancers. Eur J Radiol. 2012;81:784–8. https://doi.org/10.1016/j.ejrad.2011.01.089.
66. Gaddikeri S, Gaddikeri RS, Tailor T, Anzai Y. Dynamic contrast-enhanced MR imaging in head and neck cancer: techniques and clinical applications. Am J Neuroradiol. 2016;37:588–95. https://doi.org/10.3174/ajnr.A4458.
67. Yuan Y, Shi H, Tao X. Head and neck paragangliomas: diffusion weighted and dynamic contrast enhanced magnetic resonance imaging characteristics. BMC Med Imaging. 2016;16:12. https://doi.org/10.1186/s12880-016-0114-3.
68. Abdel Razek AA, Soliman NY, Elkhamary S, Alsharaway MK, Tawfik A. Role of diffusion-weighted MR imaging in cervical lymphadenopathy. Eur Radiol. 2006;16:1468–77. https://doi.org/10.1007/s00330-005-0133-x.
69. Holzapfel K, Duetsch S, Fauser C, Eiber M, Rummeny EJ, Gaa J. Value of diffusion-weighted MR imaging in the differentiation between benign and malignant cervical lymph nodes. Eur J Radiol. 2009;72:381–7. https://doi.org/10.1016/j.ejrad.2008.09.034.
70. Vandecaveye V, De Keyzer F, Vander Poorten V, Dirix P, Verbeken E, Nuyts S, Hermans R. Head and neck squamous cell carcinoma: value of diffusion-weighted MR imaging for nodal staging. Radiology. 2009;251:134–46. https://doi.org/10.1148/radiol.2511080128.
71. Zhang F, Chen JY. Breast cancer subtyping from plasma proteins. BMC Med Genet. 2013;6(Suppl 1):S6. https://doi.org/10.1186/1755-8794-6-S1-S6.
72. Fischbein NJ, Noworolski SM, Henry RG, Kaplan MJ, Dillon WP, Nelson SJ. Assessment of metastatic cervical adenopathy using dynamic contrast-enhanced MR imaging. Am J Neuroradiol. 2003;24:301–11.
73. Payabvash S, Chan A, Jabehdar Maralani P, Malhotra A. Quantitative diffusion magnetic resonance imaging for prediction of human papillomavirus status in head and neck squamous-cell carcinoma: a systematic review and meta-analysis. Neuroradiol J. 2019;32:232–40. https://doi.org/10.1177/1971400919849808.
74. Srinivasan A, Mohan S, Mukherji SK. Biologic imaging of head and neck cancer: the present and the future. AJNR Am J Neuroradiol. 2012;33:586–94. https://doi.org/10.3174/ajnr.A2535.

75. Ding Y, Hazle JD, Mohamed AS, Frank SJ, Hobbs BP, Colen RR, Gunn GB, Wang J, Kalpathy-Cramer J, Garden AS, et al. Intravoxel incoherent motion imaging kinetics during chemoradiotherapy for human papillomavirus-associated squamous cell carcinoma of the oropharynx: preliminary results from a prospective pilot study. NMR Biomed. 2015;28(12):1645–54. https://doi.org/10.1002/nbm.3412.
76. Kabadi SJ, Fatterpekar GM, Anzai Y, Mogen J, Hagiwara M, Patel SH. Dynamic contrast-enhanced MR imaging in head and neck cancer. Magn Reson Imaging Clin N Am. 2018;26:135–49. https://doi.org/10.1016/j.mric.2017.08.008.
77. Vaupel P, Mayer A. Hypoxia in cancer: significance and impact on clinical outcome. Cancer Metastasis Rev. 2007;26:225–39. https://doi.org/10.1007/s10555-007-9055-1.
78. Egeland TAM, Gulliksrud K, Gaustad JV, Mathiesen B, Rofstad EK. Dynamic contrast-enhanced-MRI of tumor hypoxia. Magn Reson Med. 2012;67:519–30. https://doi.org/10.1002/mrm.23014.
79. Chikui T, Kitamoto E, Kawano S, Sugiura T, Obara M, Simonetti AW, Hatakenaka M, Matsuo Y, Koga S, Ohga M, et al. Pharmacokinetic analysis based on dynamic contrast-enhanced MRI for evaluating tumor response to preoperative therapy for oral cancer. J Magn Reson Imaging. 2012;36:589–97. https://doi.org/10.1002/jmri.23704.
80. Wang P, Popovtzer A, Eisbruch A, Cao Y. An approach to identify, from DCE MRI, significant subvolumes of tumors related to outcomes in advanced head-and-neck cancer. Med Phys. 2012;39:5277–85. https://doi.org/10.1118/1.4737022.
81. Kim S, Loevner LA, Quon H, Kilger A, Sherman E, Weinstein G, Chalian A, Poptani H. Prediction of response to chemoradiation therapy in squamous cell carcinomas of the head and neck using dynamic contrast-enhanced MR imaging. AJNR Am J Neuroradiol. 2010;31:262–8. https://doi.org/10.3174/ajnr.A1817.
82. Shukla-Dave A, Hricak H, Ishill N, Moskowitz CS, Drobnjak M, Reuter VE, Zakian KL, Scardino PT, Cordon-Cardo C. Prediction of prostate cancer recurrence using magnetic resonance imaging and molecular profiles. Clin Cancer Res. 2009;15:3842–9. https://doi.org/10.1158/1078-0432.CCR-08-2453.
83. Chawla S, Loevner LA, Kim SG, Hwang WT, Wang S, Verma G, Mohan S, LiVolsi V, Quon H, Poptani H. Dynamic contrast-enhanced MRI-derived intracellular water lifetime (tau i): a prognostic marker for patients with head and neck squamous cell carcinomas. AJNR Am J Neuroradiol. 2018;39:138–44. https://doi.org/10.3174/ajnr.A5440.
84. King AD, Mo FKF, Yu KH, Yeung DKW, Zhou H, Bhatia KS, Tse GMK, Vlantis AC, Wong JKT, Ahuja AT. Squamous cell carcinoma of the head and neck: diffusion-weighted MR imaging for prediction and monitoring of treatment response. Eur Radiol. 2010;20:2213–20. https://doi.org/10.1007/s00330-010-1769-8.
85. Vandecaveye V, De Keyzer F, Nuyts S, Deraedt K, Dirix P, Hamaekers P, Vander Poorten V, Delaere P, Hermans R. Detection of head and neck squamous cell carcinoma with diffusion weighted MRI after (chemo)radiotherapy: correlation between radiologic and histopathologic findings. Int J Radiat Oncol. 2007;67:960–71. https://doi.org/10.1016/j.ijrobp.2006.09.020.
86. Furukawa M, Parvathaneni U, Maravilla K, Richards TL, Anzai Y. Dynamic contrast-enhanced MR perfusion imaging of head and neck tumors at 3 Tesla. Head Neck J Sci Spec. 2013;35:923–9. https://doi.org/10.1002/hed.23051.
87. Ishiyama M, Richards T, Parvathaneni U, Anzai Y. Dynamic contrast-enhanced magnetic resonance imaging in head and neck cancer: differentiation of new H&N cancer, recurrent disease, and benign post-treatment changes. Clin Imaging. 2015;39:566–70. https://doi.org/10.1016/j.clinimag.2015.01.016.
88. Oysu AS, Ayanoglu E, Kodalli N, Oysu C, Uneri C, Erzen C. Dynamic contrast-enhanced MRI in the differentiation of posttreatment fibrosis from recurrent carcinoma of the head and neck. Clin Imaging. 2005;29:307–12. https://doi.org/10.1016/j.clinimag.2005.01.024.
89. Tsai CJ, McBride SM, Riaz N, Lee NY. Reducing the radiation therapy dose prescription for elective treatment areas in human papillomavirus-associated oropharyngeal carcinoma being treated with primary chemoradiotherapy at Memorial Sloan Kettering Cancer Center. Pract Radiat Oncol. 2019;9:98–101. https://doi.org/10.1016/j.prro.2018.10.015.
90. Al-Ibraheem A, Buck A, Krause BJ, Scheidhauer K, Schwaiger M. Clinical applications of FDG PET and PET/CT in head and neck cancer. J Oncol. 2009;2009:208725. https://doi.org/10.1155/2009/208725.
91. Calabrese L, Jereczek-Fossa BA, Jassem J, Rocca A, Bruschini R, Orecchia R, Chiesa F. Diagnosis and management of neck metastases from an unknown primary. Acta Otorhinolaryngol Ital. 2005;25:2–12.
92. Nieder C, Gregoire V, Ang KK. Cervical lymph node metastases from occult squamous cell carcinoma: cut down a tree to get an apple? Int J Radiat Oncol Biol Phys. 2001;50:727–33. https://doi.org/10.1016/s0360-3016(01)01462-6.
93. Zhu L, Wang N. 18F-fluorodeoxyglucose positron emission tomography-computed tomography as a diagnostic tool in patients with cervical nodal metastases of unknown primary site: a meta-analysis. Surg Oncol. 2013;22:190–4.
94. Babin E, Desmonts C, Hamon M, Benateau H, Hitier M. PET/CT for assessing mandibular invasion by intraoral squamous cell carcinomas. Clin Otolaryngol. 2008;33:47–51. https://doi.org/10.1111/j.1749-4486.2007.01569.x.

95. Gordin A, Golz A, Keidar Z, Daitzchman M, Bar-Shalom R, Israel O. The role of FDG-PET/CT imaging in head and neck malignant conditions: impact on diagnostic accuracy and patient care. Otolaryngol Head Neck Surg. 2007;137:130–7. https://doi.org/10.1016/j.otohns.2007.02.001.
96. Roh JL, Yeo NK, Kim JS, Lee JH, Cho KJ, Choi SH, Nam SY, Kim SY. Utility of 2-[18F] fluoro-2-deoxy-D-glucose positron emission tomography and positron emission tomography/computed tomography imaging in the preoperative staging of head and neck squamous cell carcinoma. Oral Oncol. 2007;43:887–93. https://doi.org/10.1016/j.oraloncology.2006.10.011.
97. Senft A, Hoekstra OS, Witte BI, Leemans CR, de Bree R. Screening for distant metastases in head and neck cancer patients using FDG-PET and chest CT: validation of an algorithm. Eur Arch Otorhinolaryngol. 2016;273:2643–50. https://doi.org/10.1007/s00405-015-3773-8.
98. Peters TT, Senft A, Hoekstra OS, Castelijns JA, Witte BI, Leemans CR, de Bree R. Pretreatment screening on distant metastases and head and neck cancer patients: validation of risk factors and influence on survival. Oral Oncol. 2015;51:267–71. https://doi.org/10.1016/j.oraloncology.2014.12.006.
99. Haerle SK, Schmid DT, Ahmad N, Hany TF, Stoeckli SJ. The value of (18)F-FDG PET/CT for the detection of distant metastases in high-risk patients with head and neck squamous cell carcinoma. Oral Oncol. 2011;47:653–9. https://doi.org/10.1016/j.oraloncology.2011.05.011.
100. Krabbe CA, Pruim J, van der Laan BF, Rodiger LA, Roodenburg JL. FDG-PET and detection of distant metastases and simultaneous tumors in head and neck squamous cell carcinoma: a comparison with chest radiography and chest CT. Oral Oncol. 2009;45:234–40. https://doi.org/10.1016/j.oraloncology.2008.05.024.
101. Gupta T, Master Z, Kannan S, Agarwal JP, Ghsoh-Laskar S, Rangarajan V, Murthy V, Budrukkar A. Diagnostic performance of post-treatment FDG PET or FDG PET/CT imaging in head and neck cancer: a systematic review and meta-analysis. Eur J Nucl Med Mol Imaging. 2011;38:2083–95. https://doi.org/10.1007/s00259-011-1893-y.
102. Ong SC, Schoder H, Lee NY, Patel SG, Carlson D, Fury M, Pfister DG, Shah JP, Larson SM, Kraus DH. Clinical utility of 18F-FDG PET/CT in assessing the neck after concurrent chemoradiotherapy for locoregional advanced head and neck cancer. J Nucl Med. 2008;49:532–40. https://doi.org/10.2967/jnumed.107.044792.
103. Porceddu SV, Jarmolowski E, Hicks RJ, Ware R, Weih L, Rischin D, Corry J, Peters LJ. Utility of positron emission tomography for the detection of disease in residual neck nodes after (chemo) radiotherapy in head and neck cancer. Head Neck. 2005;27:175–81. https://doi.org/10.1002/hed.20130.
104. Mehanna H, Wong WL, McConkey CC, Rahman JK, Robinson M, Hartley AG, Nutting C, Powell N, Al-Booz H, Robinson M, et al. PET-CT surveillance versus neck dissection in advanced head and neck cancer. N Engl J Med. 2016;374:1444–54. https://doi.org/10.1056/NEJMoa1514493.
105. Gerisch G, Ecke M, Schroth-Diez B, Gerwig S, Engel U, Maddera L, Clarke M. Self-organizing actin waves as planar phagocytic cup structures. Cell Adhes Migr. 2009;3:373–82. https://doi.org/10.4161/cam.3.4.9708.
106. Huellner MW. PET/MR in head and neck cancer—an update. Semin Nucl Med. 2021;51:26–38. https://doi.org/10.1053/j.semnuclmed.2020.07.006.

Radiotherapy and Immunotherapy for Head and Neck Cancer

7

Jack M. Qian and Jonathan D. Schoenfeld

Abbreviations

BED	Biologically effective dose
bid	Twice a day
CTLA-4	Cytotoxic T-lymphocyte-associated protein 4
ENE	Extranodal extension
FDG	Fluorodeoxyglucose
fx	Fraction
HNSCC	Head and neck squamous cell carcinoma
HPV	Human papillomavirus
ICI	Immune checkpoint inhibitor
iNOS	Inducible nitric oxide synthase
irAE	Immune-related adverse effect
LA	Locally advanced
MDSC	Myeloid-derived suppressor cell
NK	Natural killer
NSCLC	Non-small-cell lung cancer
OS	Overall survival
PD-1	Programmed cell death protein 1
PD-L1	Programmed death-ligand 1
PET-CT	Positron emission tomography–computed tomography
PFS	Progression-free survival
SBRT	Stereotactic body radiotherapy
SCC	Squamous cell carcinoma
TAM	Tumor-associated macrophage
Tregs	T-regulatory cells

J. M. Qian
Harvard Radiation Oncology Program, Massachusetts General Hospital/Brigham and Women's Hospital/Dana-Farber Cancer Institute, Boston, MA, USA
e-mail: Jack_Qian@dfci.harvard.edu

J. D. Schoenfeld (✉)
Department of Radiation Oncology, Dana-Farber/Brigham and Women's Cancer Center, Harvard Medical School, Boston, MA, USA
e-mail: Jonathan_Schoenfeld@dfci.harvard.edu

Key Points

- The immune system plays a critical role in carcinogenesis.
- Radiotherapy has diverse immunomodulatory effects than can both stimulate and inhibit an antitumor immune response.
- Preclinical studies in head and neck cancer models support synergy between radiotherapy and immunotherapy and suggest additional avenues to modulate the interaction.
- Existing clinical data indicate that the combination of radiotherapy and immunotherapy is relatively safe and well-tolerated by patients, but efficacy results have not yet matched those seen in other tumor types.
- Ongoing clinical trials for both recurrent/metastatic disease and locally advanced disease will provide further insight on how to improve patient outcomes.

R. A. Chandra, R. J. Li (eds.), *Multidisciplinary Management of Head and Neck Cancer*,
https://doi.org/10.1007/978-3-031-05973-5_7

Introduction

Head and neck cancers comprise a significant portion of the global cancer burden; when aggregating subsites, they are the eighth most common cancer worldwide by both incidence and mortality [1]. Although the vast majority of head and neck cancers are squamous cell carcinomas (HNSCC) and have traditionally been associated with tobacco and alcohol use, human papillomavirus (HPV)-associated oropharyngeal squamous cell carcinoma (SCC) has emerged as a new disease entity with markedly different biological behavior [2].

Ever since the foundational work of Henri Coutard, who was the first to use X-rays to treat laryngeal cancer almost 100 years ago [3], radiation therapy has played a key role in the treatment of HNSCC. Radiation continues to be used extensively in both the curative and palliative setting, although the distinction between the two is now sometimes blurred with growing recognition of the oligometastatic state, where patients with limited numbers of metastases can achieve prolonged survival, or even cure [4, 5]. Technological advancements, in both imaging and treatment delivery, have enabled more precise radiation treatment that has reduced treatment-related morbidity and improved patient outcomes. However, even with the use of modern radiation techniques, there are still opportunities for further improvement [4].

The immune system has a critical role in tumor development, and the development of immune evasion by tumors is a key step in carcinogenesis [6, 7]. Attempts to reinvigorate an antitumor immune response have been widely integrated into practice following the development of the immune checkpoint inhibitors (ICIs) targeted against the immune checkpoint receptors cytotoxic T-lymphocyte-associated protein 4 (CTLA-4), programmed cell death protein 1 (PD-1), and programmed death-ligand 1 (PD-L1). Since the initial FDA approval of ipilimumab (a CTLA-4 inhibitor) in 2011 for the treatment of metastatic melanoma based on a proven overall survival advantage [8], antibodies blocking CTLA-4 and PD-1/PD-L1 have been tested and approved across a wide spectrum of malignancies. In HNSCC, both pembrolizumab and nivolumab (PD-1 inhibitors) have gained FDA approval for use in recurrent/metastatic HNSCC after progression through platinum-based chemotherapy [9–11]. Pembrolizumab additionally has been approved in the US for use in the first-line setting in patients with recurrent/metastatic HNSCC, either in combination with chemotherapy or alone as monotherapy depending on tumor/tumor microenvironment PD-L1 expression [12].

Unfortunately, overall response rates to PD-1 (inhibitors in unselected patients with HNSCC) remain low at approximately 10–20% [9–12], although patients who do respond can have long-lasting, durable remissions, as has been the case with other solid tumor patients who respond to PD-1 blockade [13]. The possibility of durable long-term response has been a driver of the rapid uptake in clinical practice and has invigorated efforts to develop predictive biomarkers. Tumor mutational burden, a potential surrogate for tumor neoantigens that can be recognized by the immune system, is one such biomarker, leading to the first ever histology-agnostic FDA approval of the PD-1 inhibitor pembrolizumab for mismatch repair deficient tumors of any histology [14, 15], though there is increasing recognition that the types of mutations and ability to generate neoantigens may be as important as the number of mutations present [16]. PD-L1 expression on both tumor cells and infiltrated immune cells has also been explored as a biomarker across several histologies with varying results; in HNSCC, subgroup analyses of Checkmate 141, KEYNOTE-040, and KEYNOTE-048 all suggest that higher PD-L1 expression does correlate with the likelihood of survival benefit [10–12]. It is less clear whether patients with low or no PD-L1 expression still benefit from PD-1 directed therapy; analyses of Checkmate 141 and KEYNOTE-048 show questionable benefit for the PD-L1-negative subgroup when comparing the treatment and control arms [11, 17]. Finally, for HNSCC patients, HPV-associated malignancies with relatively fewer tumor mutations as compared to tobacco-associated malignancies may also respond to immune checkpoint blockade as novel viral-associated neoantigens might be recognized by the immune system. Indeed, subgroup analyses of the Checkmate 141 and KEYNOTE-040 trials did not show any clear differences in response or clinical benefit based on p16 expression status (a surrogate for HPV-associated tumors) [10, 11, 18].

In addition to better patient selection through the use of predictive biomarkers, augmenting the antitumor immune response with other therapies could also improve immunotherapy response rates. Radiation therapy increasingly has been recognized to have diverse immunomodulatory effects, and there has consequently been interest in possible synergism between radiotherapy and immunotherapy. In the remainder of this chapter, we will summarize the preclinical data that illustrate the immune effects of radiotherapy, review the unique immune landscape of HNSCC, and finally discuss both current preclinical and clinical data relevant to the combination of radiotherapy and immunotherapy specifically in HNSCC (Fig. 7.1).

Immune Effects of Radiotherapy

Traditionally, the antitumor effects of radiotherapy have been attributed to direct cytotoxicity secondary to the induction of DNA damage, and while it was known over 40 years ago that radiotherapy also depends on an intact immune system to exert its full antitumor effect [20], the interaction between the immune system and radiotherapy has garnered more interest in the past two decades. It is now recognized that the immune effects of radiation may contribute significantly to an antitumor response; however, these immune effects are also quite complex and can be both immunostimulatory and immunosuppressive.

Radiation can induce immunogenic cell death, which gives rise to adaptive immune responses [21, 22]. Many mechanisms can be involved in this process, and a full detailed review is beyond the scope of this discussion. However, recent studies have shown radiation can promote immune activation via calreticulin-, ATP-, and HMGB-mediated pathways [22, 23]. Radiation

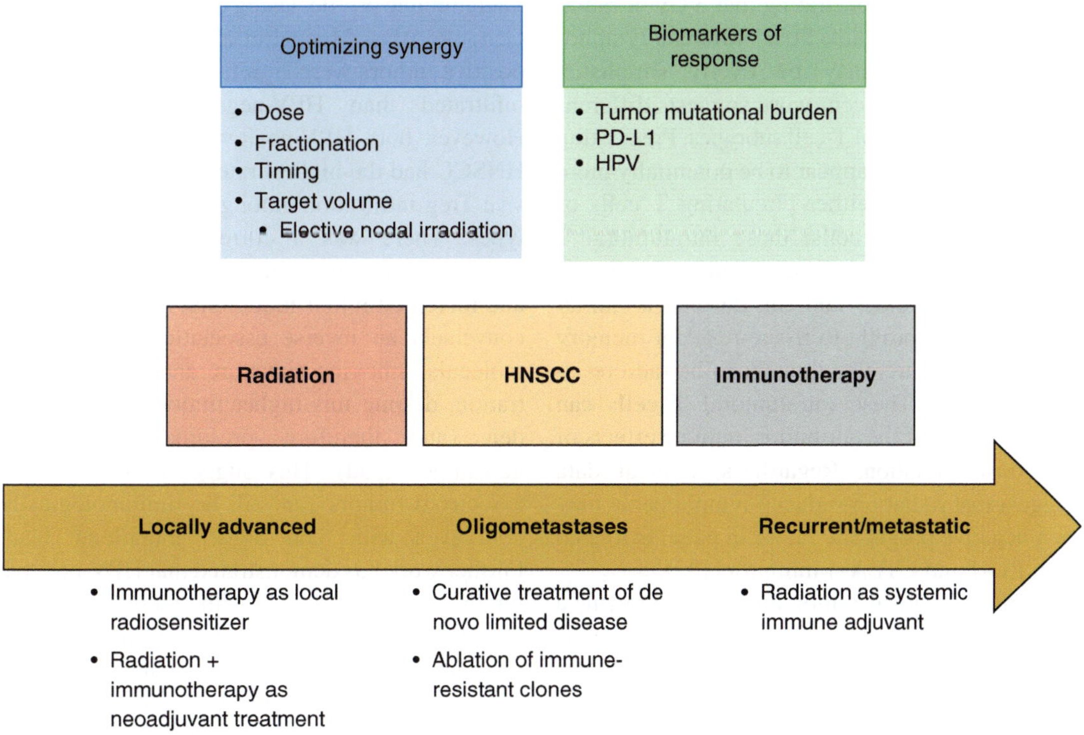

Fig. 7.1 Opportunities for radioimmunotherapy in HNSCC. (With permissions from [19])

also leads to the presence of cytosolic DNA, which triggers the cGAS/STING pathway and subsequent production of type-I interferon [24, 25]. Type-I interferon is crucial for the activation of dendritic cells, which ultimately recruit and prime T cells. These signals together are critical for the initial development of an immune response specific to tumor neoantigens.

Radiation can promote antitumor immunity through additional mechanisms. Radiation can diversify antigen presentation by tumor cells through promotion of intracellular peptide degradation as well as upregulation of MHC expression [26, 27]. This ultimately can enhance recognition and tumor cell killing by cytotoxic T cells [28]. Radiation has also been associated with increased production of other immune stimulating cytokines and chemokines, which together can promote the infiltration of T cells into tumors and modulate the function of these T cells, as well as dendritic cells and macrophages [23].

Radiation also has immunosuppressive effects that could be detrimental to an antitumor immune response. Lymphocytes are radiosensitive, with in vitro studies demonstrating that 3 Gy of radiation is enough to deplete 90% of human lymphocytes [29]. This may be overly simplistic, however, as more recent work suggests differential radiosensitivity of T-cell subtypes. Preexisting intratumoral T cells appear to be potentially more radioresistant than either circulating T cells or lymphoid tissue T cells; these intratumoral T cells survive even high doses (20 Gy) of radiation in preclinical studies and can develop a similar transcriptomic profile to tissue-resident memory T cells, which are also thought to be radioresistant [30, 31]. These intratumoral T cells can mediate some of the antitumor immune effects of high-dose radiation. Regardless, clinical data suggest that radiation-induced lymphopenia may be a negative prognostic factor in patients treated with PD-1 and CTLA-4 inhibitors [32].

Within the local tumor microenvironment, a variety of inhibitory immune cells, such as T-regulatory cells (Tregs), myeloid-derived suppressor cells (MDSCs), and tumor-associated macrophages (TAMs, and specifically M2 macrophages), are often already present. In several studies, radiation increases recruitment of these inhibitory immune cells and can also modulate their function toward an even more immunosuppressive phenotype [23]. There may also be dose-dependent effects of radiation; for instance, Vanpouille-Box et al. demonstrated that as radiation doses were escalated to 12–18 Gy, there was induction of Trex1, a DNA exonuclease which degrades cytosolic DNA and thus prevents activation of the cGAS/STING pathway [25]. The balance between competing activating and inhibitory immune responses, then, likely plays a key role in the probability of a successful antitumor immune response and provides opportunity for therapeutic intervention.

Immune Landscape of HNSCC

Work over the past decade has helped characterize the immune landscape of HNSCC. As noted above, HPV-associated oropharyngeal SCC is a different disease entity from other non-HPV-driven, tobacco-associated HNSCC, with a distinct immune profile. Using data from The Cancer Genome Atlas, Mandal et al. showed that HPV-positive tumors were significantly more immune infiltrated than HPV-negative tumors [33]. However, both HPV-positive and HPV-negative HNSCC had the highest rate of immunosuppressive Treg infiltration among ten different cancer types. There was a correlation between the molecular smoking signature of HNSCC tumors and increased tumor mutational burden, but also conversely an inverse association between the molecular smoking signature and immune infiltration, despite this higher tumor mutation burden (and therefore presumably increased neoantigen load). This suggests that tobacco-associated tumors can still be immunologically cold even with their higher mutational load. Further work has demonstrated that HPV-positive tumors are associated with increased T-cell receptor diversity, higher levels of immune cytolytic activity, and an overall enriched inflammatory response [34, 35]. The anatomic subsite where head and neck cancer develops likely plays a key role in tumor immunity as well; the oropharynx contains particularly lymphoid-rich tissue, and this unique immune environment may explain why the improved prognosis for HPV-driven HNSCC is largely limited to oropharyngeal tumors [36]. Additional work on oropharyngeal SCC has confirmed a higher degree of infiltration of CD8+ T cells in HPV-positive vs. HPV-negative tumors [37]. Overall, these studies suggest that the increased sensitivity of HPV-associated oropharyngeal SCC to chemotherapy and radiotherapy may at least in part be mediated through immune mechanisms [38, 39] and that differing immunotherapeutic approaches may be optimal for HPV-positive and HPV-negative HNSCC.

HNSCC also appears to be uniquely associated with high levels of natural killer (NK) cell infiltration, even when compared to other highly immune infiltrated cancer types [33, 37]. Patients with high levels of NK-cell infiltration were also found to have improved survival compared to those with low levels of infiltration [33]. The potential antitumor effects of NK cells is an emerging area of research and has been reviewed elsewhere [40]; currently, there is limited clinical data on their role in HNSCC, or whether opportunities for synergy between NK-directed therapies and radiation exist.

Preclinical Evidence for Radioimmunotherapy in HNSCC Models

Augmenting Antitumor Cellular Immunity

Preclinical work in HNSCC models has demonstrated synergy between radiotherapy and immunotherapy. In a poorly immunogenic orthotopic HNSCC mouse model, Oweida et al. demonstrated effective tumor cell killing when both 10 Gy of radiation and an anti-PD-L1 antibody were administered together, but not for either treatment individually [41]. Tumor control was correlated with increased tumor T-cell infiltration and was abrogated when CD4+ and CD8+ T cells were depleted. In addition, although much of research on antitumor immunity has focused on the role of T cells, work from Kim et al. in a mouse model of HPV-associated HNSCC suggests that the combination of radiation and PD-1 inhibition also promotes maturation and activation of B cells, leading to the development of memory B cells, plasma cells, and antigen-specific B cells, as well as increasing formation of B-cell germinal centers in tumor draining lymph nodes [42]. Finally, there is growing interest in harnessing additional molecular pathways to promote antitumor immunity. For instance, in a mouse model of HPV-driven carcinoma, Dillon et al. demonstrated that inhibitors of ATR, a key protein in the DNA damage response pathway, significantly sensitized tumors to radiation, and this effect was correlated with the upregulation of interferon-stimulated genes and a significant increase in innate immune cell infiltration into the tumor microenvironment [43]. Xiao et al. showed that ASTX600, an inhibitor of IAP1/2 and XIAP, proteins that modulate apoptosis and the tumor necrosis factor signaling pathway, significantly enhanced T-cell-mediated tumor cell killing when combined with radiation and PD-1 inhibition in a mouse model of oral cavity carcinoma [44].

Decreasing an Immunosuppressive Microenvironment

The immunosuppressive microenvironment remains a challenge even with combined radiotherapy and immunotherapy. Following up on their initial study demonstrating synergy between radiation/PD-1 inhibition [41], Oweida et al. demonstrated that the antitumor immune responses to combined radiation and PD-1 inhibition in their HNSCC mouse model were ultimately transient, as compensatory mechanisms of immune evasion were activated, including upregulation of another immune checkpoint, TIM-3, as well as increased tumor infiltration of Tregs [41, 45]. Adding an anti-TIM-3 antibody further delayed tumor growth, but the response was still not durable; only targeted depletion of Tregs was able to induce durable immunologic memory. Another group has explored the use of cyclophosphamide and an inhibitor of inducible nitric oxide synthase (iNOS) as immunomodulatory agents in a mouse model of HPV-associated HNSCC. When combined with traditional chemoradiation, addition of these two agents increased the CD8+ T-cell/Treg ratio and decreased immunosuppression [46]. In this particular model system, the combination of radiation with PD-1 and CTLA-4 inhibition only minimally altered the immunologically cold tumor microenvironment, but the addition of cyclophosphamide and the iNOS inhibitor shifted the balance of infiltrated immune cells away from immunosuppressive types (such as MDSCs) to those more associated with antitumor immunity (such as dendritic cells and antitumor M1 macrophages). This led to an increased CD8+ T-cell-dependent response and complete tumor rejection in more than 70% of the treated mice [47]. This is now being investigated in a clinical trial, NCT03844763, which explores the use of cyclophosphamide, avelumab (a PD-L1 inhibitor), and radiation therapy in the treatment of recurrent/metastatic HNSCC.

Radiation Dose and Fractionation Effects

Additional studies have demonstrated the importance of radiation dose and fractionation in generating an effective antitumor immune response. Consistent with work in other diseases [48], Morisada et al. showed in a syngeneic mouse oral cavity carcinoma model that hypofractionated radiation (16 Gy in two fractions) was associated with preservation of both peripheral and tumor-infiltrating lymphocytes, reduction of both peripheral and tumor-associated MDSCs, and increased expression of interferon genes, when compared to conventionally fractionated radiation (20 Gy in ten fractions) [49]. Moreover, analysis of the draining lymph nodes (which notably were included within the radiation fields) suggested that 20 Gy in ten fractions suppressed local tumor-specific T-cell responses. Consequently, only 16 Gy in two fractions demonstrated synergy with an anti-PD-1 antibody in these mice. Additional work by this group suggests a dose-dependent effect of radiation on both antigen release and T-cell priming, with 8 Gy in a single fraction enhancing these pathways compared to 2 Gy in a single fraction, resulting in increased tumor cell susceptibility to T-cell-mediated killing [50]. However, the doses used in these preclinical models differ from those used in clinical practice, as do the size of the treated tumors, and so it is uncertain how these findings might translate to the treatment of HNSCC patients.

Clinical Evidence for Radioimmunotherapy in HNSCC

Recurrent/Metastatic Setting

Despite the widespread use of ICIs in advanced malignancies, prospective clinical data on their combination with radiotherapy remain scarce, particularly in HNSCC. The unique immune-related adverse effects (irAEs) that have been observed with ICIs are now well established [51], and there have been concerns that the pro-inflammatory effects of radiation could enhance toxicities when combined with ICIs. Reassuringly, however, most of the available clinical data to date suggests that the combination of radiation and ICIs is generally well tolerated [52]. For instance, in a cohort of 133 patients with metastatic melanoma, non-small-cell lung cancer (NSCLC), or renal cell cancer who received palliative radiation to a wide range of anatomic sites, Bang et al. demonstrated numerically higher rates of irAEs when radiation was given within 14 days of immunotherapy, but the toxicities were generally mild with rates of grade 3+ toxicity less than 10% [53]. Similarly, a prospective phase I trial of pembrolizumab and stereotactic body radiotherapy (SBRT) in patients with a variety of metastatic solid tumors also demonstrated a grade 3+ toxicity rate of less than 10% [54]. Notably, this study did include four patients with HNSCC, and radiation was delivered to two distinct anatomic sites in more than 60% of the cohort. Finally, a phase 2 trial which randomized 62 patients with metastatic HNSCC to nivolumab with or without SBRT to a single metastatic site did not find a significant difference in either grade 3–5 adverse events (13% for nivolumab alone vs. 10% for nivolumab with SBRT, $p = 0.70$) or any grade adverse events (70% for nivolumab alone vs. 87% for nivolumab with SBRT, $p = 0.12$) with the addition of SBRT [55].

Nevertheless, a few key issues must be considered when interpreting these and other safety data. Just as dose and fractionation likely affect potential antitumor immunity induced by radiation (as demonstrated in preclinical work), it is probable that these parameters influence potential toxicities when combined with ICIs. The relative timing of radiation and immunotherapy is likely to be important as well; notably, radiation recall, a relatively rare, unpredictable, and poorly understood phenomenon wherein an inflammatory reaction can develop in previously irradiated tissue following administration of a new systemic agent [56] has now been reported following ICI administration [57, 58]. Additionally, the anatomic site treated with radiation could influence the side effect profile of combination treatment; for instance, the landmark PACIFIC trial, which demonstrated a significant overall survival benefit to adjuvant durvalumab (an anti-PD-L1 antibody) after definitive chemoradiation for stage III NSCLC, also showed an increase in any-grade pneumonitis with the addition of durvalumab (although rates of clinically relevant pneumonitis, i.e., grade 3+, were similar between treatment groups and low overall) [59]. Within the brain, there is a potential increased risk of developing radiation necrosis after treatment of brain metastases with combined ICIs and radiation [60, 61]. Finally, as discussed earlier, in certain settings, radiation can induce lymphopenia, which could ultimately interfere with the efficacy of ICIs [32]. These data highlight the importance of collecting robust radiation treatment and toxicity data to facilitate future analyses as we study combination radiation and immunotherapy treatments.

There are very few efficacy data relevant to the addition of radiation to ICIs in patients with recurrent or metastatic HNSCC. In general, the primary rationale for radiation in this setting is to help stimulate a systemic antitumor immune response or abscopal effect. This is particularly difficult to study retrospectively, as disentangling a true abscopal effect from a delayed response to immunotherapy is challenging [62]. The only available prospective data for HNSCC comes from the randomized phase 2 trial noted above, in which 62 patients with metastatic HNSCC were randomized to nivolumab with or without SBRT to a single metastatic site (9 Gy × 3 fractions, between the first and second doses of nivolumab). Ultimately, there was no improvement in overall response rate (34.5% for nivolumab alone vs. 29.0% for nivolumab with SBRT, $p = 0.86$) [55]. In NSCLC, a similarly designed phase 2 trial of pembrolizumab with or without SBRT to a single metastatic site in patients with advanced NSCLC also failed to meet its primary endpoint, although it did demonstrate a doubling of overall response rate with the addition of SBRT that was not statistically significant (18% for pembrolizumab alone vs. 36% for pembrolizumab with SBRT, $p = 0.07$) [63]. Differences between the designs of these two studies include the anti-PD-1 agent used (nivolumab vs. pembrolizumab), the type of cancer (HNSCC vs. NSCLC), timing of SBRT (between first and second dose of nivolumab vs. prior to starting pembrolizumab), and dose of SBRT (9 Gy × 3 fractions vs. 8 Gy × 3 fractions). Given the results of these trials, further research is clearly needed; Table 7.1 summarizes ongoing trials that will help address these questions specifically in patients with recurrent/metastatic HNSCC. Notably, however, only a few of these studies are randomized, and so any efficacy data will require confirmation in larger, phase 3 trials.

Finally, as noted above, there is growing recognition of an oligometastatic disease state. Contrary to previous conceptualization of metastatic disease as inevitably widespread and thus incurable, the oligometastatic hypothesis suggests that there is a wide range of metastatic potential that varies among different cancers and from patient to patient and that an intermediate state likely exists between purely localized disease and widely metastatic disease, wherein a

Table 7.1 Ongoing trials evaluating combinations of ICIs and radiation in the management of recurrent/metastatic HNSCC (with permissions from [19])

NCT#	Title	Inclusion criteria	Treatment arms	Timing	Phase
NCT03539198	Study of proton SBRT and immunotherapy for recurrent/progressive locoregional or metastatic head and neck cancer	Recurrent/metastatic HNSCC, ≥2 metastatic sites	1: Nivolumab given every 2 weeks, with proton SBRT to one metastatic site administered with cycle 3	Concurrent	N/A
NCT03283605	Immunotherapy and SBRT for metastatic head and neck carcinomas	Metastatic HNSCC, ≥2 metastatic sites	1: Durvalumab + tremelimumab for 4 cycles (4 weeks each), SBRT between cycles 2–3	Concurrent	1/2
NCT03844763	CONFRONT: Targeting the tumor microenvironment in R/M SCCHN	Recurrent/metastatic HNSCC	1: Avelumab, cyclophosphamide, and radiation (8 Gy/1 fx) to a single site 1 week after first dose of avelumab	Concurrent	1/2
NCT03522584	Durvalumab, tremelimumab and hypofractionated radiation therapy in treating patients with recurrent or metastatic head and neck squamous cell carcinoma	Recurrent/metastatic HNSCC; progression through prior PD-1/PD-L1 inhibitor	1: Durvalumab + tremelimumab for 4 cycles (4 weeks each) followed by durvalumab alone for 9 cycles; SBRT during week 3 in 3 fractions, every other day	Concurrent	1/2
NCT03474497	UCDCC#272: IL-2, radiotherapy, and pembrolizumab in patients refractory to checkpoint blockade	Recurrent/metastatic HNSCC; progression through prior PD-1/PD-L1 inhibitor	1: One cycle of pembrolizumab, then SBRT (24 Gy/3 fx) and intratumoral injection of interleukin-2 during cycle 2, then additional pembrolizumab	Concurrent	1/2
NCT03317327	REPORT: REirradiation and programmed cell death protein 1 (PD-1) blockade on recurrent squamous cell head and neck tumors	Recurrent HNSCC after prior radiation or second primary HNSCC	1: Nivolumab with re-irradiation to 60 Gy (in 1.5 Gy bid fx), followed by nivolumab for up to 12 months	Concurrent	1/2
NCT04340258	Trial combining pembrolizumab and cesium 131 brachytherapy with salvage surgery in HNSCC	Resectable recurrent HNSCC after prior surgery or radiation	1: One dose of pembrolizumab, then salvage surgery with implantation of Cesium-131 brachytherapy seeds (60–70 Gy), followed by adjuvant pembrolizumab for 6 months	Concurrent	1/2
NCT04454489	Quad shot radiotherapy in combination with immune checkpoint inhibition	Recurrent/metastatic HNSCC	1: Pembrolizumab given every 3 weeks; quad-shot radiation (14.8 Gy in 4 bid fx) between cycles 2 and 3	Concurrent	2
NCT03313804	Priming immunotherapy in advanced disease with radiation	Recurrent/metastatic HNSCC	1: Nivolumab, pembrolizumab, or atezolimuab, with either SBRT (BED >100 Gy) or 30 Gy fractionated RT	Concurrent	2

NCT03386357	Radiotherapy with pembrolizumab in metastatic HNSCC	Recurrent/metastatic HNSCC, ≥2 metastatic sites, progression through platinum-based therapy	1: Radiation to 1–3 metastases (36 Gy/12 fx), with pembrolizumab starting between fraction 3 and 4	Concurrent	2
			2: Pembrolizumab alone		
NCT03511391	CHEERS: CHEckpoint inhibition in combination with an immunoboost of external body radiotherapy in solid tumors	Recurrent/metastatic HNSCC, progression through platinum-based therapy	1: Two cycles of nivolumab, then SBRT to 1–3 metastases (24 Gy/3 fx) prior to cycle 3	Concurrent	2
			2: Nivolumab alone		
NCT03085719	Targeting PD-1 therapy resistance with focused high or high and low dose radiation in SCCHN	Metastatic HNSCC, progression through prior PD-1 inhibition, ≥3 metastatic sites	1: Pembrolizumab and high-dose SBRT (3 fx) to 1 metastatic site	Concurrent	2
			2: Pembrolizumab and high-dose SBRT (3 fx) to 1 metastatic site, and low-dose radiation (2 fx) to another site		
NCT03546582	KEYSTROKE: SBRT ± pembrolizumab in patients with local-regionally recurrent or second primary head and neck carcinoma	Recurrent HNSCC after prior radiation or second primary HNSCC	1: Re-irradiation with SBRT over 2 weeks, then pembrolizumab every 3 weeks for up to 2 years	Sequential	2
			2: Re-irradiation with SBRT over 2 weeks		
NCT03521570	Intensity-modulated radiation therapy & nivolumab for recurrent or second primary head & neck squamous cell cancer	Recurrent HNSCC after prior radiation or second primary HNSCC	1: One dose of nivolumab, then radiation with concurrent nivolumab, then adjuvant nivolumab for 5 months	Concurrent + sequential	2
NCT02289209	Reirradiation with pembrolizumab in locoregional inoperable recurrence or second primary squamous cell CA of the head and neck	Unresectable recurrent HNSCC after prior radiation or second primary HNSCC	1: Pembrolizumab with re-irradiation to 60 Gy (in 1.2 Gy bid fx), followed by pembrolizumab for 3 months	Concurrent + sequential	2
NCT02684253	Screening trial of nivolumab with image guided, stereotactic body radiotherapy (SBRT) versus nivolumab alone in patients with metastatic head and neck squamous cell carcinoma (HNSCC)	Metastatic HNSCC, ≥2 metastatic sites	1: One cycle of nivolumab, then SBRT (27 Gy/3 fx) with the second cycle, followed by additional nivolumab	Concurrent	2
			2: Nivolumab alone		

BED biologically effective dose, *bid* twice a day, *fx* fraction, *HNSCC* head and neck squamous cell carcinoma, *ICIs* immune checkpoint inhibitors, *SBRT* stereotactic body radiotherapy limited number of metastases might develop with limited further metastatic potential [64]. Aggressive local treatment of patients with limited metastases would thus potentially offer a significant survival benefit. Results from several randomized phase 2 trials have supported this hypothesis (though notably HNSCC was not represented in any of these studies) [65–69]. Consequently, there is interest in the addition of ICIs to radiation in this population of patients to improve outcomes [70]. In this setting, radiation would be administered at ablative doses to all metastatic sites, and so the addition of ICIs would also be intended to augment the local effects of radiation at each treatment site. To our knowledge, no prospective clinical data has yet been published on the combination of radiation and ICIs in patients with oligometastatic HNSCC, though there is at least one ongoing clinical trial (NCT03283605, which examines the use of durvalumab, tremelimumab [a CTLA-4 inhibitor], and SBRT in patients with HNSCC with fewer than 10 metastases).

Related to the overall concept of oligometastases is oligoprogression, or the development of a limited number of progressive metastatic lesions after a period of stability on systemic therapy [71]. In the context of ICIs, oligoprogression may herald general immune escape in patients who had previously been responding to treatment. However, in certain cases oligoprogression may develop as a result of resistant tumor clones that lack particular tumor antigens or antigen presentation, or because of differences in the underlying immune microenvironment of the anatomic site that permit localized immune escape (e.g., brain) [72, 73]. If this is the case, local treatment such as radiation to these oligoprogressive sites may enable the patient to continue to derive benefit from ICIs [74–76]. We are testing this paradigm prospectively in SCCHN (NCT03085719).

Locally Advanced/Definitive Setting

ICIs are being investigated in the setting of curative treatment of earlier stages of disease across all cancer types, including HNSCC. Addition of ICIs to radiation in this setting would be intended to potentially augment the local effects of radiation (i.e., as a radiosensitizer) and address micrometastatic disease. Several possible combinations are under investigation—immunotherapy added to a chemoradiation regimen to intensify therapy (for patients with currently poor outcomes), immunotherapy given concurrently with radiation instead of chemotherapy or with a lower dose of radiation (potentially as a way to reduce treatment morbidity while maintaining overall efficacy), or immunotherapy administered adjuvantly and/or as induction (i.e., sequential therapy). To date adjuvant immunotherapy has proven successful in NSCLC; as noted earlier, the PACIFIC trial demonstrated a significant and meaningful overall survival benefit for adjuvant durvalumab starting within 6 weeks of completing standard chemoradiation for unresectable stage III NSCLC, with an increase in 2-year overall survival from 55.6% to 66.3% [77]. Of note, the magnitude of benefit was greater for patients who were randomized within 2 weeks of completing chemoradiation. Adjuvant immunotherapy also has newly demonstrated success in esophagogastric cancer; Checkmate-577 demonstrated a doubling of median disease-free survival (22.4 vs. 11.0 months) with the administration of adjuvant nivolumab compared to placebo following neoadjuvant chemoradiation and surgical resection in patients with esophageal and gastroesophageal cancer, though full trial results have yet to be published [78].

As shown in Table 7.2, ongoing trials are evaluating various combinations of radiation and ICIs for HNSCC in the definitive setting, and several have now reported safety data. In general, combinations of PD-1/PD-L1 inhibitors with definitive radiation appear well tolerated with no unexpected toxicities. KEYCHAIN is a randomized phase 2 study of radiation combined with concurrent and adjuvant pembrolizumab compared with radiation and concurrent cisplatin in intermediate-risk p16-positive HNSCC; the safety lead-in phase of the study found only one dose-limiting toxicity (grade 4 adrenal insufficiency) among eight patients in the pembrolizumab arm, and so the trial has proceeded to its

Table 7.2 Ongoing trials evaluating combinations of ICIs and radiation in the definitive management of locally advanced HNSCC (with permissions from [19])

NCT#	Title	Inclusion criteria	Treatment arms	Timing	Phase
NCT02819752	PEmbrolizumab combined with chemoradiotherapy in squamous cell carcinoma of the head and neck (PEACH)	LA HNSCC	1: Pembrolizumab added to standard chemoradiation, 3 doses concurrently, 4 doses adjuvantly	Concurrent + sequential	1
NCT04477759	Dose-escalated hypofractionated adaptive radiotherapy for head and neck cancer (DEHART)	LA HNSCC, cisplatin-ineligible, or primary metastatic HNSCC	1: MR-guided hypofractionated radiation (50–60 Gy/15 fx); atezolizumab given with fraction 1 and 11 of radiation, then every 4 weeks for up to 1 year	Concurrent + sequential	1
NCT03509012	CLOVER: Immunotherapy in combination with chemoradiation in patients with advanced solid tumors	LA HNSCC	1: Durvalumab concurrent with standard radiation and cisplatin	Concurrent	1
NCT02764593	RTOG 3504: Safety testing of adding nivolumab to chemotherapy in patients with intermediate and high-risk local-regionally advanced head and neck cancer	LA HNSCC, intermediate or high risk	1: One dose of nivolumab as induction, then radiation (70 Gy/35 fx) and nivolumab with weekly cisplatin, then adjuvant nivolumab for 7 doses	Concurrent + sequential	1
			2: One dose of nivolumab as induction, then radiation (70 Gy/35 fx) and nivolumab with bolus cisplatin, then adjuvant nivolumab for 7 doses		
			3: One dose of nivolumab as induction, then radiation (70 Gy/35 fx) and nivolumab with weekly cetuximab, then adjuvant nivolumab for 7 doses		
			4: One dose of nivolumab as induction, then radiation (70 Gy/35 fx) with nivolumab, then adjuvant nivolumab for 7 doses		

(continued)

Table 7.2 (continued)

NCT#	Title	Inclusion criteria	Treatment arms	Timing	Phase
NCT03051906	DUCRO-HN: Durvalumab, cetuximab and radiotherapy in head neck cancer	LA HNSCC	1: Durvalumab every 4 weeks, cetuximab weekly, and radiation to 69.96 Gy/33 fx, followed by adjuvant durvalumab for 6 months	Concurrent + sequential	1/2
NCT03247712	Neoadjuvant immunoradiotherapy in head & neck cancer	Resectable LA HNSCC	1: Neoadjuvant SBRT (24–40 Gy/3–5 fx) and nivolumab, followed by surgery, followed by adjuvant nivolumab	Concurrent + sequential	1/2
NCT02296684	Immunotherapy with MK-3475 in surgically resectable head and neck squamous cell carcinoma	Resectable LA HNSCC, except p16-positive oropharyngeal SCC	1: Two doses of pembrolizumab neoadjuvantly followed by surgery and standard risk-adapted adjuvant (chemo)radiation	Sequential	2
			2: One dose of pembrolizumab neoadjuvantly, followed by surgery and standard risk-adapted adjuvant (chemo)radiation, followed by adjuvant pembrolizumab for up to 6 doses for patients with ENE or positive margins		
NCT03894891	Induction TPN followed by nivolumab with radiation in locoregionally advanced laryngeal and hypopharyngeal cancer	LA p16-negative SCC of larynx or hypopharynx	1: Induction cisplatin, docetaxel, and nivolumab, followed by concurrent radiation and nivolumab	Concurrent + sequential	2
NCT03708224	Phase II study of perioperative immunotherapy in patients with advanced non-virally associated squamous cell carcinoma	Resectable LA HNSCC, except p16-positive oropharyngeal SCC	1: One dose of atezolizumab neoadjuvantly, followed by surgery and standard risk-adapted adjuvant (chemo)radiation, followed by atezolizumab every 3 weeks for up to 12 cycles	Sequential	2
			2: One dose of atezolizumab and tocilizumab neoadjuvantly, followed by surgery and standard risk-adapted adjuvant (chemo) radiation, followed by atezolizumab every 3 weeks for up to 12 cycles		

NCT03426657	Radiotherapy with double checkpoint blockade of locally advanced HNSCC	LA HNSCC	1: One cycle of induction cisplatin, docetaxel, durvalumab, and tremelimumab; patients with increased CD8+ T-cell infiltration on interval biopsy then receive durvalumab, tremelimumab, and radiation, followed by adjuvant durvalumab for 8 months	Concurrent + sequential	2
NCT03532737	Concomitant immune check point inhibitor with radiochemotherapy in head and neck cancer	LA HNSCC, non-nasopharynx	1: Pembrolizumab for 6 cycles (3 weeks each), and chemoradiation starting with cycle 2, with either bolus cisplatin or cetuximab, and radiation to 66–70 Gy/30–35 fx	Concurrent + sequential	2
NCT02892201	Pembrolizumab in HNSCC with residual disease after radiation	LA HNSCC with residual disease after definitive radiation	1: Pembrolizumab for 4 cycles, followed by evaluation for salvage surgery; unresectable patients continue pembrolizumab for up to 1 year	Sequential	2
NCT03721757	CA209-891: Neoadjuvant and adjuvant nivolumab as immune checkpoint inhibition in oral cavity cancer (NICO)	LA oral cavity SCC	1: One dose of neoadjuvant nivolumab followed by surgery, then one dose of nivolumab, then standard postoperative radiation or chemoradiation (60 Gy/30 fx), then 6 months of adjuvant nivolumab	Sequential	2
NCT03944915	De-escalation therapy for human papillomavirus negative disease (DEPEND)	LA p16-negative HNSCC	1: Induction carboplatin, paclitaxel, and nivolumab, followed by response-adapted chemoradiation (66–75 Gy)	Sequential	2
NCT04405154	A study of concomitant camrelizumab with chemoradiation for locally advanced head and neck cancer	LA HNSCC	1: Camrelizumab for 8 cycles (2 weeks each), with standard chemoradiation (bolus cisplatin and radiation [66 Gy/33 fx]) starting with cycle 2	Concurrent + sequential	2

(continued)

Table 7.2 (continued)

NCT#	Title	Inclusion criteria	Treatment arms	Timing	Phase
NCT02777385	Pembrolizumab in combination with cisplatin and intensity modulated radiotherapy (IMRT) in head and neck cancer	LA HNSCC, intermediate or high risk	1: Pembrolizumab for one initial dose, then concurrent with radiation and weekly cisplatin, then adjuvant pembrolizumab for a total of 8 doses	Concurrent + sequential	2
			2: Radiation and weekly cisplatin, followed by adjuvant pembrolizumab for 8 doses	Sequential	
NCT03383094	KEYCHAIN: Chemoradiation vs immunotherapy and radiation for head and neck cancer	LA HNSCC, p16-positive, intermediate risk	1: Pembrolizumab and standard radiation to 70 Gy/33–35 fx, followed by adjuvant pembrolizumab for up to 20 cycles (3 weeks each)	Concurrent	2
			2: Standard chemoradiation to 70 Gy/33–35 fx with bolus cisplatin		
NCT02707588	PembroRad: Tolerance and efficacy of pembrolizumab or cetuximab combined with RT in patients with locally advanced HNSCC	LA HNSCC	1: Radiation (69.96 Gy/33 fx) with concurrent pembrolizumab	Concurrent	2
			2: Radiation (69.96 Gy/33 fx) with concurrent cetuximab		
NCT02609503	Pembrolizumab + radiation for locally Adv SCC of the head and neck (SCCHN) not eligible cisplatin	LA HNSCC, cisplatin-ineligible	1: Radiation (70 Gy/35 fx) with 3 concurrent cycles of pembrolizumab, then 3 adjuvant cycles	Concurrent + sequential	2
NCT03258554	NRG-HN004: Radiation therapy with durvalumab or cetuximab in treating patients with Locoregionally advanced head and neck cancer who cannot take cisplatin	LA HNSCC, cisplatin-ineligible	1: Durvalumab for 7 cycles (4 weeks each); radiation to 70 Gy/35 fx starting week 2	Concurrent + sequential	2/3
			2: Cetuximab for 8 cycles (weekly); radiation to 70 Gy/35 fx starting week 2		
NCT01810913	RTOG 1216: Testing docetaxel-cetuximab or the addition of an immunotherapy drug, atezolizumab, to the usual chemotherapy and radiation therapy in high-risk head and neck cancer	Resected LA HNSCC, except p16-positive oropharyngeal SCC, with pathologic ENE or positive margins	1: Atezolizumab for 8 cycles (3 weeks each) following surgery, with standard chemoradiation (to 60 Gy/30 fx with weekly cisplatin) starting week 2	Concurrent + sequential	2/3

NCT03811015	EA3161: Testing immunotherapy versus observation in patients with HPV throat cancer	p16-positive oropharyngeal SCC, intermediate risk	1: Radiation (70 Gy/35 fx) and concurrent weekly cisplatin, then adjuvant nivolumab for 12 months	Sequential	2/3
			2: Radiation (70 Gy/35 fx) and concurrent weekly cisplatin, then observation		
NCT03452137	IMvoke010: A study of atezolizumab (anti−Pd-L1 antibody) as adjuvant therapy after definitive local therapy in patients with high-risk locally advanced squamous cell carcinoma of the head and neck	LA HNSCC after definitive local therapy (chemoradiation or surgery + (chemo)radiation)	1: Adjuvant atezolizumab for 1 year	Sequential	3
			2: Placebo for 1 year		
NCT03576417	NIVOPOSTOP: A trial evaluating the addition of nivolumab to cisplatin-RT for treatment of cancers of the head and neck	Resected LA HNSCC, with ENE, positive margins, or multiple positive nodes	1: One dose of nivolumab, then nivolumab concurrent with radiation (66 Gy/33 fx) and bolus cisplatin	Concurrent + sequential	3
			2: Radiation (66 Gy/33 fx) with bolus cisplatin		
NCT03673735	Maintenance immune check-point inhibitor following post-operative chemo-radiation in subjects with HPV-negative HNSCC (ADHERE)	Surgically resected p16-negative HNSCC with pathologic ENE or positive margins	1: One dose of induction durvalumab followed by standard chemoradiation (bolus cisplatin and radiation [66 Gy/33 fx]), followed by 6 months of adjuvant durvalumab	Sequential	3
			2: Standard chemoradiation (bolus cisplatin and radiation [66 Gy/33 fx])		

(continued)

Table 7.2 (continued)

NCT#	Title	Inclusion criteria	Treatment arms	Timing	Phase
NCT03700905	IMSTAR-HN: Study of nivolumab alone or in combination with ipilimumab as immunotherapy vs standard follow-up in surgical resectable HNSCC after adjuvant therapy	Resectable LA HNSCC, except p16-positive oropharyngeal SCC	1: One dose of neoadjuvant nivolumab followed by surgery, followed by standard risk-adapted adjuvant (chemo)radiation, followed by either adjuvant nivolumab or adjuvant nivolumab + ipilimumab for 6 months	Sequential	3
			2: Surgical resection followed by standard risk adapted adjuvant (chemo)radiation		
NCT03765918	Study of pembrolizumab given prior to surgery and in combination with radiotherapy given post-surgery for advanced head and neck squamous cell carcinoma (MK-3475-689)	Resectable LA HNSCC	1: Two doses of neoadjuvant pembrolizumab, then surgery, then pembrolizumab with adjuvant radiation or chemoradiation, then adjuvant pembrolizumab for 12 additional doses	Concurrent + sequential	3
			2: Surgery followed by adjuvant radiation or chemoradiation		
NCT03673735	ADHERE: Maintenance immune check-point inhibitor following post-operative chemo-radiation in subjects with HPV-negative HNSCC	Resected LA HNSCC, except p16-positive oropharyngeal SCC, with pathologic ENE or positive margins	1: Following surgery, one dose of durvalumab, then standard radiation (66 Gy/33 fx) with bolus cisplatin, then adjuvant durvalumab for 6 doses	Sequential	3
			2: Following surgery, standard radiation (66 Gy/33 fx) with bolus cisplatin		
NCT03040999	KEYNOTE-412: Study of pembrolizumab (MK-3475) or placebo with chemoradiation in participants with locally advanced head and neck squamous cell carcinoma	LA HNSCC	1: One dose of induction pembrolizumab, then pembrolizumab with radiation (70 Gy/35 fx) and bolus cisplatin, then adjuvant pembrolizumab for a total of 17 doses	Concurrent + sequential	3
			2: Standard radiation (70 Gy/35 fx) with bolus cisplatin		

NCT02999087	REACH: Randomized trial of avelumab-cetuximab-radiotherapy versus SOCs in LA SCCHN	LA HNSCC, both cisplatin eligible and ineligible	1: Cetuximab and avelumab, 1 dose prior to radiation, then concurrent during radiation (69.96 Gy/33 fx), then adjuvant avelumab for 12 months	Concurrent + sequential	3
			2: Standard radiation (69.96 Gy/33 fx) with concurrent bolus cisplatin for cisplatin-eligible patients		
			3: Standard radiation (69.96 Gy/33 fx) with concurrent cetuximab for cisplatin-ineligible patients		
NCT02952586	Javelin 100: Study to compare avelumab in combination with standard of care chemoradiotherapy (SoC CRT) versus SoC CRT for definitive treatment in patients with locally advanced squamous cell carcinoma of the head and neck	LA HNSCC	1: One dose of induction avelumab, then avelumab with radiation (70 Gy/35 fx) and bolus cisplatin, then adjuvant avelumab for 12 months	Concurrent + sequential	3
			2: Radiation (70 Gy/35 fx) and bolus cisplatin		

ENE extranodal extension, *fx* fraction, *HNSCC* head and neck squamous cell carcinoma, *ICIs* immune checkpoint inhibitors, *LA* locally advanced, *SBRT* stereotactic body radiotherapy phase 2 component [79]. A single-arm phase 2 trial of radiation administered with concurrent and adjuvant pembrolizumab in cisplatin-ineligible patients with locally advanced HNSCC similarly demonstrated relatively low toxicity in the first 12 enrolled patients, and 11 of 12 patients received all planned cycles of pembrolizumab [80]. Finally, PembroRad is a randomized phase 2 trial of radiation combined with concurrent pembrolizumab versus radiation combined with concurrent cetuximab, again in cisplatin-ineligible patients with locally advanced HNSCC. There have been 133 patients randomized in a 1:1 fashion, and the pembrolizumab arm was found to have significantly less mucositis or dermatitis within the radiation field than the cetuximab arm [81].

Early results also suggest that intensification of existing chemoradiation regimens with the addition of ICIs is reasonably safe. In a small phase 1 trial of concurrent and adjuvant avelumab added to standard cetuximab/radiation in 10 cisplatin-ineligible patients with locally advanced HNSCC, no grade 4–5 toxicities were observed, and only one of eight evaluable patients discontinued avelumab for toxicity [82]. REACH is a phase 3 trial that is also comparing concurrent avelumab, cetuximab, and radiation, followed by 12 months of adjuvant avelumab, against either standard bolus cisplatin with radiation or cetuximab with radiation (depending on if the patient is judged to be fit for cisplatin or not) in patients with locally advanced HNSCC; results for the 82 patients randomized during the safety phase of the trial suggested that addition of avelumab was tolerable, with 88% of patients completing concurrent avelumab as per protocol, and rates of grade 4+ events similar between control and experimental arms [83]. Similarly, a single-arm phase 1b study of the addition of concurrent and adjuvant pembrolizumab to standard radiation and weekly cisplatin in patients with locally advanced HNSCC demonstrated in 59 patients that concurrent pembrolizumab did not prevent patients from completing chemoradiation, and only 5 of 59 patients ultimately discontinued treatment because of irAEs [84]. Finally, RTOG 3504 is a four-arm phase 1 trial in patients with intermediate- or high-risk HNSCC that is examining the addition of concurrent and adjuvant nivolumab to either radiation alone or radiation with weekly cisplatin, bolus cisplatin, or cetuximab; safety results from the latter three arms again demonstrated that nivolumab did not prevent timely completion of chemoradiation, and rates of dose-limiting toxicities were low [85].

Efficacy data are now just starting to be reported from some of these ongoing trials. One of the single-arm phase 2 trials noted above [80] of radiation with concurrent and adjuvant pembrolizumab in cisplatin-ineligible patients with locally advanced HNSCC ultimately enrolled 29 patients, and reported 1-year progression-free survival (PFS) and overall survival (OS) of 76% and 86%, respectively [86]. More recently, efficacy results from PembroRad were presented, with oncologic outcomes found to be not significantly different between the pembrolizumab vs. cetuximab arms (2-year PFS 42% vs. 40%, $p = 0.41$; 2-year OS 62% vs. 55%, $p = 0.49$) [87]. Finally, Javelin 100 was a double-blind, placebo-controlled phase 3 trial that randomized 697 patients with locally advanced HNSCC to standard of care cisplatin-based chemoradiation with or without concurrent and adjuvant (for 12 months) avelumab, with PFS as the primary endpoint. The trial was terminated early for futility following a planned interim analysis, in which PFS was found to favor the placebo + chemoradiation arms (hazard ratio 1.21, $p = 0.92$), and rates of grade 3 or higher adverse events were also slightly higher in the avelumab arm compared to placebo (88% vs. 82%) [88]. Exploratory analyses did not reveal any improvement for either time to locoregional failure or distant metastatic failure, and the PFS results were generally consistent across subgroups as well. One possible exception was the results for the PD-L1 high subgroup (defined as $\geq$25%), where avelumab seemed to confer a PFS benefit compared to placebo; however, the number of patients was small, and the study did not stratify on the basis of PD-L1 status, so this observation remains purely hypothesis generating.

The disappointing results of Javelin 100 invite comparison to the successful incorporation of

PD-L1 blockade into the treatment of locally advanced NSCLC as evidenced by the PACIFIC study. Given the high risk of lymph node metastases in patients with locally advanced HNSCC, standard radiation generally entails elective treatment of the draining cervical lymph node chains (in contrast to NSCLC, where elective lymph nodes are not intentionally irradiated). These draining lymph nodes are precisely where antigen-presenting cells migrate to for T-cell priming, following radiation to the primary tumor [23, 27]. Correlative positron emission tomography–computed tomography (PET-CT) studies from a recently published clinical trial of neoadjuvant ICIs (nivolumab or nivolumab and ipilimumab) prior to surgery in patients with oral cavity SCC provides further support for the importance of the draining lymph nodes; following initiation of neoadjuvant ICIs, there was a high rate of increased fluorodeoxyglucose (FDG) uptake in the draining cervical lymph nodes on an interval PET-CT, which ultimately on surgical pathology demonstrated only reactive findings without any evidence of cancer. This observed increase in FDG uptake may therefore represent radiographic evidence of a mounting immune response [89]. Given the radiosensitivity of lymphocytes, then, it seems possible that radiation (particularly longer conventionally fractionated regimens) that electively treats the draining lymph nodes following the receipt of ICI could actually hinder T-cell priming. Indeed, as noted above, there is some preclinical data to support this, as Morisada et al. demonstrated in an syngeneic mouse model of oral cavity cancer that 20 Gy in ten fractions compared to 16 Gy in two fractions to both the primary tumor and the draining lymph nodes blunted tumor-specific CD8+ T-cell responses within those draining lymph nodes (although notably tumors were implanted in the mice legs, and thus, this is not a perfect model for head and neck lymphatics) [49]. The phase 2 trial reported by Weiss et al. also noted a rate of grade 3+ lymphopenia of 58.6% [86]. Another notable issue is that the design of Javelin 100, as well as many of the other trials described above, incorporated both concurrent and adjuvant ICIs in the experimental arm, whereas PACIFIC (and Checkmate-577) only tested the value of adjuvant immunotherapy. Timing and sequencing of ICIs and radiation remains a critical issue that requires further study, although the concerns regarding radiation-induced T-cell death may be particularly problematic when ICI is administered concurrently as compared with sequentially [90]. Finally, as demonstrated in the preclinical work above, radiation dose and fractionation are also likely critical to successful synergy between radiation and ICIs; however, the hypofractionated regimens that appear to have the greatest immunologic potential in preclinical models differ tremendously from the long conventionally fractionated regimens (1.8–2 Gy/fraction) used in the current standard management of HNSCC. PACIFIC did also employ conventional fractionation, though standard total doses for NSCLC are somewhat lower than for HNSCC (54–66 Gy versus 70 Gy). Overall, given the years of experience supporting the current standard radiation regimen and fields used in the definitive management of HNSCC, careful studies will be required to determine what kinds of modifications to elective nodal irradiation, timing/sequencing, dose, and/or fractionation are required to maximize synergy with ICIs and ultimately improve patient outcomes. There is already significant heterogeneity among the ongoing trials in Tables 7.1 and 7.2 with regard to these parameters, and so examining the results collectively will hopefully be informative.

Conclusions/Future Directions

There remains excitement for the possibility of combining radiotherapy and immunotherapy to improve outcomes for patients with HNSCC. Ongoing trials will help advance this emerging field, and the developing paradigm of oligometastatic disease provides further opportunity to integrate improving systemic and local therapies. Biomarker studies conducted in parallel will also inform optimal patient selection for combined treatment approaches. Moreover, while this chapter has largely focused on ICIs (and PD-1/PD-L1 targeted therapies in particular) given their widespread use, immunotherapeutic agents targeting other checkpoints and pathways are in development as well [91], as are trials testing their combination with radiation (e.g., NCT04220775). Nevertheless, significant work remains to be done in both the preclinical and clinical space to determine the dose, fractionation, timing, target, and field size of radiation that will be the most synergistic with immunotherapies. Finding the optimal balance between the immunostimulatory and immunosuppressive effects of radiotherapy is key and hopefully will herald continued improvement in outcomes for patients with HNSCC.

Acknowledgments Portions of this chapter were reproduced from prior work by the authors: Qian JM and Schoenfeld JD (2021) Radiotherapy and Immunotherapy for Head and Neck Cancer: Current Evidence and Challenges. *Front. Oncol.* 10:608772. https://doi.org/10.3389/fonc.2020.608772.

References

1. Bray F, Ferlay J, Soerjomataram I, Siegel RL, Torre LA, Jemal A. Global cancer statistics 2018: GLOBOCAN estimates of incidence and mortality worldwide for 36 cancers in 185 countries. CA Cancer J Clin. 2018;68(6):394–424.
2. Taberna M, Mena M, Pavón MA, Alemany L, Gillison ML, Mesía R. Human papillomavirus-related oropharyngeal cancer. Ann Oncol. 2017;28(10):2386–98.
3. Coutard H. Principles of X ray therapy of malignant diseases. Lancet. 1934;224(5784):1–8.
4. Leeman JE, Li J-G, Pei X, Venigalla P, Zumsteg ZS, Katsoulakis E, et al. Patterns of treatment failure and postrecurrence outcomes among patients with locally advanced head and neck squamous cell carcinoma after chemoradiotherapy using modern radiation techniques. JAMA Oncol. 2017;3(11):1487–94.
5. Vengaloor Thomas T, Packianathan S, Bhanat E, Albert A, Abraham A, Gordy X, et al. Oligometastatic head and neck cancer: comprehensive review. Head Neck. 2020;42(8):2194–201.
6. Mascaux C, Angelova M, Vasaturo A, Beane J, Hijazi K, Anthoine G, et al. Immune evasion before tumour invasion in early lung squamous carcinogenesis. Nature. 2019;571(7766):570–5.
7. Galon J, Bruni D. Tumor immunology and tumor evolution: intertwined histories. Immunity. 2020;52(1):55–81.
8. Hodi FS, O'Day SJ, McDermott DF, Weber RW, Sosman JA, Haanen JB, et al. Improved survival with ipilimumab in patients with metastatic melanoma. N Engl J Med. 2010;363(8):711–23.
9. Seiwert TY, Burtness B, Mehra R, Weiss J, Berger R, Eder JP, et al. Safety and clinical activity of pembrolizumab for treatment of recurrent or metastatic squamous cell carcinoma of the head and neck (KEYNOTE-012): an open-label, multicentre, phase 1b trial. Lancet Oncol. 2016;17(7):956–65.
10. Cohen EEW, Soulières D, Le Tourneau C, Dinis J, Licitra L, Ahn M-J, et al. Pembrolizumab versus methotrexate, docetaxel, or cetuximab for recurrent or metastatic head-and-neck squamous cell carcinoma (KEYNOTE-040): a randomised, open-label, phase 3 study. Lancet. 2019;393(10167):156–67.
11. Ferris RL, Blumenschein G, Fayette J, Guigay J, Colevas AD, Licitra L, et al. Nivolumab for recurrent squamous-cell carcinoma of the head and neck. N Engl J Med. 2016;375(19):1856–67.
12. Burtness B, Harrington KJ, Greil R, Soulières D, Tahara M, de Castro G, et al. Pembrolizumab alone or with chemotherapy versus cetuximab with chemotherapy for recurrent or metastatic squamous cell carcinoma of the head and neck (KEYNOTE-048): a randomised, open-label, phase 3 study. Lancet. 2019;394(10212):1915–28.
13. Alexander BM, Schoenfeld JD, Trippa L. Hazards of hazard ratios—deviations from model assumptions in immunotherapy. N Engl J Med. 2018;378(12):1158–9.
14. Le DT, Uram JN, Wang H, Bartlett BR, Kemberling H, Eyring AD, et al. PD-1 blockade in tumors with mismatch-repair deficiency. N Engl J Med. 2015;372(26):2509–20.
15. Marabelle A, Fakih MG, Lopez J, Shah M, Shapira-Frommer R, Nakagawa K, et al. 1192O—association of tumour mutational burden with outcomes in patients with select advanced solid tumours treated with pembrolizumab in KEYNOTE-158. Ann Oncol. 2019 Oct;1(30):v477–8.
16. Chen DS, Mellman I. Elements of cancer immunity and the cancer-immune set point. Nature. 2017;541(7637):321–30.
17. Schoenfeld JD, Fell G, Haddad RI, Trippa L. Keynote 48: is it really for everyone? Oral Oncol. 2020;105:104762.
18. Ferris RL, Blumenschein G, Fayette J, Guigay J, Colevas AD, Licitra L, et al. Nivolumab vs investigator's choice in recurrent or metastatic squamous cell carcinoma of the head and neck: 2-year long-term survival update of CheckMate 141 with analyses by tumor PD-L1 expression. Oral Oncol. 2018;81:45–51.
19. Qian JM, Schoenfeld JD. Radiotherapy and immunotherapy for head and neck cancer: current evidence and challenges. Front Oncol. 2021;10:3307.
20. Stone HB, Peters LJ, Milas L. Effect of host immune capability on radiocurability and subsequent transplantability of a murine fibrosarcoma. J Natl Cancer Inst. 1979;63(5):1229–35.
21. Galluzzi L, Vitale I, Warren S, Adjemian S, Agostinis P, Martinez AB, et al. Consensus guidelines for the definition, detection and interpretation of immunogenic cell death. J Immunother Cancer. 2020;8(1):e000337.
22. Golden EB, Frances D, Pellicciotta I, Demaria S, Helen Barcellos-Hoff M, Formenti SC. Radiation

fosters dose-dependent and chemotherapy-induced immunogenic cell death. Onco Targets Ther. 2014;3:e28518.
23. Weichselbaum RR, Liang H, Deng L, Fu Y-X. Radiotherapy and immunotherapy: a beneficial liaison? Nat Rev Clin Oncol. 2017;14(6):365–79.
24. Deng L, Liang H, Xu M, Yang X, Burnette B, Arina A, et al. STING-dependent cytosolic DNA sensing promotes radiation-induced type I interferon-dependent antitumor immunity in immunogenic tumors. Immunity. 2014;41(5):843–52.
25. Vanpouille-Box C, Alard A, Aryankalayil MJ, Sarfraz Y, Diamond JM, Schneider RJ, et al. DNA exonuclease Trex1 regulates radiotherapy-induced tumour immunogenicity. Nat Commun. 2017;8:15618.
26. Reits EA, Hodge JW, Herberts CA, Groothuis TA, Chakraborty M, Wansley EK, et al. Radiation modulates the peptide repertoire, enhances MHC class I expression, and induces successful antitumor immunotherapy. J Exp Med. 2006;203(5):1259–71.
27. Sharabi AB, Nirschl CJ, Kochel CM, Nirschl TR, Francica BJ, Velarde E, et al. Stereotactic radiation therapy augments antigen-specific PD-1-mediated antitumor immune responses via cross-presentation of tumor antigen. Cancer Immunol Res. 2015;3(4):345–55.
28. Garnett CT, Palena C, Chakraborty M, Chakarborty M, Tsang K-Y, Schlom J, et al. Sublethal irradiation of human tumor cells modulates phenotype resulting in enhanced killing by cytotoxic T lymphocytes. Cancer Res. 2004;64(21):7985–94.
29. Nakamura N, Kusunoki Y, Akiyama M. Radiosensitivity of CD4 or CD8 positive human T-lymphocytes by an in vitro colony formation assay. Radiat Res. 1990;123(2):224–7.
30. Arina A, Beckett M, Fernandez C, Zheng W, Pitroda S, Chmura SJ, et al. Tumor-reprogrammed resident T cells resist radiation to control tumors. Nat Commun. 2019;10(1):3959.
31. Grayson JM, Harrington LE, Lanier JG, Wherry EJ, Ahmed R. Differential sensitivity of naive and memory CD8+ T cells to apoptosis in vivo. J Immunol. 2002;169(7):3760–70.
32. Pike LRG, Bang A, Mahal BA, Taylor A, Krishnan M, Spektor A, et al. The impact of radiation therapy on lymphocyte count and survival in metastatic cancer patients receiving PD-1 immune checkpoint inhibitors. Int J Radiat Oncol Biol Phys. 2019;103(1):142–51.
33. Mandal R, Şenbabaoğlu Y, Desrichard A, Havel JJ, Dalin MG, Riaz N, et al. The head and neck cancer immune landscape and its immunotherapeutic implications. JCI Insight. 2016;1(17):e89829.
34. Wang J, Sun H, Zeng Q, Guo X-J, Wang H, Liu H-H, et al. HPV-positive status associated with inflamed immune microenvironment and improved response to anti-PD-1 therapy in head and neck squamous cell carcinoma. Sci Rep. 2019;9(1):13404.
35. Chen Y-P, Wang Y-Q, Lv J-W, Li Y-Q, Chua MLK, Le Q-T, et al. Identification and validation of novel microenvironment-based immune molecular subgroups of head and neck squamous cell carcinoma: implications for immunotherapy. Ann Oncol. 2019;30(1):68–75.
36. Chakravarthy A, Henderson S, Thirdborough SM, Ottensmeier CH, Su X, Lechner M, et al. Human papillomavirus drives tumor development throughout the head and neck: improved prognosis is associated with an immune response largely restricted to the oropharynx. J Clin Oncol. 2016;34(34):4132–41.
37. Schoenfeld JD, Gjini E, Rodig SJ, Tishler RB, Rawal B, Catalano PJ, et al. Evaluating the PD-1 axis and immune effector cell infiltration in oropharyngeal squamous cell carcinoma. Int J Radiat Oncol Biol Phys. 2018;102(1):137–45.
38. Spanos WC, Nowicki P, Lee DW, Hoover A, Hostager B, Gupta A, et al. Immune response during therapy with cisplatin or radiation for human papillomavirus-related head and neck cancer. Arch Otolaryngol Head Neck Surg. 2009;135(11):1137–46.
39. Özcan-Wahlbrink M, Schifflers C, Riemer AB. Enhanced radiation sensitivity of human papillomavirus-driven head and neck cancer: focus on immunological aspects. Front Immunol. 2019;10:2831.
40. Shimasaki N, Jain A, Campana D. NK cells for cancer immunotherapy. Nat Rev Drug Discov. 2020;19(3):200–18.
41. Oweida A, Lennon S, Calame D, Korpela S, Bhatia S, Sharma J, et al. Ionizing radiation sensitizes tumors to PD-L1 immune checkpoint blockade in orthotopic murine head and neck squamous cell carcinoma. Onco Targets Ther. 2017;6(10):e1356153.
42. Kim SS, Shen S, Miyauchi S, Sanders PD, Franiak-Pietryga I, Mell L, et al. B cells improve overall survival in HPV-associated squamous cell carcinomas and are activated by radiation and PD-1 blockade. Clin Cancer Res. 2020;26(13):3345–59.
43. Dillon MT, Bergerhoff KF, Pedersen M, Whittock H, Crespo-Rodriguez E, Patin EC, et al. ATR inhibition potentiates the radiation-induced inflammatory tumor microenvironment. Clin Cancer Res. 2019;25(11):3392–403.
44. Xiao R, Allen CT, Tran L, Patel P, Park S-J, Chen Z, et al. Antagonist of cIAP1/2 and XIAP enhances antitumor immunity when combined with radiation and PD-1 blockade in a syngeneic model of head and neck cancer. Onco Targets Ther. 2018;7(9):e1471440.
45. Oweida A, Hararah MK, Phan A, Binder D, Bhatia S, Lennon S, et al. Resistance to radiotherapy and PD-L1 blockade is mediated by TIM-3 upregulation and regulatory T-cell infiltration. Clin Cancer Res. 2018;24(21):5368–80.
46. Hanoteau A, Newton JM, Krupar R, Huang C, Liu H-C, Gaspero A, et al. Tumor microenvironment modulation enhances immunologic benefit of chemoradiotherapy. J Immunother Cancer. 2019;7(1):10.
47. Newton JM, Hanoteau A, Liu H-C, Gaspero A, Parikh F, Gartrell-Corrado RD, et al. Immune microenvironment modulation unmasks therapeutic benefit of radiotherapy and checkpoint inhibition. J Immunother Cancer. 2019;7(1):216.

48. Dewan MZ, Galloway AE, Kawashima N, Dewyngaert JK, Babb JS, Formenti SC, et al. Fractionated but not single-dose radiotherapy induces an immune-mediated abscopal effect when combined with anti-CTLA-4 antibody. Clin Cancer Res. 2009;15(17):5379–88.
49. Morisada M, Clavijo PE, Moore E, Sun L, Chamberlin M, Van Waes C, et al. PD-1 blockade reverses adaptive immune resistance induced by high-dose hypofractionated but not low-dose daily fractionated radiation. Onco Targets Ther. 2018;7(3):e1395996.
50. Morisada M, Moore EC, Hodge R, Friedman J, Cash HA, Hodge JW, et al. Dose-dependent enhancement of T-lymphocyte priming and CTL lysis following ionizing radiation in an engineered model of oral cancer. Oral Oncol. 2017;71:87–94.
51. Martins F, Sofiya L, Sykiotis GP, Lamine F, Maillard M, Fraga M, et al. Adverse effects of immune-checkpoint inhibitors: epidemiology, management and surveillance. Nat Rev Clin Oncol. 2019;16(9):563–80.
52. Hwang WL, Pike LRG, Royce TJ, Mahal BA, Loeffler JS. Safety of combining radiotherapy with immune-checkpoint inhibition. Nat Rev Clin Oncol. 2018;15(8):477–94.
53. Bang A, Wilhite TJ, Pike LRG, Cagney DN, Aizer AA, Taylor A, et al. Multicenter evaluation of the tolerability of combined treatment with PD-1 and CTLA-4 immune checkpoint inhibitors and palliative radiation therapy. Int J Radiat Oncol Biol Phys. 2017;98(2):344–51.
54. Luke JJ, Lemons JM, Karrison TG, Pitroda SP, Melotek JM, Zha Y, et al. Safety and clinical activity of pembrolizumab and multisite stereotactic body radiotherapy in patients with advanced solid tumors. J Clin Oncol. 2018;36(16):1611–8.
55. McBride S, Sherman E, Tsai CJ, Baxi S, Aghalar J, Eng J, et al. Randomized phase II trial of nivolumab with stereotactic body radiotherapy versus nivolumab alone in metastatic head and neck squamous cell carcinoma. J Clin Oncol. 2020;39(1):30–7.
56. Azria D, Magné N, Zouhair A, Castadot P, Culine S, Ychou M, et al. Radiation recall: a well recognized but neglected phenomenon. Cancer Treat Rev. 2005;31(7):555–70.
57. Shibaki R, Akamatsu H, Fujimoto M, Koh Y, Yamamoto N. Nivolumab induced radiation recall pneumonitis after two years of radiotherapy. Ann Oncol. 2017;28(6):1404–5.
58. Schoenfeld JD, Nishino M, Severgnini M, Manos M, Mak RH, Hodi FS. Pneumonitis resulting from radiation and immune checkpoint blockade illustrates characteristic clinical, radiologic and circulating biomarker features. J Immunother Cancer. 2019;7(1):112.
59. Antonia SJ, Villegas A, Daniel D, Vicente D, Murakami S, Hui R, et al. Durvalumab after chemoradiotherapy in stage III non-small-cell lung cancer. N Engl J Med. 2017;377(20):1919–29.
60. Colaco RJ, Martin P, Kluger HM, Yu JB, Chiang VL. Does immunotherapy increase the rate of radiation necrosis after radiosurgical treatment of brain metastases? J Neurosurg. 2016;125(1):17–23.
61. Martin AM, Cagney DN, Catalano PJ, Alexander BM, Redig AJ, Schoenfeld JD, et al. Immunotherapy and symptomatic radiation necrosis in patients with brain metastases treated with stereotactic radiation. JAMA Oncol. 2018;4(8):1123–4.
62. Reynders K, Illidge T, Siva S, Chang JY, De Ruysscher D. The abscopal effect of local radiotherapy: using immunotherapy to make a rare event clinically relevant. Cancer Treat Rev. 2015;41(6):503–10.
63. Theelen WSME, Peulen HMU, Lalezari F, van der Noort V, de Vries JF, Aerts JGJV, et al. Effect of pembrolizumab after stereotactic body radiotherapy vs pembrolizumab alone on tumor response in patients with advanced non-small cell lung cancer: results of the PEMBRO-RT phase 2 randomized clinical trial. JAMA Oncol. 2019;5(9):1276–82.
64. Hellman S, Weichselbaum RR. Oligometastases. J Clin Oncol. 1995;13(1):8–10.
65. Palma DA, Olson R, Harrow S, Gaede S, Louie AV, Haasbeek C, et al. Stereotactic ablative radiotherapy versus standard of care palliative treatment in patients with oligometastatic cancers (SABR-COMET): a randomised, phase 2, open-label trial. Lancet. 2019;393(10185):2051–8.
66. Gomez DR, Blumenschein GR, Lee JJ, Hernandez M, Ye R, Camidge DR, et al. Local consolidative therapy versus maintenance therapy or observation for patients with oligometastatic non-small-cell lung cancer without progression after first-line systemic therapy: a multicentre, randomised, controlled, phase 2 study. Lancet Oncol. 2016;17(12):1672–82.
67. Iyengar P, Wardak Z, Gerber DE, Tumati V, Ahn C, Hughes RS, et al. Consolidative radiotherapy for limited metastatic non-small-cell lung cancer: a phase 2 randomized clinical trial. JAMA Oncol. 2018;4(1):e173501.
68. Ost P, Reynders D, Decaestecker K, Fonteyne V, Lumen N, De Bruycker A, et al. Surveillance or metastasis-directed therapy for oligometastatic prostate cancer recurrence: a prospective, randomized, multicenter phase II trial. J Clin Oncol. 2018;36(5):446–53.
69. Phillips R, Shi WY, Deek M, Radwan N, Lim SJ, Antonarakis ES, et al. Outcomes of observation vs stereotactic ablative radiation for oligometastatic prostate cancer: the ORIOLE phase 2 randomized clinical trial. JAMA Oncol. 2020;6(5):650–9.
70. Pitroda SP, Chmura SJ, Weichselbaum RR. Integration of radiotherapy and immunotherapy for treatment of oligometastases. Lancet Oncol. 2019;20(8):e434–42.
71. Guckenberger M, Lievens Y, Bouma AB, Collette L, Dekker A, deSouza NM, et al. Characterisation and classification of oligometastatic disease: a European Society for Radiotherapy and Oncology and European Organisation for Research and Treatment of Cancer consensus recommendation. Lancet Oncol. 2020;21(1):e18–28.

72. Pike LRG, Bang A, Ott P, Balboni T, Taylor A, Catalano P, et al. Radiation and PD-1 inhibition: favorable outcomes after brain-directed radiation. Radiother Oncol. 2017;124(1):98–103.
73. Schoenfeld JD. We are all connected: modeling the tumor-immune ecosystem. Trends Cancer. 2018;4(10):655–7.
74. Bang A, Schoenfeld JD. Immunotherapy and radiotherapy for metastatic cancers. Ann Palliat Med. 2019;8(3):312–25.
75. Klemen ND, Wang M, Feingold PL, Cooper K, Pavri SN, Han D, et al. Patterns of failure after immunotherapy with checkpoint inhibitors predict durable progression-free survival after local therapy for metastatic melanoma. J Immunother Cancer. 2019;7(1):196.
76. Basler L, Kroeze SGC, Guckenberger M. SBRT for oligoprogressive oncogene addicted NSCLC. Lung Cancer. 2017;106:50–7.
77. Antonia SJ, Villegas A, Daniel D, Vicente D, Murakami S, Hui R, et al. Overall survival with durvalumab after chemoradiotherapy in stage III NSCLC. N Engl J Med. 2018;379(24):2342–50.
78. Kelly RJ, Ajani JA, Kuzdzal J, Zander T, Van Cutsem E, Piessen G, et al. LBA9_PR Adjuvant nivolumab in resected esophageal or gastroesophageal junction cancer (EC/GEJC) following neoadjuvant chemoradiation therapy (CRT): first results of the CheckMate 577 study. Ann Oncol. 2020;31:S1193–4.
79. Sacco AG, Sharabi A, Jing Z, Pittman E, Gold KA, Sumner W, et al. Radiotherapy with concurrent and adjuvant pembrolizumab in patients with P16-positive locoregionally advanced head and neck cancer: KEYCHAIN trial lead-in results. Int J Radiat Oncol. 2019;105(1, Suppl):E363–4.
80. Weiss J, Bauman JR, Deal AM, Sheth S, Chera BS, Shen C, et al. Preliminary toxicity data from the combination of pembrolizumab and definitive-dose radiotherapy for locally advanced head and neck cancer with contraindication to cisplatin therapy. J Clin Oncol. 2018;36(15_suppl):6069.
81. Sun XS, Sire C, Tao Y, Martin L, Alfonsi M, Prevost JB, et al. A phase II randomized trial of pembrolizumab versus cetuximab, concomitant with radiotherapy (RT) in locally advanced (LA) squamous cell carcinoma of the head and neck (SCCHN): first results of the GORTEC 2015-01 "PembroRad" trial. J Clin Oncol. 2018;36(15_suppl):6018.
82. Elbers JBW, Al-Mamgani A, Tesseslaar MET, van den Brekel MWM, Lange CAH, van der Wal JE, et al. Immuno-radiotherapy with cetuximab and avelumab for advanced stage head and neck squamous cell carcinoma: results from a phase-I trial. Radiother Oncol. 2020;142:79–84.
83. Tao Y, Auperin A, Sun XS, Sire C, Martin L, Bera G, et al. Avelumab-cetuximab-radiotherapy (RT) versus standards of care (SoC) in locally advanced squamous cell carcinoma of the head and neck (SCCHN): safety phase of the randomized trial GORTEC 2017-01 (REACH). J Clin Oncol. 2018;36(15_suppl):6076.
84. Powell SF, Gold KA, Gitau MM, Sumey CJ, Lohr MM, McGraw SC, et al. Safety and efficacy of pembrolizumab with chemoradiotherapy in locally advanced head and neck squamous cell carcinoma: a phase IB study. J Clin Oncol. 2020;38(21):2427–37.
85. Ferris RL, Gillison ML, Harris J, Colevas AD, Mell LK, Kong C, et al. Safety evaluation of nivolumab (Nivo) concomitant with cetuximab-radiotherapy for intermediate (IR) and high-risk (HR) local-regionally advanced head and neck squamous cell carcinoma (HNSCC): RTOG 3504. J Clin Oncol. 2018;36(15_suppl):6010.
86. Weiss J, Vincent B, Deal A, Grilley-Olson J, Patel S, Hackman T, et al. Progression-free survival, overall survival and immunophenotyping outcomes for patients with stage III-IV head and neck cancer and cisplatin contraindication treated with definitive radiotherapy plus pembrolizumab. Int J Radiat Oncol Biol Phys. 2020;106(5):1221.
87. Bourhis J, Sire C, Tao Y, Martin L, Alfonsi M, Prevost JB, et al. LBA38 Pembrolizumab versus cetuximab, concomitant with radiotherapy (RT) in locally advanced head and neck squamous cell carcinoma (LA-HNSCC): results of the GORTEC 2015-01 "PembroRad" randomized trial. Ann Oncol. 2020;31:S1168.
88. Lee NY, Ferris RL, Psyrri A, Haddad RI, Tahara M, Bourhis J, et al. Avelumab plus standard-of-care chemoradiotherapy versus chemoradiotherapy alone in patients with locally advanced squamous cell carcinoma of the head and neck: a randomised, double-blind, placebo-controlled, multicentre, phase 3 trial. Lancet Oncol. 2021;22(4):450–62.
89. Schoenfeld JD, Hanna GJ, Jo VY, Rawal B, Chen Y-H, Catalano PS, et al. Neoadjuvant nivolumab or nivolumab plus ipilimumab in untreated oral cavity squamous cell carcinoma: a phase 2 open-label randomized clinical trial. JAMA Oncol. 2020;6(10):1563–70.
90. Buchwald ZS, Wynne J, Nasti TH, Zhu S, Mourad WF, Yan W, et al. Radiation, immune checkpoint blockade and the abscopal effect: a critical review on timing, dose and fractionation. Front Oncol. 2018;8:612.
91. Marin-Acevedo JA, Dholaria B, Soyano AE, Knutson KL, Chumsri S, Lou Y. Next generation of immune checkpoint therapy in cancer: new developments and challenges. J Hematol Oncol. 2018;11(1):39.

8 Epidemiology and Genomics of Head and Neck Squamous Cell Carcinoma

Katherine Wai and Hyunseok Kang

Epidemiology of Head and Neck Cancer

Overview

Cancers of the head and neck are a group of heterogeneous diseases that can involve multiple sites, including the nasal cavity, nasopharynx, paranasal sinuses, oral cavity, pharynx, larynx, and salivary gland. In 2020, there were over 900,000 new diagnoses of head and neck cancer worldwide, with greater than 450,000 deaths [1]. While there are several different histologic subtypes, more than 90% of head and neck cancers are squamous cell carcinomas (SCC). Among oral cavity/oropharynx cancers, the overall incidence has increased approximately 1% per year between 2008 and 2017, and an estimated 54,010 new cases will be diagnosed in 2021 in the United States alone [2].

K. Wai
Department of Otolaryngology and Head and Neck Surgery, University of California, San Francisco, CA, USA

H. Kang (✉)
Department of Medicine, University of California, San Francisco, CA, USA
e-mail: Hyunseok.kang@ucsf.edu

Risk Factors

Tobacco smoking and alcohol use are well-known risk factors for the development of HNSCC. There is a correlation between length and amount of tobacco smoking and risk of cancer, which can largely be attributed to common genetic alterations [3]. Although alcohol consumption has been shown to be an independent risk factor for HNSCC, it is most commonly thought to have a synergistic effect in combination with tobacco use, especially in oral and pharyngeal SCC [4]. Betel nut chewing is also an independent risk factor for HNSCC, and its effects appear to be synergistic with tobacco and alcohol [5].

Briefly, although some observational studies have been designed to study the potential role of marijuana as a carcinogen for HNSCC, to date, none have shown or suggested that marijuana use confers a greater risk of these cancers. There is some data to suggest that marijuana use may be associated with different tumor subsite, with a greater proportion having oropharyngeal cancer compared to nonmarijuana users [6, 7]. However, this effect may be confounded by HPV exposure, rather than a causal association.

There is a growing body of literature to study the relationship between the oral microbiome and the risk of HNSCC. Chronic periodontitis, an inflammatory infection associated with Gram-

R. A. Chandra, R. J. Li (eds.), *Multidisciplinary Management of Head and Neck Cancer*,
https://doi.org/10.1007/978-3-031-05973-5_8

negative anaerobic bacteria in the dental biofilm, has been shown to be an independent risk factor for incident HNSCC cases [8]. Furthermore, there are differences in the underlying bacterial diversity between tumor and nontumor oral mucosal tissue. However, it remains unclear whether these changes reflect the fact that certain bacteria simply develop and grow better within the tumor microenvironment, or if these bacteria are actively involved in a mechanism driving tumor growth [9–11]. Additional studies will be required to better characterize these observed associations.

Human Papillomavirus

Over a decade ago, there was an apparent rise in the incidence of oropharyngeal SCC, particularly among men younger than 60 years of age [12]. Although this trend was observed worldwide, it was particularly notable within North America and Europe [12–15]. The overall prevalence of human papillomavirus (HPV)-positive oropharyngeal cancer was 40.5% prior to 2000 compared to 72.2% after 2000, with the majority of reported data weighted toward Western European and North American cases [16]. High-risk subtypes of HPV are known to be tumorigenic in cervical cancer and are accepted to be largely similar in oropharyngeal cancer [17]. HPV16, and to a lesser extent HPV18, are the predominant subtypes implicated in tumors of the head and neck.

Risk factors for HPV-positive oropharyngeal (OP) SCC are different than non-HPV-positive cases and may explain these changing epidemiologic trends. Oral HPV infection is a main risk factor for HPV-positive OPSCC and oral HPV infection is predominantly sexually transmitted [18, 19]. Both the number of lifetime sexual partners and the practice of oral sex have been showed to be associated with an increased risk for oral HPV infection, including both low-risk and high-risk HPV subtypes [19, 20].

Evolving Changes in Management of HNSCC

The treatment of HNSCC is varied given the heterogeneity of disease and often involves upfront surgery, followed by adjuvant radiation and/or chemotherapy. Despite aggressive multimodal treatment, the morbidity and mortality remain high [21].

The introduction of immunotherapy has begun to change the landscape of treatment for HNSCC and is currently approved for recurrent, unresectable, or metastatic disease [22]. Tumor cells are known to both exploit the immune system to encourage progression by promoting growth and survival and simultaneously evade it using several mechanisms, including secretion of tumor-derived immunosuppressive factors [23, 24]. Cancer immunotherapies aim to treat cancer via modulation of the immune system. In recurrent/metastatic (R/M) HNSCC, several key clinical trials studying the effects of immune checkpoint inhibitors (including pembrolizumab and nivolumab, both anti-PD-1 checkpoint blockade therapies) induced responses in only 13–18% of patients and prolonged overall survival [25–27]. Beyond R/M HNSCC, several clinical trials are currently underway to evaluate the role of immunotherapy in the neoadjuvant and adjuvant setting, though to date, there is no consensus on which patients would benefit most from immunotherapy and when those therapies should be employed [28].

Another evolving change in the management of HNSCC is among patients with HPV-positive OPSCC. Unlike the HPV-negative OPSCC counterparts, those with HPV-positive cancers have a much better prognosis and overall survival, even with similar disease burden [29]. The difference in outcome between these two subgroups was so significant that the eighth edition tumor-node-metastasis (TNM) staging system was revised to account for this. As a result, there is great interest in de-intensification treatment strategies for this population in order to minimize the toxicities associated with standard, traditional approaches to OPSCC without compromising overall survival [30]. A wide variety of strategies have been, or are actively, undergoing investigation, including the use of cetuximab or immunotherapy instead of platinum-based chemotherapy, reduced dose of adjuvant radiation, and induction chemotherapy with adjuvant radiation and/or chemotherapy. Early results from phase II clinical trials for de-escalated radiation dosing have shown comparable outcomes in select patients [31]. However, additional phase III clinical trial studies are needed before a new standard of care protocol for HPV-positive OPSCC can be implemented.

Genomic Landscape of Head and Neck Cancer

Genomic Characteristics: Overview

Recent development in genomics research of HNSCC opened up a possibility of improving patient survival and defining personalized treatment approaches. A number of efforts to elucidate genomics of head and neck cancer have sought to characterize a comprehensive profile of somatic genomic alterations. The Cancer Genome Atlas (TCGA) was the large effort and included 279 samples from patients with HNSCC [32]. Most HNSCC tumors showed copy number alterations, including losses and gains of chromosomal regions similar to SCC of the lung. Through these endeavors, new insights into the genomic characteristics of HNSCC and differences between HPV-positive and HPV-negative HNSCC were delineated (Fig. 8.1).

Genomic Landscape of HPV-Negative HNSCC

Multiple studies have described the genetic makeup of HPV-negative tumors. Although this is not a comprehensive list, common notable genes seen in several large cohorts of HPV-negative HNSCC include *TP53*, *CDKN2A*, and *NOTCH* genetic alterations [32–34]. Among some of the findings specific to HPV-negative HNSCC, the TCGA study reported novel co-amplifications of chromosome 11q13 (*CCND1*), and 11q22 (*BIRC2* and *YAP1*), novel focal deletions in tumor suppressor genes (*FAT1*, *NOTCH1*, CDKN2A), and further identified a subset of HPV-negative tumors with a three gene constellation of wild-type *TP53* with concurrent mutations of *CASP8* with *HRAS* [32]. Some of these genes will be discussed in detail next.

TP53 is the most commonly mutated gene in HNSCC, upwards of 85% of the time among HPV-negative tumors [32]. In contrast, mutations in *TP53* are rarely seen in HPV-positive tumors [33, 35]. The primary function of p53 is to initiate cell cycle arrest, thereby allowing DNA repair enzymes to repair genomic insults. If cellular repair cannot occur, p53 initiates cellular apoptosis. Mutations in *TP53* render p53 nonfunctional—without functional p53, accumulation of the genomic insult and cell growth can continue unchecked and lead to cancer. The majority of *TP53* gene alterations are missense mutations, though inactivating mutations including nonsense, insertion or deletions, and splice site mutations have also been observed [35].

In addition to *TP53*, *CDKN2A* is also involved with cell cycle and DNA repair pathways. Similar to *TP53*, almost all HPV-negative HNSCC contain deletions in *CDKN2A* on chromosome 9p, which encodes $p16^{INK4A}$ and p14ARF and leads to inactivation of the protein [32, 33, 35, 36]. The role of $p16^{INK4A}$ is to inhibit cyclin D1-CDK4 and cyclin D1-CDK6 complexes, which mediates cellular senescence. Inhibition of these complexes prevents the transcription factor E2F from initiating transcription of S-phase genes. *CCND1* on chromosome 11q13 encodes for cyclin D1 and has also been shown to have frequent amplification mutations in roughly 30% of HPV-negative tumors [32]. Amplification of *CCND1* or deletion of *CDKN2A* occurs in approximately 60% of all HNSCC [32] and in 95% of oral cavity SCC tumors [37]. Tumors with both the mutations are

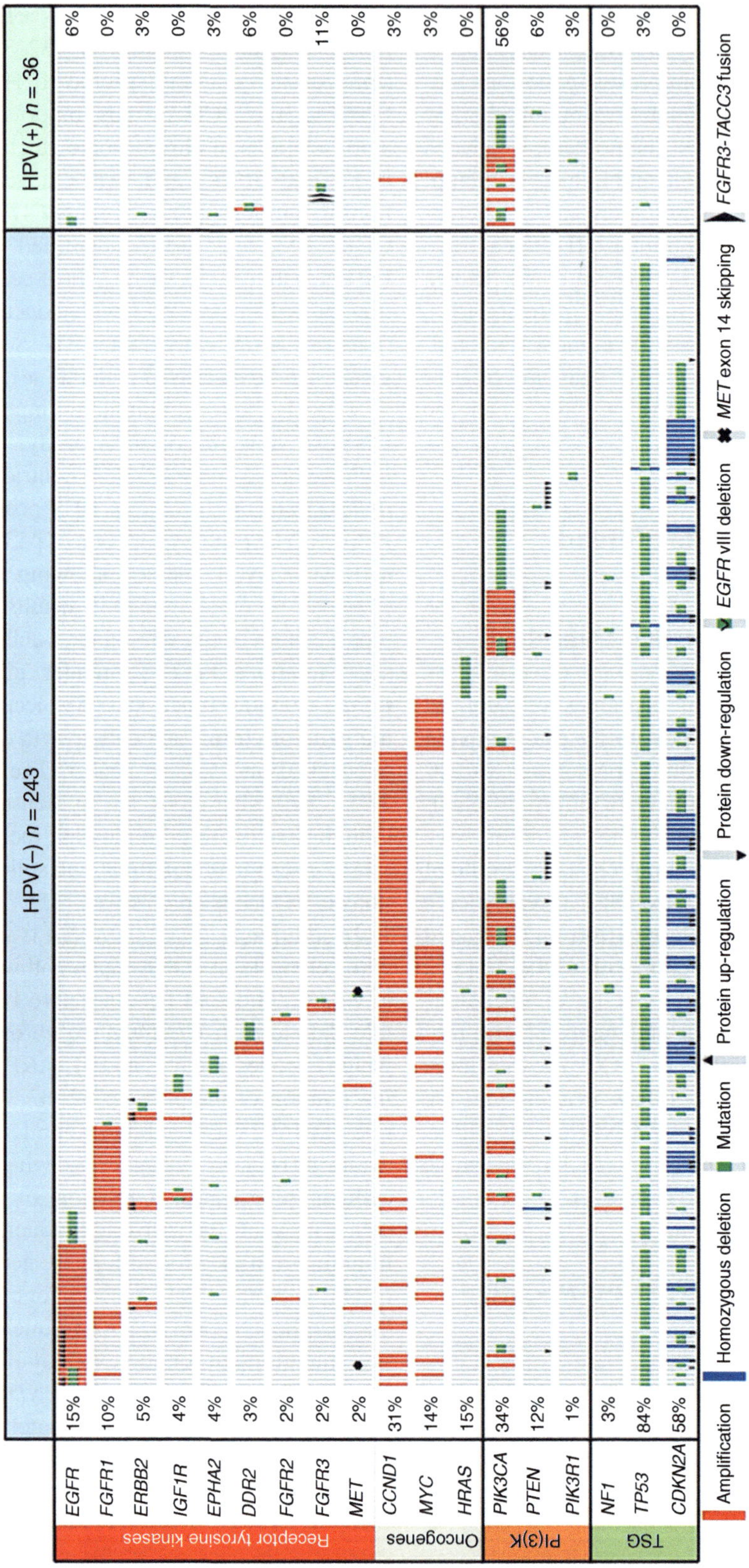

Fig. 8.1 Genomic landscape of head and neck cancer. (MS Lawrence *et al. Nature* **517**, 576–582 (2015) https://doi.org/10.1038/nature14129. This work is licensed under a Creative Commons Attribution-NonCommercial-ShareAlike 3.0 Unported licence)

associated with worse prognosis than their wild-type counterparts [38].

The NOTCH signaling pathway regulates multiple cellular processes, including proliferation, differentiation and apoptosis. The NOTCH proteins are transmembrane receptors that activate via ligand binding, releasing the NOTCH intracellular domain, which translocates to the nucleus and acts as a transcription factor. *NOTCH1* mutations are seen in approximately 10–15% of cases [33, 35]. A variety of mutations are observed, though most commonly, *NOTCH1* is inactivated via a missense mutation or mutations that truncate the protein product [35]. Nonsense mutations lead to truncated proteins that lack a C-terminal ankyrin repeat domain, thereby preventing downstream transactivation of target genes; other mutations occur in highly conserved residues which lead to poor ligand binding [33]. Interestingly, the spectrum of *NOTCH1* mutations in HNSCC are fundamentally different than those seen in hematopoietic tumors and provide evidence that *NOTCH1* has a tumor suppressor role in HNSCC rather than an oncogenic role as seen in some leukemias. For example, loss of heterozygosity was seen in 33% of patients with *NOTCH1* mutations [35]. The role of *NOTCH1* as a biomarker remains preliminary. However, there is preclinical data that showed *NOTCH1* loss-of-function mutations in HNSCC xenograft models reduced overall tumor size when treated with PI3K/mTOR inhibitors [39].

Among HNSCC, a frequent number of inactivating mutations have been observed in the *FAT1* gene, with mutations in 8–12% of cases or deletions in 6% of cases [32, 33, 40]. HPV-negative tumors have a significantly higher prevalence of *FAT1* inactivation compared to HPV-negative tumors [32]. Typically, *FAT1* encodes a cadherin-like protein and functions as a tumor suppressor, binding to β-catenin and blocking nuclear localization. The mechanism by which *FAT1* affects oncogenesis is less understood than other tumor suppressor genes such as *TP53*. In general, it is thought that loss of function of *FAT1*, through mutation, deletion, or inactivating structural rearrangements, allows translocation of β-catenin, and downstream promotes Wnt signaling and transcription of oncogenes to promote tumorigenesis [40]. *FAT1* mutations were correlated with increased expression of several cell surface receptors including VEGFR2, HER3, and PD-L1, though there is no causal mechanism elucidated at this time [41].

Another mechanism by with tumorigenesis occurs is through evasion of apoptosis. Caspase-8 plays a pivotal role in the initiation and propagation of the apoptotic signal. Inactivating mutations of *CASP8* have been reported, varying from 8% of all tumors by TCGA, to 10% in oral cavity SCC [32, 37]. Mutations displayed clustered missense and other inactivation mutations in the first death effector intron and caspase peptidase domains [32]. Among tumors analyzed in the TCGA study, most *CASP8* mutations were in oral cavity cancer. This is consistent with prior studies which have defined a novel molecular subtype of oral cavity SCC that is characterized by frequent *CASP8* mutations [37, 42]. *CASP8* and *HRAS* co-mutations have been observed and seem to be exclusive to HPV-negative HNSCC [32, 37]. Further, a three gene profile of wild-type *TP53* with mutant *HRAS* and *CASP8* has been associated with favorable clinical outcomes, which may imply an alternative pathogenesis for tumor development that is independent of *TP53* [32].

Genomic Landscape of HPV-Positive HNSCC

High-risk HPV subtypes are known to undergo genetic changes that are tumorigenic in cervical cancer, and although the exact mechanism in HNSCC remains an area of active research, it is accepted to be largely similar [17]. HPV16, and to a lesser extent HPV18, can integrate into the human genome. In doing so, viral oncoproteins E6 and E7 are overexpressed and promote tumor growth by inactivating tumor suppressor proteins [43, 44]. E6 binds to p53, which ultimately leads to degradation of p53, which causes aberrant propagation of the cell cycle and prevents apoptosis. In addition, overexpression of the oncoprotein E7 will bind to the retinoblastoma protein (pRb), inactivating pRb, leading to downstream expression of gene products that allow a cell to transition from the G1 to S phase, promoting proliferation [45].

Recent data have shown that the oncogene *PIK3CA* is the most commonly mutated gene in HPV-positive HNSCC [32, 34, 46]. In large dataset studies, *PIK3CA* is altered in approximately 30–50% of cases analyzed. Briefly, *PIK3CA* encodes p110α, a catalytic subunit of phosphoinositol 3-kinase (PI3K). The family of kinases plays several integral roles in cellular signaling, including cell growth, proliferation, and survival. The majority of *PIK3CA* alterations are localized to mutations in the helical/kinase domain in patients with HPV-positive HNSCC, which is a different mutation hotspot than *PIK3CA* alterations seen in HPV-negative cases [32, 47]. When altered, the most frequently affected pathway is the PI3K-PTEN-AKT signaling pathway (Fig. 8.2). Activation of the PI3K pathway triggers downstream effects on transcription, protein synthesis, and proliferation, overall contributing to tumor progression and survival. Interestingly, there is data supporting the fact that effects on downstream signaling are different [32] between HPV-positive versus HPV-negative tumors. In HPV-positive tumors, mutant *PIK3CA* was correlated with the activation of downstream mTOR but not AKT, which leads to constitutive activation of this signaling pathway [48].

A subset of HPV-positive tumors also show a loss of *tumor necrosis factor (TNF) receptor-associated factor 3 (TRAF3)* [32]. *TRAF3* copy loss, mutation, and/or decreased expression have all been observed in HPV-positive tumors. The receptors that signal through TRAF proteins, including TNF, toll-like receptors, RIG-1-like receptors, and interleukin-1 receptors, ultimately lead to activation of innate immune responses and nuclear factor-κB (NF-κB), which regulates apoptosis and inflammation. *TRAF3* negatively regulates NF-κB pathways, and thus mutations in *TRAF3* lead to unopposed activation of NF-κB, a finding also induced by carcinogens and oncogenic viruses [49]. Interestingly, an in-depth analysis of the TCGA dataset found that mutations in *TRAF3* and *cylindromatosis lysine 63 deubiquitinase (CYLD)*, another gene involved in NF-κB inhibition, were associated with the absence of HPV genome integration and improved overall survival [50]. These data suggest that episomal HPV that lacks integration into the genome has different gene expression profiles and that carcinogenesis in these cases may partially depend on gene defects in *TRAF3* and/or *CYLD* [50]. *CYLD* mutation is reported to be enriched in

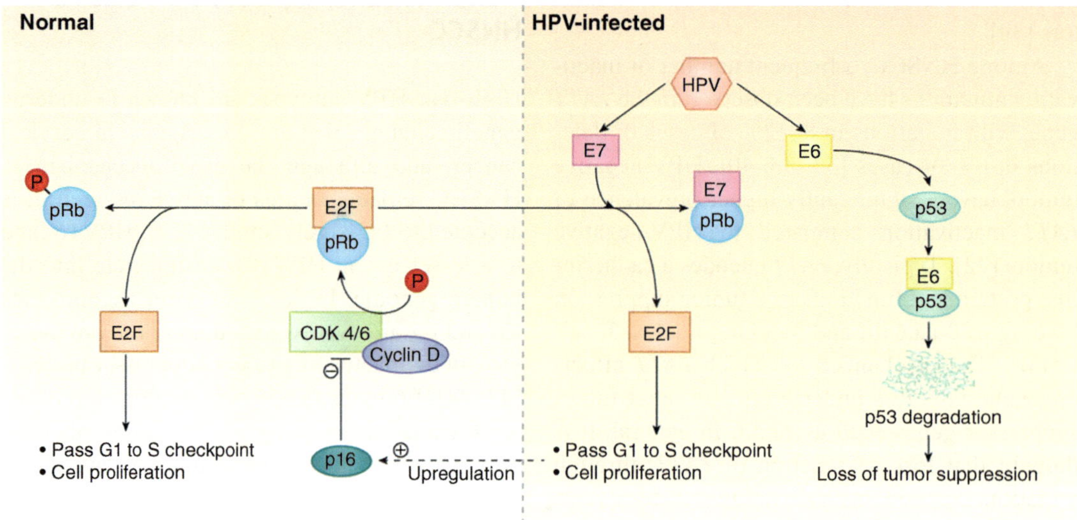

Fig. 8.2 Biology of HPV-negative and -positive HNSCC

HPV-positive HNSCC and has been associated with distinct phenotypic features with lower tumor mutational burden and less frequent alterations in *PIK3CA* mutations [51].

Molecular Profiling and Therapeutic Implications

Receptor Tyrosine Kinases

The family of receptor tyrosine kinases is diverse and includes a broad range of receptors required for cell regulation and survival. They are activated via dimerization of transmembrane receptors in the presence of different ligands. This interaction results in kinase activation and autophosphorylation, which leads to downstream activation of signal transduction pathways, resulting in transcription of genes required for cell growth and survival (Fig. 8.3).

The epidermal growth factor receptor (EGFR) has been a therapeutic target of interest, as the vast majority (>90%) of HNSCC overexpress EGFR [52]. However, EGFR expression, as quantified by immunohistochemistry techniques, has not been shown to be predictive of response to EGFR-targeted therapy, such as cetuximab, a monoclonal IgG antibody-directed against EGFR [53]. Additionally, clinical trials with erlotinib, a small molecule inhibitor of EGFR, have also shown modest response rates at best [54, 55]. Genomic sequencing has begun to restructure the role of EFGR. The TCGA group reported that only 15% of HPV-negative HNSCC and 6% of HPV-positive HNSCC have mutations or amplifications of *EGFR* [32]. In contrast to other cancer subtypes, *EGFR* mutations in HNSCC are more

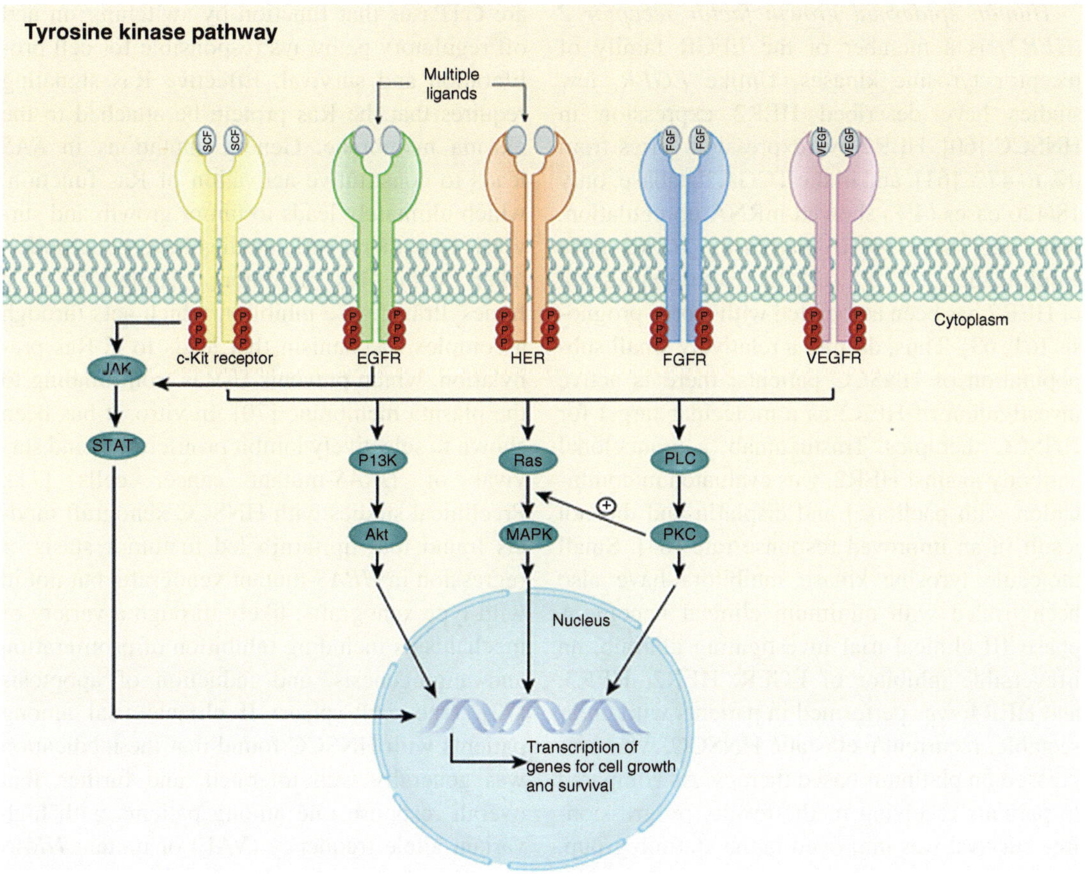

Fig. 8.3 Signaling pathways in HNSCC

dispersed across the gene, rather than clustered within specific exons [56]. Furthermore, while *HRAS* and *PIK3CA* have mutation hotspots in HNSCC tumors, *EGFR* does not show frequently mutated residues [57]. Given the lack of driver mutations but overall overexpression of EGFR, it is likely that co-activation of other pathways could explain the lack of predictable response to EGFR targeted therapy. For example, in patients with HNSCC with a *KRAS*-variant, treatment with cetuximab and definitive chemoradiation showed improved overall survival, possibly because cetuximab bolstered antitumor immunity that had otherwise been inhibited in *KRAS*-variant patients [58]. Given that the *KRAS*-variant has been shown to be associated with worse prognosis with cisplatin and definitive radiation [59], improved response in this subpopulation might suggest that it may serve as a predictive biomarker for response to EGFR inhibition.

Human epidermal growth factor receptor 2 (HER2) is a member of the EFGR family of receptor tyrosine kinases. Unlike *EGFR*, few studies have described HER2 expression in HNSCC [60]. HER2 overexpression varies from 0% to 47% [61], and in the TCGA database, only 18/426 cases (4%) showed mRNA upregulation, with only 8/18 (44%) harboring *HER2* gene amplification [32, 60, 62]. Increased expression of HER2 has been associated with worse prognosis [61, 63]. Thus, despite a relatively small subpopulation of HNSCC patients, there is active investigation of HER2 as a molecular target for HNSCC therapies. Trastuzumab, a monoclonal antibody against HER2, was evaluated in combination with paclitaxel and cisplatin and did not result in an improved response rate [64]. Small molecule tyrosine kinase inhibitors have also been trialed with minimum clinical benefit. A phase III clinical trial investigating afatinib, an irreversible inhibitor of EGFR, HER2, HER3, and HER4, was performed in patients with unresectable, recurrent/metastatic HNSCC, who progressed on platinum-based therapy. As compared to patients receiving methotrexate, progression-free survival was improved in the afatinib group, though there was no difference in overall survival [65]. In another clinical trial studying the use of afatinib as an adjunct after definitive treatment, addition of afatinib did not confer improved disease-free survival [66]. Similarly disappointing results have been seen with other HER family tyrosine kinase inhibitors, including lapatinib, gefitnib, and dacomitinib [67–69]. These results should be interpreted with caution, as the majority were investigated in recurrent/metastatic disease. However, future identification of biomarkers may predict clinical responses to HER family blockade.

HRAS

The TCGA study reported that approximately 6% of HNSCC tumors contain a *HRAS* mutation, and this prevalence may even be more frequent in oral cavity cancers [32, 42]. Ras proteins are oncogenes, encoded by three ubiquitous genes *HRAS*, *KRAS*, and *NRAS*. In short, Ras proteins are GTPases that function by switching on and off regulatory pathways responsible for cell proliferation and survival. Effective Ras signaling requires that the Ras protein be attached to the plasma membrane. Genetic mutations in *RAS* leads to constitutive activation of Ras function, which ultimately leads to tumor growth and survival. Few therapeutic studies have been undertaken with Ras inhibitors. Tipifarnib is a farnesyltransferase inhibitor, which acts through a complex mechanism that leads to H-Ras prenylation, which prevents H-Ras from binding to the plasma membrane [70]. In vitro, it has been shown to selectively inhibit proliferation and survival of *HRAS*-mutant cancer cells [71]. Preclinical studies with HNSCC xenograft models found that tipifarnib led to tumor stasis or regression in *HRAS*-mutant xenografts but not in wild-type xenografts, likely through a variety of mechanisms including inhibition of proliferation and angiogenesis, and induction of apoptosis [72]. One early phase II clinical trial among patients with HNSCC found that the medication was generally well tolerated, and further, that overall response rate among patients with high variant allele frequency (VAF) of mutant *HRAS* was 55% [73]. A pivotal study is currently ongoing to further explore this approach as a viable therapeutic option (NCT03719690).

TP53

TP53 is the most commonly mutated gene in all cancers, often seen in HPV-negative HNSCC tumors, and thus is a compelling therapeutic target. Traditionally, however, tumor suppressor genes are regarded as difficult targets. The concept of synthetic lethality, in which a combination of mutations in two or more separate genes leads to cell death, has started to gain attention. Therapeutic approaches can therefore exploit maladaptive genetic changes in cancer cells that are not present in normal cells, unlike traditional chemotherapeutic strategies. High-risk *TP53* mutations can cause drug resistance to cytotoxic therapy [74]. Tumors with loss of p53 function require activation of S and G_2 phase checkpoints to mediate growth arrest. Using a high-throughput RNA interference kinome screen, WEE1, a G_2-M cell-cycle regulatory kinase was identified as a promising therapeutic target [75]. Inhibition of WEE1 impaired growth of p53-mutant HNSCC through synthetic lethal interactions. When used in vitro and in vivo, a WEE1 inhibitor, MK-1775, has been shown to sensitize platinum-resistant p53-mutant HNSCC cells to cisplatin [76]. A different WEE1 inhibitor, AZD1775, showed similar results in HPV-positive HNSCC both in vitro and in vivo [77]. In a phase I clinical trial, patients treated with the triple combination of AZD1775, cisplatin, and docetaxel, partial response was seen in 5/10 patients, and all patients carried either *TP53* mutations or a deficient p53 pathway [78]. A similar approach can be applied to other cell-cycle regulatory proteins, employing the idea of synthetic lethality. For example, CHK1 is another G_2-M cell-cycle regulating protein. Using AZD7762, a Chk inhibitor has also been shown to sensitize p53-mutant HNSCC cells to cisplatin [79].

PIK3CA

As discussed previously in this chapter, although *PIK3CA* alterations are commonly seen in HNSCC, it is more frequently observed in HPV-positive tumors versus HPV-negative tumors. Given that HPV-positive tumors tend to activate the PI3K-mTOR signaling pathway, there is a mechanistic basis that inhibition of either PI3K or mTOR could have therapeutic benefit. Preclinical studies have shown that HNSCC tumors with activating *PIK3CA* mutations may be more sensitive to dual PI3K/mTOR inhibition [47, 48]. Several phase I and phase II clinical trials have been performed with small molecule inhibitors of PI3K and mTOR either as monotherapy or in combination with cetuximab or cytotoxic therapies, all with minimal clinical benefit at best [80].

STAT

The signal transducer and activator of transcription (STAT) family of proteins is another protein that is activated downstream of receptor tyrosine kinases. Signaling through receptor tyrosine kinases, STAT proteins can be activated, translocated to the nucleus, and bind gene-regulatory elements that promote transcription of both oncogenes and tumor suppressor genes, thereby playing a diverse role in tumorigenesis [81]. In HNSCC, there are higher levels of STAT3 expression and activation compared to noncancer patients and has become a possible therapeutic target [82, 83]. A phase 0 clinical trial showed that injection of a STAT3 decoy could downregulate STAT3 target genes in tumors, suggesting this strategy may derive clinical benefit [84]. Early clinical trials are underway (NCT03153982, NCT03195699), and at this time, it remains unknown whether STAT inhibition will have a clinical role in the treatment of HNSCC.

Immunologic Characteristics of HNSCC

Cancer immunotherapies aim to initiate and potentiate immune responses. Thus, the underlying immunologic profile of a patient with HNSCC will likely dictate the type and magnitude of the immune response that transpires, ultimately resulting in differential efficacy and benefit. New research has started to focus on improving our understanding and defining novel immunotherapeutic approaches and identifying corresponding biomarkers, especially since only a small proportion of HNSCC patients respond to immune checkpoint inhibitors (ICI) [85].

Tumor expression of the programmed death-ligand 1 (PD-L1) has been extensively studied as a biomarker in ICI therapy. Results have been variable, though the consensus from the Society for Immunotherapy of Cancer concluded that tumor PD-L1 expression generally correlated with improved efficacy of anti PD-1/PD-L1 ICI, with increased predictive value when including PD-L1 expression on tumor infiltrating immune cells [22]. However, the definition of what it means to have PD-L1 positive expression has not yet been standardized across tumor types. For practical purposes, combined positive score (CPS) which is defined by the number of PD-L1 staining cells (tumor cells, lymphocytes, macrophages) divided by the total number of viable tumor cells, multiplied by 100, is used to guide treatment of first-line recurrent/metastatic HNSCC patients. In a large phase 3 study, single-agent pembrolizumab has demonstrated overall survival benefit in patients with CPS $\geq$ 1 [86].

Several studies have sought to characterize the immune landscape within the tumor microenvironment (TME) as a strategy to predict clinical response. These investigations are based on the understanding that different immune cell subsets may dominate in different cancer types, or even the same cancer type in different patients, which would likely affect ICI efficacy [87]. Chemokine gene signatures representing high versus low CD8+ T-cell infiltration into the TME was evaluated in a cohort of 134 HNSCC from the University of Chicago, and a cohort of 424 HNSCC from TCGA database. The presence of a high T-cell inflamed phenotype was associated with increased levels of multiple immune checkpoints including PD-L1 expression, supporting the hypothesis that this chemokine signature could represent sensitivity to anti-PD-1/PD-L1 therapies [88]. Following the same logic regarding the importance of the TME, immune gene signatures related to IFN-γ signaling and activated T-cell biology have been delineated and modeled. Immune gene expression profiles, comprised of 18 genes indicative of a T-cell inflamed TME phenotype, have been shown to predict response to pembrolizumab in several different cancer types, including HNSCC [89, 90]. A similar finding of a six IFN-γ gene panel was also associated with improved overall response and progression-free survival in R/M HNSCC [27].

Conclusion

The overall epidemiology of HNSCC has transformed in the last few decades, with an increasing number of HPV-positive tumors and decreasing number of HPV-negative cases, classically associated with tobacco smoking and alcohol. Improving genomic technologies have allowed for comprehensive characterization of hundreds of HNSCC and thus have contributed to a greater understanding of the landscape of somatic genomic alterations, and the genetic changes that distinguish HPV-negative tumors (*TP53*, *CDKN2A*, *NOTCH1*, *CASP8*, *FAT1*) as compared to HPV-positive tumors (*PIK3CA*, *TRAF3*). Although many of these genetic alterations have yet to generate new therapies, several targeted clinical trials that take advantage of our understanding of the molecular profile of HNSCC have been performed or are currently underway, in hopes of delivering personalized, more efficacious therapies. Beyond genomics, novel immunotherapeutic strategies have created an urgent need for the identification of biomarkers to predict clinical benefit. Future studies combining our understanding of the molecular and immunologic landscape of HNSCC will be pivotal to optimizing and delivering personalized treatment.

References

1. Sung H, Ferlay J, Siegel RL, et al. Global Cancer Statistics 2020: GLOBOCAN estimates of incidence and mortality worldwide for 36 cancers in 185 countries. CA Cancer J Clin. 2021;71(3):209–49. https://doi.org/10.3322/caac.21660.
2. American Cancer Society. Cancer facts and figures. Washington, DC: American Cancer Society; 2021.
3. Brennan JA, Boyle JO, Koch WM, et al. Association between cigarette smoking and mutation of the p53 gene in squamous-cell carcinoma of the head and

neck. N Engl J Med. 1995;332(11):712–7. https://doi.org/10.1056/NEJM199503163321104.
4. Hashibe M, Brennan P, Chuang SC, et al. Interaction between tobacco and alcohol use and the risk of head and neck cancer: pooled analysis in the International Head and Neck Cancer Epidemiology Consortium. Cancer Epidemiol Biomark Prev. 2009;18(2):541–50. https://doi.org/10.1158/1055-9965.EPI-08-0347.
5. Guha N, Warnakulasuriya S, Vlaanderen J, Straif K. Betel quid chewing and the risk of oral and oropharyngeal cancers: a meta-analysis with implications for cancer control. Int J Cancer. 2014;135(6):1433–43. https://doi.org/10.1002/ijc.28643.
6. Marks MA, Chaturvedi AK, Kelsey K, et al. Association of marijuana smoking with oropharyngeal and oral tongue cancers: pooled analysis from the INHANCE consortium. Cancer Epidemiol Biomark Prev. 2014;23(1):160–71. https://doi.org/10.1158/1055-9965.EPI-13-0181.
7. Xie M, Gupta MK, Archibald SD, Stanley Jackson B, Young JEM, Zhang H. Marijuana and head and neck cancer: an epidemiological review. J Otolaryngol Head Neck Surg. 2018;47(1):73. https://doi.org/10.1186/s40463-018-0319-2.
8. Tezal M, Sullivan MA, Hyland A, et al. Chronic periodontitis and the incidence of head and neck squamous cell carcinoma. Cancer Epidemiol Biomark Prev. 2009;18(9):2406–12. https://doi.org/10.1158/1055-9965.EPI-09-0334.
9. Pushalkar S, Ji X, Li Y, et al. Comparison of oral microbiota in tumor and non-tumor tissues of patients with oral squamous cell carcinoma. BMC Microbiol. 2012;12:144. https://doi.org/10.1186/1471-2180-12-144.
10. Schmidt BL, Kuczynski J, Bhattacharya A, et al. Changes in abundance of oral microbiota associated with oral cancer. PLoS One. 2014;9(6):e98741. https://doi.org/10.1371/journal.pone.0098741.
11. Hayes RB, Ahn J, Fan X, et al. Association of oral microbiome with risk for incident head and neck squamous cell cancer. JAMA Oncol. 2018;4(3):358–65. https://doi.org/10.1001/jamaoncol.2017.4777.
12. Chaturvedi AK, Engels EA, Anderson WF, Gillison ML. Incidence trends for human papillomavirus-related and -unrelated oral squamous cell carcinomas in the United States. J Clin Oncol. 2008;26(4):612–9. https://doi.org/10.1200/JCO.2007.14.1713.
13. Mork J, Moller B, Dahl T, Bray F. Time trends in pharyngeal cancer incidence in Norway 1981-2005: a subsite analysis based on a reabstraction and recoding of registered cases. Cancer Causes Control. 2010;21(9):1397–405. https://doi.org/10.1007/s10552-010-9567-9.
14. Blomberg M, Nielsen A, Munk C, Kjaer SK. Trends in head and neck cancer incidence in Denmark, 1978-2007: focus on human papillomavirus associated sites. Int J Cancer. 2011;129(3):733–41. https://doi.org/10.1002/ijc.25699.
15. Forte T, Niu J, Lockwood GA, Bryant HE. Incidence trends in head and neck cancers and human papillomavirus (HPV)-associated oropharyngeal cancer in Canada, 1992-2009. Cancer Causes Control. 2012;23(8):1343–8. https://doi.org/10.1007/s10552-012-0013-z.
16. Mehanna H, Beech T, Nicholson T, et al. Prevalence of human papillomavirus in oropharyngeal and nonoropharyngeal head and neck cancer--systematic review and meta-analysis of trends by time and region. Head Neck. 2013;35(5):747–55. https://doi.org/10.1002/hed.22015.
17. Nulton TJ, Olex AL, Dozmorov M, Morgan IM, Windle B. Analysis of The Cancer Genome Atlas sequencing data reveals novel properties of the human papillomavirus 16 genome in head and neck squamous cell carcinoma. Oncotarget. 2017;8(11):17684–99. https://doi.org/10.18632/oncotarget.15179.
18. D'Souza G, Kreimer AR, Viscidi R, et al. Case-control study of human papillomavirus and oropharyngeal cancer. N Engl J Med. 2007;356(19):1944–56. https://doi.org/10.1056/NEJMoa065497.
19. Gillison ML, Broutian T, Pickard RK, et al. Prevalence of oral HPV infection in the United States, 2009-2010. JAMA. 2012;307(7):693–703. https://doi.org/10.1001/jama.2012.101.
20. Smith EM, Ritchie JM, Summersgill KF, et al. Age, sexual behavior and human papillomavirus infection in oral cavity and oropharyngeal cancers. Int J Cancer. 2004;108(5):766–72. https://doi.org/10.1002/ijc.11633.
21. Cheraghlou S, Schettino A, Zogg CK, Judson BL. Changing prognosis of oral cancer: an analysis of survival and treatment between 1973 and 2014. Laryngoscope. 2018;128(12):2762–9. https://doi.org/10.1002/lary.27315.
22. Cohen EEW, Bell RB, Bifulco CB, et al. The Society for Immunotherapy of Cancer consensus statement on immunotherapy for the treatment of squamous cell carcinoma of the head and neck (HNSCC). J Immunother Cancer. 2019;7(1):184. https://doi.org/10.1186/s40425-019-0662-5.
23. Ferris RL. Immunology and immunotherapy of head and neck cancer. J Clin Oncol. 2015;33(29):3293–304. https://doi.org/10.1200/jco.2015.61.1509.
24. Horton JD, Knochelmann HM, Day TA, Paulos CM, Neskey DM. Immune evasion by head and neck cancer: foundations for combination therapy. Trends Cancer. 2019;5(4):208–32. https://doi.org/10.1016/j.trecan.2019.02.007.
25. Le X, Ferrarotto R, Wise-Draper T, Gillison M. Evolving role of immunotherapy in recurrent metastatic head and neck cancer. J Natl Compr Cancer Netw. 2020;18(7):899–906. https://doi.org/10.6004/jnccn.2020.7590.
26. Ferris RL, Blumenschein G Jr, Fayette J, et al. Nivolumab for recurrent squamous-cell carcinoma of the head and neck. N Engl J Med. 2016;375(19):1856–67. https://doi.org/10.1056/NEJMoa1602252.
27. Seiwert TY, Burtness B, Mehra R, et al. Safety and clinical activity of pembrolizumab for treat-

ment of recurrent or metastatic squamous cell carcinoma of the head and neck (KEYNOTE-012): an open-label, multicentre, phase 1b trial. Lancet Oncol. 2016;17(7):956–65. https://doi.org/10.1016/S1470-2045(16)30066-3.
28. Farlow JL, Birkeland AC, Swiecicki PL, Brenner JC, Spector ME. Window of opportunity trials in head and neck cancer. J Cancer Metastasis Treat. 2019;5:18. https://doi.org/10.20517/2394-4722.2018.100.
29. Fakhry C, Westra WH, Li S, et al. Improved survival of patients with human papillomavirus-positive head and neck squamous cell carcinoma in a prospective clinical trial. J Natl Cancer Inst. 2008;100(4):261–9. https://doi.org/10.1093/jnci/djn011.
30. Strohl MP, Wai KC, Ha PK. De-intensification strategies in HPV-related oropharyngeal squamous cell carcinoma-a narrative review. Ann Transl Med. 2020;8(23):1601. https://doi.org/10.21037/atm-20-2984.
31. Bigelow EO, Seiwert TY, Fakhry C. Deintensification of treatment for human papillomavirus-related oropharyngeal cancer: current state and future directions. Oral Oncol. 2020;105:104652. https://doi.org/10.1016/j.oraloncology.2020.104652.
32. Cancer Genome Atlas N. Comprehensive genomic characterization of head and neck squamous cell carcinomas. Nature. 2015;517(7536):576–82. https://doi.org/10.1038/nature14129.
33. Stransky N, Egloff AM, Tward AD, et al. The mutational landscape of head and neck squamous cell carcinoma. Science. 2011;333(6046):1157–60. https://doi.org/10.1126/science.1208130.
34. Chung CH, Guthrie VB, Masica DL, et al. Genomic alterations in head and neck squamous cell carcinoma determined by cancer gene-targeted sequencing. Ann Oncol. 2015;26(6):1216–23. https://doi.org/10.1093/annonc/mdv109.
35. Agrawal N, Frederick MJ, Pickering CR, et al. Exome sequencing of head and neck squamous cell carcinoma reveals inactivating mutations in NOTCH1. Science. 2011;333(6046):1154–7. https://doi.org/10.1126/science.1206923.
36. Lechner M, Frampton GM, Fenton T, et al. Targeted next-generation sequencing of head and neck squamous cell carcinoma identifies novel genetic alterations in HPV+ and HPV- tumors. Genome Med. 2013;5(5):49. https://doi.org/10.1186/gm453.
37. Pickering CR, Zhang J, Yoo SY, et al. Integrative genomic characterization of oral squamous cell carcinoma identifies frequent somatic drivers. Cancer Discov. 2013;3(7):770–81. https://doi.org/10.1158/2159-8290.CD-12-0537.
38. Beck TN, Kaczmar J, Handorf E, et al. Phospho-T356RB1 predicts survival in HPV-negative squamous cell carcinoma of the head and neck. Oncotarget. 2015;6(22):18863–74. https://doi.org/10.18632/oncotarget.4321.
39. Sambandam V, Frederick MJ, Shen L, et al. PDK1 mediates NOTCH1-mutated head and neck squamous carcinoma vulnerability to therapeutic PI3K/mTOR inhibition. Clin Cancer Res. 2019;25(11):3329–40. https://doi.org/10.1158/1078-0432.CCR-18-3276.
40. Morris LG, Kaufman AM, Gong Y, et al. Recurrent somatic mutation of FAT1 in multiple human cancers leads to aberrant Wnt activation. Nat Genet. 2013;45(3):253–61. https://doi.org/10.1038/ng.2538.
41. Chen Z, Zhang C, Chen J, et al. The proteomic landscape of growth factor signaling networks associated with FAT1 mutations in head and neck cancers. Cancer Res. 2021;81:4402. https://doi.org/10.1158/0008-5472.CAN-20-3659.
42. Su SC, Lin CW, Liu YF, et al. Exome sequencing of oral squamous cell carcinoma reveals molecular subgroups and novel therapeutic opportunities. Theranostics. 2017;7(5):1088–99. https://doi.org/10.7150/thno.18551.
43. Werness BA, Levine AJ, Howley PM. Association of human papillomavirus types 16 and 18 E6 proteins with p53. Science. 1990;248(4951):76–9. https://doi.org/10.1126/science.2157286.
44. Psyrri A, DeFilippis RA, Edwards AP, Yates KE, Manuelidis L, DiMaio D. Role of the retinoblastoma pathway in senescence triggered by repression of the human papillomavirus E7 protein in cervical carcinoma cells. Cancer Res. 2004;64(9):3079–86. https://doi.org/10.1158/0008-5472.can-03-3739.
45. Serrano M, Hannon GJ, Beach D. A new regulatory motif in cell-cycle control causing specific inhibition of cyclin D/CDK4. Nature. 1993;366(6456):704–7. https://doi.org/10.1038/366704a0.
46. Seiwert TY, Zuo Z, Keck MK, et al. Integrative and comparative genomic analysis of HPV-positive and HPV-negative head and neck squamous cell carcinomas. Clin Cancer Res. 2015;21(3):632–41. https://doi.org/10.1158/1078-0432.CCR-13-3310.
47. Lui VW, Hedberg ML, Li H, et al. Frequent mutation of the PI3K pathway in head and neck cancer defines predictive biomarkers. Cancer Discov. 2013;3(7):761–9. https://doi.org/10.1158/2159-8290.CD-13-0103.
48. Sewell A, Brown B, Biktasova A, et al. Reverse-phase protein array profiling of oropharyngeal cancer and significance of PIK3CA mutations in HPV-associated head and neck cancer. Clin Cancer Res. 2014;20(9):2300–11. https://doi.org/10.1158/1078-0432.CCR-13-2585.
49. Zhang PL, Pellitteri PK, Law A, et al. Overexpression of phosphorylated nuclear factor-kappa B in tonsillar squamous cell carcinoma and high-grade dysplasia is associated with poor prognosis. Mod Pathol. 2005;18(7):924–32. https://doi.org/10.1038/modpathol.3800372.
50. Hajek M, Sewell A, Kaech S, Burtness B, Yarbrough WG, Issaeva N. TRAF3/CYLD mutations identify a distinct subset of human papillomavirus-associated head and neck squamous cell carcinoma. Cancer. 2017;123(10):1778–90. https://doi.org/10.1002/cncr.30570.
51. Williams EA, Montesion M, Alexander BM, et al. CYLD mutation characterizes a subset of HPV-

positive head and neck squamous cell carcinomas with distinctive genomics and frequent cylindroma-like histologic features. Mod Pathol. 2021;34(2):358–70. https://doi.org/10.1038/s41379-020-00672-y.
52. Dassonville O, Formento JL, Francoual M, et al. Expression of epidermal growth factor receptor and survival in upper aerodigestive tract cancer. J Clin Oncol. 1993;11(10):1873–8. https://doi.org/10.1200/JCO.1993.11.10.1873.
53. Licitra L, Storkel S, Kerr KM, et al. Predictive value of epidermal growth factor receptor expression for first-line chemotherapy plus cetuximab in patients with head and neck and colorectal cancer: analysis of data from the EXTREME and CRYSTAL studies. Eur J Cancer. 2013;49(6):1161–8. https://doi.org/10.1016/j.ejca.2012.11.018.
54. Soulieres D, Senzer NN, Vokes EE, Hidalgo M, Agarwala SS, Siu LL. Multicenter phase II study of erlotinib, an oral epidermal growth factor receptor tyrosine kinase inhibitor, in patients with recurrent or metastatic squamous cell cancer of the head and neck. J Clin Oncol. 2004;22(1):77–85. https://doi.org/10.1200/JCO.2004.06.075.
55. William WN Jr, Tsao AS, Feng L, et al. Single arm, phase II study of cisplatin, docetaxel, and erlotinib in patients with recurrent and/or metastatic head and neck squamous cell carcinomas. Oncologist. 2018;23(5):526–e49. https://doi.org/10.1634/theoncologist.2017-0661.
56. Lawrence MS, Stojanov P, Mermel CH, et al. Discovery and saturation analysis of cancer genes across 21 tumour types. Nature. 2014;505(7484):495–501. https://doi.org/10.1038/nature12912.
57. Chang MT, Asthana S, Gao SP, et al. Identifying recurrent mutations in cancer reveals widespread lineage diversity and mutational specificity. Nat Biotechnol. 2016;34(2):155–63. https://doi.org/10.1038/nbt.3391.
58. Weidhaas JB, Harris J, Schaue D, et al. The KRAS-variant and cetuximab response in head and neck squamous cell cancer: a secondary analysis of a randomized clinical trial. JAMA Oncol. 2017;3(4):483–91. https://doi.org/10.1001/jamaoncol.2016.5478.
59. Chung CH, Lee JW, Slebos RJ, et al. A 3′-UTR KRAS-variant is associated with cisplatin resistance in patients with recurrent and/or metastatic head and neck squamous cell carcinoma. Ann Oncol. 2014;25(11):2230–6. https://doi.org/10.1093/annonc/mdu367.
60. Pollock NI, Grandis JR. HER2 as a therapeutic target in head and neck squamous cell carcinoma. Clin Cancer Res. 2015;21(3):526–33. https://doi.org/10.1158/1078-0432.CCR-14-1432.
61. Cavalot A, Martone T, Roggero N, Brondino G, Pagano M, Cortesina G. Prognostic impact of HER-2/neu expression on squamous head and neck carcinomas. Head Neck. 2007;29(7):655–64. https://doi.org/10.1002/hed.20574.
62. Birkeland AC, Yanik M, Tillman BN, et al. Identification of targetable ERBB2 aberrations in head and neck squamous cell carcinoma. JAMA Otolaryngol Head Neck Surg. 2016;142(6):559–67. https://doi.org/10.1001/jamaoto.2016.0335.
63. Xia W, Lau YK, Zhang HZ, et al. Strong correlation between c-erbB-2 overexpression and overall survival of patients with oral squamous cell carcinoma. Clin Cancer Res. 1997;3(1):3–9.
64. Gillison ML, Glisson BS, O'Leary E, et al. Phase II trial of trastuzumab (T), paclitaxel (P) and cisplatin (C) in metastatic (M) or recurrent (R) head and neck squamous cell carcinoma (HNSCC): response by tumor EGFR and HER2/neu status. J Clin Oncol. 2006;24(18 Suppl):5511. https://doi.org/10.1200/jco.2006.24.18_suppl.5511.
65. Machiels JP, Haddad RI, Fayette J, et al. Afatinib versus methotrexate as second-line treatment in patients with recurrent or metastatic squamous-cell carcinoma of the head and neck progressing on or after platinum-based therapy (LUX-Head & Neck 1): an open-label, randomised phase 3 trial. Lancet Oncol. 2015;16(5):583–94. https://doi.org/10.1016/S1470-2045(15)70124-5.
66. Burtness B, Haddad R, Dinis J, et al. Afatinib vs placebo as adjuvant therapy after chemoradiotherapy in squamous cell carcinoma of the head and neck: a randomized clinical trial. JAMA Oncol. 2019;5(8):1170–80. https://doi.org/10.1001/jamaoncol.2019.1146.
67. Stewart JS, Cohen EE, Licitra L, et al. Phase III study of gefitinib compared with intravenous methotrexate for recurrent squamous cell carcinoma of the head and neck [corrected]. J Clin Oncol. 2009;27(11):1864–71. https://doi.org/10.1200/JCO.2008.17.0530.
68. Harrington K, Temam S, Mehanna H, et al. Postoperative adjuvant lapatinib and concurrent chemoradiotherapy followed by maintenance lapatinib monotherapy in high-risk patients with resected squamous cell carcinoma of the head and neck: a phase III, randomized, double-blind, placebo-controlled study. J Clin Oncol. 2015;33(35):4202–9. https://doi.org/10.1200/JCO.2015.61.4370.
69. Prawira A, Brana-Garcia I, Spreafico A, et al. Phase I trial of dacomitinib, a pan-human epidermal growth factor receptor (HER) inhibitor, with concurrent radiotherapy and cisplatin in patients with locoregionally advanced squamous cell carcinoma of the head and neck (XDC-001). Investig New Drugs. 2016;34(5):575–83. https://doi.org/10.1007/s10637-016-0367-2.
70. Whyte DB, Kirschmeier P, Hockenberry TN, et al. K- and N-Ras are geranylgeranylated in cells treated with farnesyl protein transferase inhibitors. J Biol Chem. 1997;272(22):14459–64. https://doi.org/10.1074/jbc.272.22.14459.
71. End DW, Smets G, Todd AV, et al. Characterization of the antitumor effects of the selective farnesyl protein transferase inhibitor R115777 in vivo and in vitro. Cancer Res. 2001;61(1):131–7.
72. Gilardi M, Wang Z, Proietto M, et al. Tipifarnib as a precision therapy for HRAS-mutant head and neck squamous cell carcinomas. Mol Cancer Ther.

2020;19(9):1784–96. https://doi.org/10.1158/1535-7163.MCT-19-0958.
73. Ho AL, Brana I, Haddad R, et al. Tipifarnib in head and neck squamous cell carcinoma with HRAS mutations. J Clin Oncol. 2021;39(17):1856–64. https://doi.org/10.1200/JCO.20.02903.
74. Hientz K, Mohr A, Bhakta-Guha D, Efferth T. The role of p53 in cancer drug resistance and targeted chemotherapy. Oncotarget. 2017;8(5):8921–46. https://doi.org/10.18632/oncotarget.13475.
75. Moser R, Xu C, Kao M, et al. Functional kinomics identifies candidate therapeutic targets in head and neck cancer. Clin Cancer Res. 2014;20(16):4274–88. https://doi.org/10.1158/1078-0432.CCR-13-2858.
76. Osman AA, Monroe MM, Ortega Alves MV, et al. Wee-1 kinase inhibition overcomes cisplatin resistance associated with high-risk TP53 mutations in head and neck cancer through mitotic arrest followed by senescence. Mol Cancer Ther. 2015;14(2):608–19. https://doi.org/10.1158/1535-7163.MCT-14-0735-T.
77. Tanaka N, Patel AA, Wang J, et al. Wee-1 kinase inhibition sensitizes high-risk HPV+ HNSCC to apoptosis accompanied by downregulation of MCl-1 and XIAP antiapoptotic proteins. Clin Cancer Res. 2015;21(21):4831–44. https://doi.org/10.1158/1078-0432.CCR-15-0279.
78. Mendez E, Rodriguez CP, Kao MC, et al. A phase I clinical trial of AZD1775 in combination with neoadjuvant weekly docetaxel and cisplatin before definitive therapy in head and neck squamous cell carcinoma. Clin Cancer Res. 2018;24(12):2740–8. https://doi.org/10.1158/1078-0432.CCR-17-3796.
79. Gadhikar MA, Sciuto MR, Alves MV, et al. Chk1/2 inhibition overcomes the cisplatin resistance of head and neck cancer cells secondary to the loss of functional p53. Mol Cancer Ther. 2013;12(9):1860–73. https://doi.org/10.1158/1535-7163.MCT-13-0157.
80. Marquard FE, Jucker M. PI3K/AKT/mTOR signaling as a molecular target in head and neck cancer. Biochem Pharmacol. 2020;172:113729. https://doi.org/10.1016/j.bcp.2019.113729.
81. Calo V, Migliavacca M, Bazan V, et al. STAT proteins: from normal control of cellular events to tumorigenesis. J Cell Physiol. 2003;197(2):157–68. https://doi.org/10.1002/jcp.10364.
82. Grandis JR, Drenning SD, Chakraborty A, et al. Requirement of Stat3 but not Stat1 activation for epidermal growth factor receptor- mediated cell growth In vitro. J Clin Invest. 1998;102(7):1385–92. https://doi.org/10.1172/JCI3785.
83. Grandis JR, Drenning SD, Zeng Q, et al. Constitutive activation of Stat3 signaling abrogates apoptosis in squamous cell carcinogenesis in vivo. Proc Natl Acad Sci U S A. 2000;97(8):4227–32. https://doi.org/10.1073/pnas.97.8.4227.
84. Sen M, Thomas SM, Kim S, et al. First-in-human trial of a STAT3 decoy oligonucleotide in head and neck tumors: implications for cancer therapy. Cancer Discov. 2012;2(8):694–705. https://doi.org/10.1158/2159-8290.CD-12-0191.
85. Ferris RL, Blumenschein G Jr, Fayette J, et al. Nivolumab vs investigator's choice in recurrent or metastatic squamous cell carcinoma of the head and neck: 2-year long-term survival update of CheckMate 141 with analyses by tumor PD-L1 expression. Oral Oncol. 2018;81:45–51. https://doi.org/10.1016/j.oraloncology.2018.04.008.
86. Burtness B, Harrington KJ, Greil R, et al. Pembrolizumab alone or with chemotherapy versus cetuximab with chemotherapy for recurrent or metastatic squamous cell carcinoma of the head and neck (KEYNOTE-048): a randomised, open-label, phase 3 study. Lancet. 2019;394(10212):1915–28. https://doi.org/10.1016/S0140-6736(19)32591-7.
87. Mandal R, Senbabaoglu Y, Desrichard A, et al. The head and neck cancer immune landscape and its immunotherapeutic implications. JCI Insight. 2016;1(17):e89829. https://doi.org/10.1172/jci.insight.89829.
88. Saloura V, Izumchenko E, Zuo Z, et al. Immune profiles in primary squamous cell carcinoma of the head and neck. Oral Oncol. 2019;96:77–88. https://doi.org/10.1016/j.oraloncology.2019.06.032.
89. Ayers M, Lunceford J, Nebozhyn M, et al. IFN-gamma-related mRNA profile predicts clinical response to PD-1 blockade. J Clin Invest. 2017;127(8):2930–40. https://doi.org/10.1172/JCI91190.
90. Cristescu R, Mogg R, Ayers M, et al. Pan-tumor genomic biomarkers for PD-1 checkpoint blockade-based immunotherapy. Science. 2018;362(6411):eaar3593. https://doi.org/10.1126/science.aar3593.

Biomarkers in Head and Neck Cancer

9

Zachary A. Oaks, Colette J. Shen, Siddharth H. Sheth, Gaorav P. Gupta, and Bhishamjit S. Chera

Introduction

Prior to the incorporation of prognostic and predictive biomarkers into clinical decision making, TNM staging was the mainstay of patient prognostication and would often guide treatment decisions. In early biomarker studies, physicians and researchers would need primary tissue to evaluate biomarkers including p53, human papilloma virus (HPV), and proliferative rate. The limitation of this approach is that it requires significant tissue for analysis, and in some cases may require invasive procedures or surgical resection to obtain adequate tumor tissue. With the advent of highly sensitive technologies such as next generation sequencing and digital PCR, much less tissue is required to evaluate tissue biomarkers. Genomic and transcriptional profiling of head and neck cancers, such as in The Cancer Genome Atlas, has generated a comprehensive molecular portrait of this constellation of malignant diseases, and in the process revealed new classes of biomarkers. There has also been the development of less invasive testing of biomarkers for certain head and neck cancer patient subsets, who can now have circulating tumor DNA (ctDNA) levels analyzed from blood samples which can inform the treating physician of response to treatment as well as detect recurrences earlier than conventional physical exam and radiographic studies.

Z. A. Oaks · C. J. Shen · G. P. Gupta · B. S. Chera (✉)
Department of Radiation Oncology, University of North Carolina, Chapel Hill, NC, USA
e-mail: zachary.oaks@unchealth.unc.edu; colette_shen@med.unc.edu; gaorav_gupta@med.unc.edu; bchera@med.unc.edu

S. H. Sheth
Division of Oncology, Department of Medicine, University of North Carolina, Chapel Hill, NC, USA
e-mail: siddharth.sheth@med.unc.edu

R. A. Chandra, R. J. Li (eds.), *Multidisciplinary Management of Head and Neck Cancer*,
https://doi.org/10.1007/978-3-031-05973-5_9

P53

p53 is colloquially referred to as the "guardian of the genome" as it functions as a critical tumor suppressor gene regulating the cell cycle, especially at the G1/S transition in the cell cycle. Genetic alterations in p53 are common in cancer in general, and is frequently altered in head and neck squamous cell carcinoma (HNSCC) [1, 2]. Early biomarker studies in head and neck cancer were based on direct studies of biopsied or resected tissue, with PCR and immunohistochemistry being common techniques. Tissue-based studies included p53 mutations and expression levels, HPV status, and expression profiles of infiltrating lymphocytes [3]. Early studies found that p53 was overexpressed in 78% of head and neck tumors of patients who smoked at least 20 cigarettes per day [2]. Conversely, in the same study only 1 of 7 non-smokers had p53 overexpression [2]. Interestingly, in tumor samples from patients that had quit smoking for at least 5 years, there was still elevated p53 levels in 90% of patients [2]. The Cancer Genome Atlas Network (TCGA) showed that in smoking-related HNSCC, inactivity p53 mutations were present in 84% of the cases, whereas very few (3%) of patients with HPV-positive HNSCC had p53 mutations (Fig. 9.1) [4]. In a study of 420 patients who underwent surgical management of HNSCC, 53% had mutations in p53 [1]. Patients with p53 mutations had a significantly lower 5-year overall survival of 40.7%, compared to 54.8% in those with wild-type p53 [1]. The authors also analyzed whether the mutations in p53 would disrupt its function based on the type of mutation and its location within the p53 gene. They showed that patients with p53 mutations that would likely disrupt its function had significantly worse overall survival compared to patients with wild-type p53, while those with "non-disruptive" mutations did not have a significantly worse overall survival [1]. Thus, the presence of p53 mutations, and more specifically disruptive mutations, are prognostic of poorer outcomes.

In the setting of locoregionally recurrent HNSCC, p53 mutations were found in 45% of recurrences and predictive of which patients would have recurrence, while p53 expression was unrelated to likelihood of recurrence [5]. In a similar study of locoregional recurrence of HNSCC following either treatment with radiation therapy in the primary or adjuvant setting, 44% patients were found to have p53 mutations, and the presence of a p53 mutation was correlated with higher risk of recurrent disease and earlier time to recurrence [6]. When applied to a specific disease site, p53 expression in >5% of cells in nasopharyngeal cancer was linked to significantly decreased overall survival, whereas the proliferation marker Ki-67 and anti-apoptotic protein bcl-2 were not associated with worse overall survival [7]. In laryngeal cancer, patients with >75% of their tumor cells expressing mutant p53 had higher likelihood of disease recurrence within 10 years at 48%, compared to 81% disease free survival at 10 years in patients with low p53 mutant expression [8].

Though P53 is commonly altered and is prognostic, its role as a predictive biomarker is unclear, and there are ongoing clinical trials to answer this question. Of note, a randomized phase III trial of adenoviral p53 gene therapy showed that patients with wild-type p53 or low mutant p53 expression had better treatment response to gene therapy than those with highly expressed mutant p53 [9]. Because p53 is a relatively unstable protein, its predictive value is limited to experimental therapy, and requires primary tumor tissue for analysis, alternative biomarkers have taken primacy in the clinic.

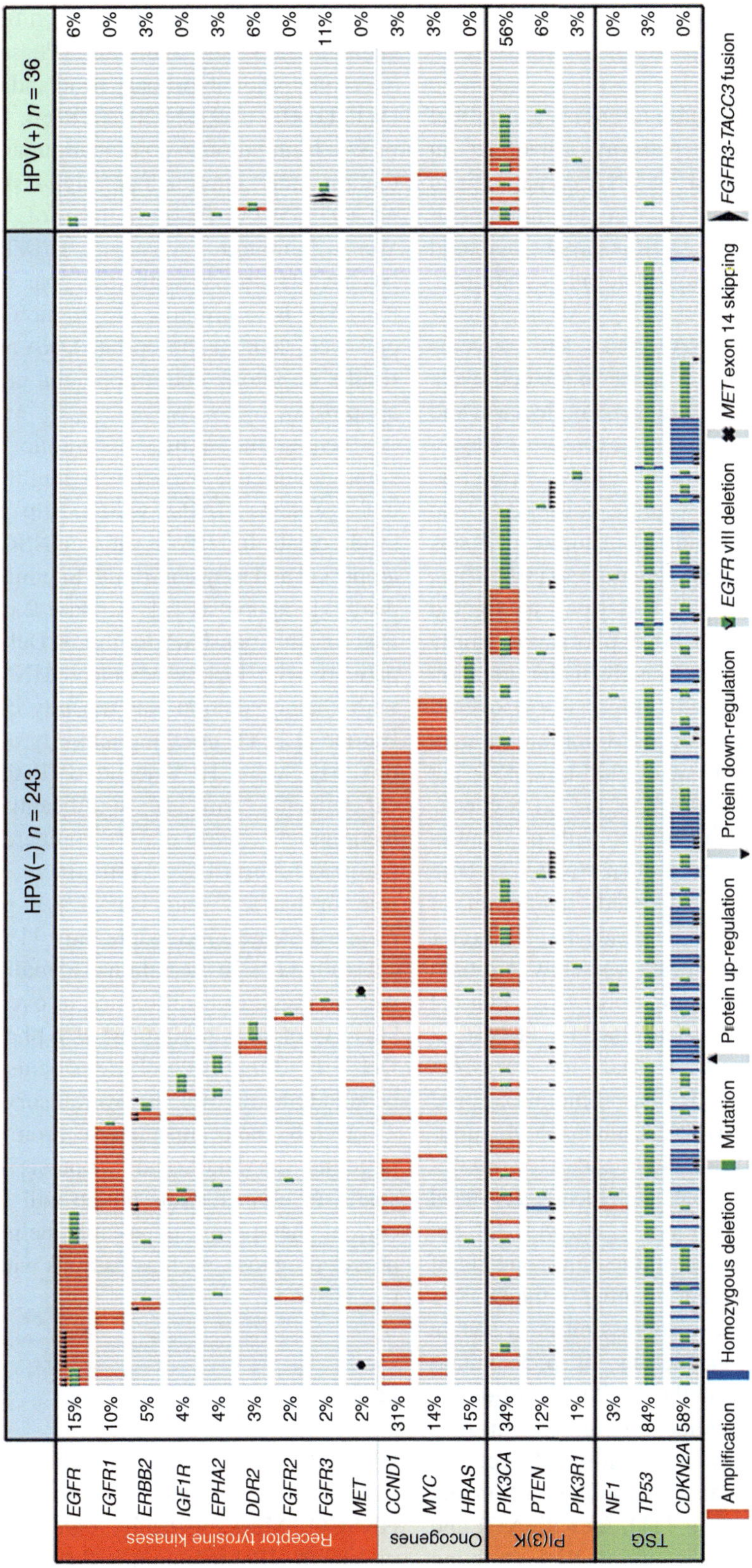

Fig. 9.1 Alteration events for key genes are displayed by sample ($n = 279$). *TSG* tumor suppressor gene. (Reprinted with permission. MS Lawrence *et al. Nature* **517**, 576–582 (2015) https://doi.org/10.1038/nature14129)

DNA Damage Response and Apoptosis

The DNA damage response (DDR) is a term that describes intracellular pathways involved in responding to and repairing DNA damage, or in the setting of significant damage, promoting cell death via apoptosis. The DDR is largely mediated by p53 in normal cells under normal conditions. Ataxia telangiectasia mutated (ATM) is a sensor of double-stranded DNA breaks and prevents progression from G1 to S phase in the cell cycle via stabilization of p53. Ataxia telangiectasia and Rad3 related protein (ATR) is a sensor of single-strand DNA breaks and regulates G2/M progression in the cell cycle. In tumor cells, ATM and ATR are upregulated and aid in tumor cell survival in the setting of replicative stress [10]. Thus, inhibition of ATM and ATR is ripe for targeted therapies. As described above, p53 is commonly mutated in HNSCC which results in deregulation of G1/S progression and confers worse overall survival outcomes. Thus, ATM inhibition may have benefit in patients with wild-type p53, but not in those with mutations, as the G1/S checkpoint is unregulated in tumors with p53 mutations. Conversely, in p53 mutant HNSCC, where the G1/S phase is deregulated, regulation of G2/M progression by ATR is essential for tumor cell survival [10]. There are currently multiple clinical trials studying numerous ATR (Phase I trial for locally advanced HNSCC: NCT02567422; Phase I trial for recurrent HNSCC: NCT04576091; Phase I biomarker trial for newly diagnosed HNSCC amenable to surgical resection: NCT03022409) and ATM inhibitors (Phase I trial in patients with locally advanced and metastatic solid tumors: NCT02588105), but none are available for use in the clinic outside of trials at this time. Another member of the DDR is Wee1, which regulates progression at G2/M in the cell cycle. In the setting of p53 mutant HNSCC, regulation of the G2/M transition by Wee1 becomes essential for tumor survival. Mechanistically, in vitro inhibition of Wee1 in the setting of DNA damage results in a fourfold increase in cell apoptosis [11]. As with ATR and ATM inhibitors, there are multiple trials studying the role of Wee1 inhibition in combination with chemotherapy and radiation (Phase I trial of Wee1 inhibition in primary head and neck cancer (WISTERIA trial): NCT03028766; Phase I trial of Wee1 inhibition in combination with chemoradiation for intermediate-/high-risk HNSCC: NCT02585973; Phase I trial of neoadjuvant Wee1 inhibition in stage III/IVB HNSCC: NCT02508246). Overall, the goal of inhibiting the DDR is to promote progression through the cell cycle without appropriate DNA repair, which results in mitotic catastrophe and cell death via apoptosis.

Even in the setting of mitotic catastrophe, some tumor cells overexpress inhibitor of apoptosis proteins (IAPs) which help tumor cells evade death. In HPV-negative HNSCC IAPs are amplified and may contribute to inhibition of cell death via apoptosis [4]. A phase 2 double-blind trial of the pan-IAP inhibitor Debio 1143 in patients with locally advanced HNSCC treated with chemoradiation was recently completed in Europe [12]. Of note, the majority of patients in this trial were HPV-negative. Patients in the experimental arm had significantly improved locoregional control of 54% compared to 33% in the placebo arm [12]. Furthermore, progression free survival at 2 years was 72% in patients who received Debio 1143, while only 41% in the placebo arm [12]. In this relatively small study, there was no significant difference in overall survival [12]. As of July 2020, there is a phase 3 trial of Debio 1143 that is enrolling patients which may be able to better evaluate overall survival.

Overall, there are multiple promising agents targeting the DDR and apoptosis in different stages of clinical trials which will complement current therapies and possibly help those with poorer prognosis such as patients with p53 negative HNSCC.

HPV

HPV infection causes predilection to anogenital cancer, cervical cancer, and HNSCC. Mechanistically, carcinogenesis secondary to HPV infection is through the E6 protein, which inhibits p53 activity, and the E7 protein which binds the tumor suppressor retinoblastoma (Rb) protein [13]. Rb is responsible for cell cycle regulation at the G1/S interface, thus in the setting of E7 binding of Rb which inhibits normal Rb function and promotes proteolysis, the cell cycle proceeds unregulated. In addition to its inhibition of Rb, E7 also binds p21 and p27 and inhibits their normal function, which includes cell cycle arrest through negative regulation of CDK2 [14–16]. Under normal conditions, the proliferative activity caused by E7 inhibition of Rb, p21, and p27 would activate p53 regulated cell cycle arrest and subsequent apoptosis. As a result of Rb inhibition and degradation secondary to HPV infection, there is upregulation of p16, a tumor suppressor that inhibits G1/S phase transition. The upregulation of p16 in HPV infection makes it a useful surrogate in detecting HPV-positive HNSCC [17]. Intracellularly, the E6 protein recruits the E6AP protein, which ubiquitinates and degrades p53. Thus, the combination of increased proliferative capacity and loss of regulatory function from p53 results in tumorigenesis. Interestingly, not all E6 proteins have the same affinity for E6AP. HPV-16 E6 has the highest affinity for E6AP, and thus is effective in degrading p53 [13]. HPV-18 E6 is relatively less effective at E6AP recruitment, while HPV-11 has very low E6AP recruitment, and thus minimal effect on p53 degradation [13]. Hence, detection of HPV-16 E6 and E7 DNA in either tumor specimens or in the peripheral blood can be a useful biomarker for a subset of HNSCC.

HPV proteins were first identified in oral SCC in 1983 using serum-derived antibodies from guinea pigs immunized with papilloma antigen [18]. Not long after, HPV-11 and HPV-16 viral DNA was detected in human HNSCC tumors, thus establishing HPV infection as a possible underlying mechanism of tumorigenesis in head and neck cancers [19]. In an early retrospective IHC study of HPV-positive tumors, they were found to be less likely to harbor p53 mutations relative to HPV-negative HNSCC, 10% vs. 67% respectively [20]. In the same study, it was shown that survival is improved in HPV-positive HNSCC, with a hazard ratio of 0.57 for HPV-positive HNSCC when compared to HPV-negative HNSCC [20]. In 2000, HPV E7 DNA was found in the peripheral blood of HPV-positive HNSCC, which was thought to be a result of tumor shedding HPV DNA into the bloodstream, as HPV is not typically found in the peripheral blood because its complete lifecycle is within epithelial cells [21].

In a meta-analysis from 2005 of over 5000 HNSCCs, one quarter of these tumors were HPV-positive, and when looking at site-specific data there appeared to be enrichment of HPV-positive cancers in the oropharynx (36%) versus 24% of oral cavity and 24% of laryngeal cancers [22]. Of further interest, different HPV subtypes seemed to cluster to different HNSCC sites. Within HPV-positive HNSCC, HPV-16 was the predominant infection of the oropharynx (87%), whereas HPV-18 was present in 68% and 69% of oral and laryngeal HPV-positive SCCs, respectively [22]. Over time, the prevalence of HPV-associated oropharyngeal SCC has significantly increased, for example, the rate of HPV-positive oropharyngeal SCC was 51% in the United States and 35% in Europe prior to 2000 [23]. Patients with oropharyngeal cancer diagnosed after 2005 were even more likely to be HPV-positive with similar rates of 70% and 73% in the United States and Europe, respectively [23]. Thus, as the rate of HPV-positive HNSCC increases, the role of HPV-related biomarkers is becoming more important over time.

Clinically, identifying HPV-positive HNSCC is important, because prognostically they have improved survival and better response to radiation and chemotherapy. Over a decade ago, a prospective analysis of patients enrolled in ECOG 2399 showed patients with HPV-positive HNSCC had significantly better clinical response to chemoradiation (84% versus 57% in HPV-negative HNSCC) and improved 2-year overall survival (95% versus 62% in HPV-negative HNSCC) [24]. Similarly, the post hoc analysis of RTOG 0129 found that 3-year overall survival of HPV-positive oropharyngeal SCC was 82.4% and 57.1% in the HPV-negative cohort [25]. Progression free survival at 3 years was also significantly improved in the HPV-positive cohort (73.3%) versus 43.4% in the HPV-negative group [25]. The TROG 02.02 trial which studied stage III and IV oropharyngeal SCC response to chemoradiation with or without tirapazamine, used p16 as a surrogate for HPV positivity [17]. The authors found that patients with p16 positive oropharyngeal cancers had improved 2-year overall survival of 91% compared to 74% in the p16 negative/HPV-negative cohort [17]. Further bolstering the use of p16 as a HPV surrogate was the post hoc analysis of RTOG 0129 with p16 IHC and presence of HPV DNA in fixed tissues with similar overall survival and progression free survival at 3 years as the patients with HPV-positive tumors, as there is significant overlap between the two groups with kappa of 0.8 [25]. A review of using p16 status alone as a surrogate for HPV status found sensitivity of p16 IHC ranged from 46% to 98%, with specificity ranging from 50% to 97% [26]. When IHC staining for p16 was combined with HPV DNA PCR, sensitivity and specificity had significant improvement ranging from 97% to 100% and 88–100%, respectively [26]. A study of 150 HNSCC patients in Sweden found that HPV positivity conferred better long-term outcomes as well. HPV-positive patients had significantly better 5-year disease free survival of 81% compared to 36% in HPV-negative patients [27]. Conversely, when viral load within tumors is measured by the number of viral genome copies, it was found that viral load does not impact survival [27]. Thus, it is the presence of HPV, not the viral load that impacts patient outcomes. Regarding p16 staining and HPV positivity, the American Society of Clinical Oncology and College of American Pathologist came to a consensus that "Pathologists should report p16 IHC positivity as a surrogate for [high risk] HR-HPV in tissue specimens (i.e., noncytology) when there is at least 70% nuclear and cytoplasmic expression with at least moderate to strong intensity" [28].

Even in the setting of recurrent disease, HPV-positive disease confers better overall survival compared to those with HPV-negative disease. Using p16 as a surrogate for HPV positivity, patients from RTOG 0129 and 0522 who had disease progression were stratified based on p16 status [29]. The authors found that the 2-year overall survival was significantly improved, 54.6%, in patients with p16 positive disease compared to those that were p16 negative (27.6%), and median survival was 2.6 years versus 0.8 years, respectively [29]. The trend of improved survival in p16 positive patients was significant, independent of site of recurrence or whether or not the patient had salvage surgery [29]. Clinically, the ability to identify patients with HPV-positive disease is important as there are numerous trials showing excellent tumor control with significant reduction in treatment toxicity [30–32].

Circulating HPV DNA is an emerging biomarker that can be utilized to predict treatment response and detect early recurrence in oropharyngeal SCC [33–38]. In two phase II prospective trials studying deintensification of chemoradiotherapy, circulating HPV DNA (ctHPVDNA) was analyzed prior to treatment, weekly during treatment, and after treatment in patients with p16 positive oropharyngeal SCC [34, 35]. In 103 patients with p16 positive oropharyngeal SCC, digital PCR quantification of the E7 gene from HPV16 identified 84 patients with pre-treatment serum positive for ctHPVDNA from HPV16 [34]. Of the 19 patients negative for HPV16 circulating tumor DNA, another testing for ctHPVDNA from HPV 18, 31, 33, and 35 identified eight more HPV-positive cases. In concert with data from healthy controls, it was determined that ctHPVDNA had a specificity of 97% and sensitivity of 89% for detecting non-metastatic oropharyngeal SCC (Fig. 9.2) [34]. Interestingly, tumor burden with regard to both higher T-stage and N-stage did not positively correlate with ctHPVDNA copy number (Fig. 9.2) [34]. Conversely, this pattern is not seen in patients with distant disease, there is a significant positive correlation between distant disease and ctHPVDNA load [33]. During treatment of these patients with deintensified chemoradiation, those with clearance of >95% of ctHPVDNA by 4 weeks of treatment and favorable clinical factors including less than 10 pack-years of tobacco use and <T4 tumors had 100% regional disease free survival, whereas those with initially low ctHPVDNA levels, T4 tumors, or >10 pack-years of tobacco use had regional disease free survival of 65% at

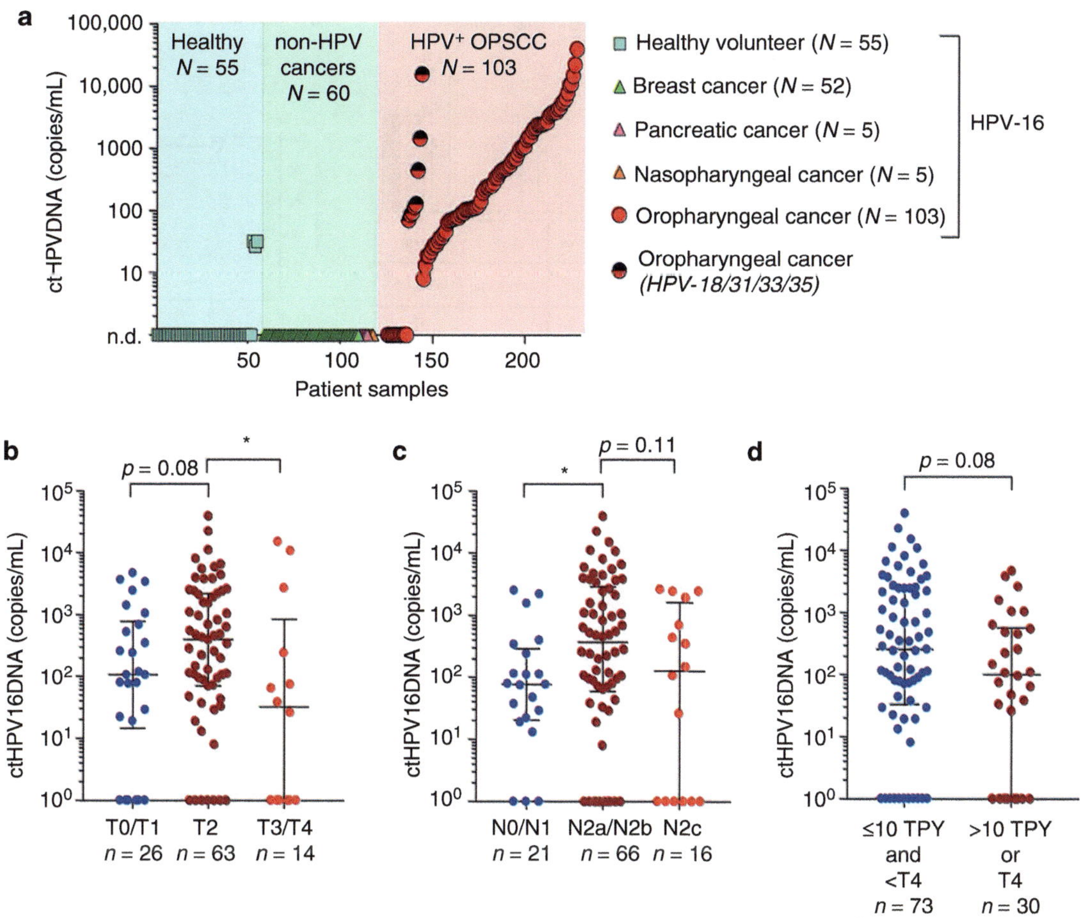

Fig. 9.2 Detection of plasma ctHPVDNA in HPV-associated oropharyngeal cancer patients. (**a**) Measurement of ctHPVDNA copies/mL plasma in 55 healthy volunteers, 60 patients with non-HPV-associated cancers, and 103 patients with HPV-associated OPSCC. Two-toned (red-black) circles denote patients who were negative for ctHPV16DNA but positive for ctHPVDNA from an alternative high-risk HPV strain (-18/31/33/35). (**b–d**) Baseline levels of ctHPV16DNA stratified by (**b**) tumor stage, (**c**) nodal stage, and (**d**) clinical risk factors. Median and interquartile range is shown for each graph. $*p < 0.05$; p values based on a two-sided Mann–Whitney test. (Reprinted with permission. *BS Chera et al. Clin Cancer Res* 2019;25:4682–4690)

18 months (Fig. 9.3) [34]. In analysis of ctHPVDNA levels during follow-up for monitoring recurrence, 115 patients were analyzed for ctHPVDNA, 76% had undetectable levels during follow-up, and zero of these patients had been diagnosed with recurrent disease at the time of publication, corresponding to a negative predictive value of ctHPVDNA of 100% (Fig. 9.4) [35]. Conversely, 24% of patients had a detectable amount of ctHPVDNA following completion of treatment [35]. Not all of these patients developed recurrent disease, and further parsing of the data found that of those with two consecutively abnormal ctHPVDNA levels, which in this case was 16 patients, 15 had biopsy-proven recurrences [35]. In addition to the negative predictive value of 100% for ctHPVDNA, two consecutive positive tests have a positive predictive value of 94%, and test sensitivity and specificity were 100% and 99%, respectively (Fig. 9.5) [35]. Clinically, patients with two consecutive positive tests had 2-year recurrence free survival of only 5%, whereas those with undetectable ctHPVDNA levels had a 2-year recurrence free survival of

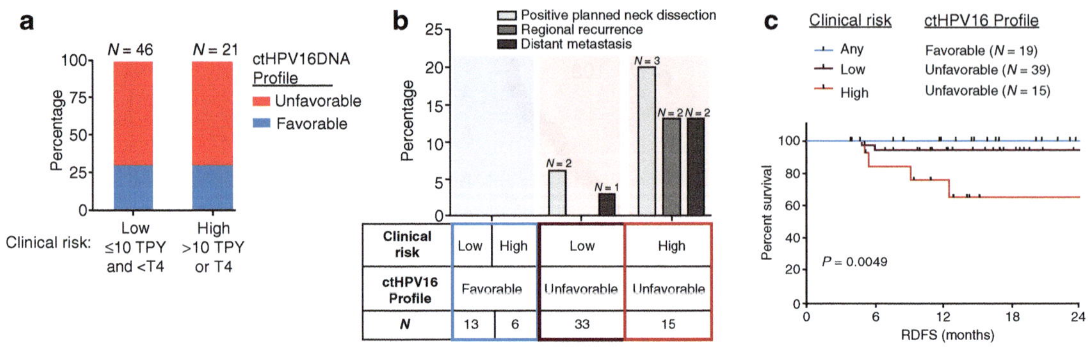

Fig. 9.3 A favorable ctHPV16DNA clearance profile correlates with disease control in OPSCC patients treated with CRT. (**a**) Percentages of clinical low-risk and clinical high-risk patients with a favorable ctHPV16DNA profile, which is defined as baseline ctHPV16DNA >200 copies/mL and >95% clearance by Week 4. (**b**) Percentage of patients within each subgroup who experience a positive neck dissection, regional recurrence, or distant metastasis. (**c**) Kaplan–Meier analysis of RDFS stratified by clinical risk and ctHPV16DNA profile. *p* value calculated using a two-tailed log-rank test for trend. (Reprinted with permission. *BS Chera et al. Clin Cancer Res* 2019;25:4682–4690)

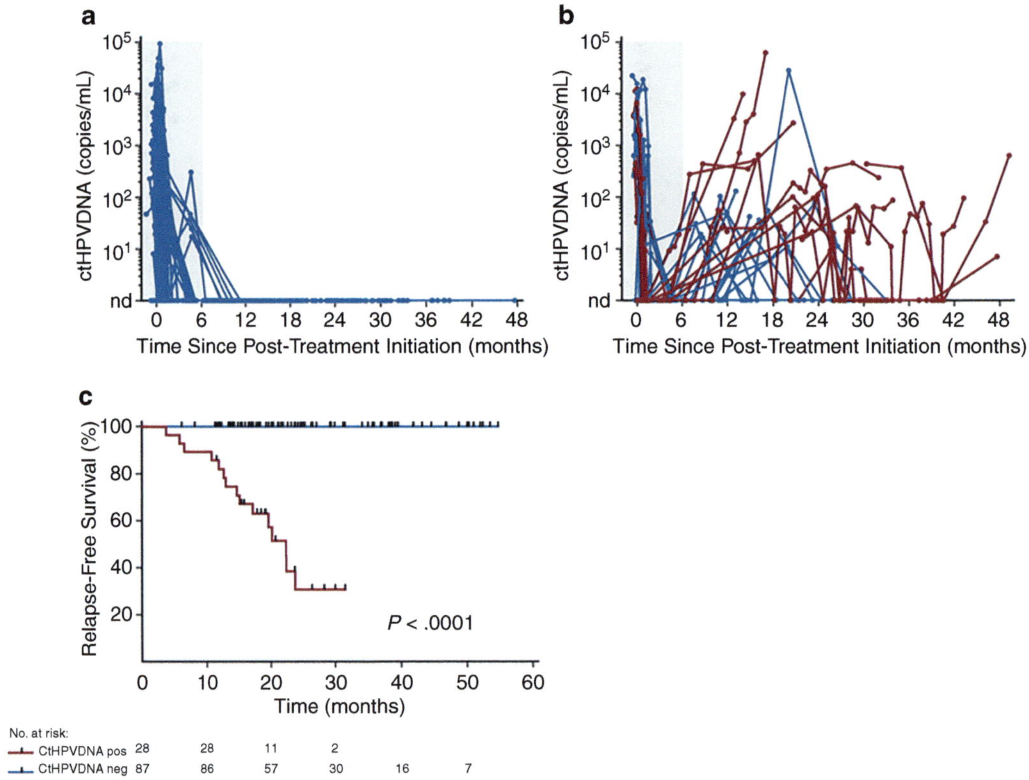

Fig. 9.4 Longitudinal ctHPVDNA surveillance identifies patients at high risk of disease recurrence. (**a**) Eighty-seven out of 115 patients (76%) had undetectable ctHPVDNA levels at all post-treatment surveillance time points (starting at 6 months after start of CRT). (**b**) Twenty-eight patients had a positive ctHPVDNA test result during post-treatment surveillance. Red lines indicate patients who have been diagnosed with biopsy-proven disease recurrence. The blue shaded region indicates the interval for initial treatment with CRT and assessment of response by positron emission tomography/computed tomography, which was not included in the post-treatment surveillance period. (**c**) Kaplan–Meier relapse-free survival of patients with undetectable ctHPVDNA at all surveillance time points versus patients with at least one abnormal ctHPVDNA blood test. *p* value was calculated using a two-tailed log-rank test. *ctHPVDNA* circulating tumor human papillomavirus DNA, *neg* negative, *pos* positive. (Reprinted with permission. *BS Chera et al. Journal of Clinical Oncology* 2020;38:1050–1058)

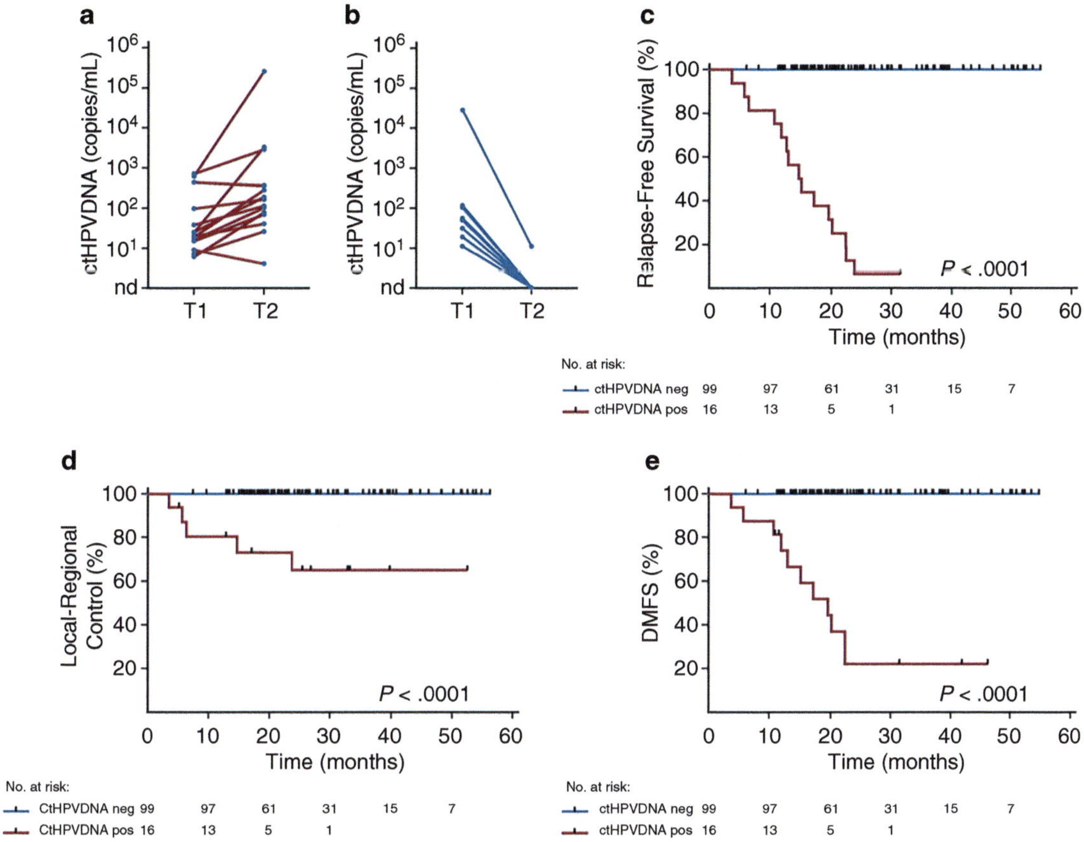

Fig. 9.5 Patients with two consecutively abnormal ctHPVDNA surveillance tests have a higher risk for disease recurrence. ctHPVDNA levels at the first abnormal surveillance time point (T1) and a subsequent time point (T2) for patients with (**a**) biopsy-proven disease recurrence or (**b**) no clinically evident disease recurrence. Kaplan–Meier estimates for (**c**) relapse free survival, (**d**) locoregional control, and (**e**) distant metastasis-free survival for patients who had two consecutively abnormal ctHPVDNA surveillance tests (ctHPVDNA positive) versus those without two consecutively abnormal ctHPVDNA surveillance tests (ctHPVDNA negative). *p* values were calculated using a two-tailed log-rank test. *ctHPVDNA* circulating tumor human papillomavirus DNA, *neg* negative, *pos* positive. (Reprinted with permission. *BS Chera et al. Journal of Clinical Oncology* 2020;38:1050–1058)

100% (Fig. 9.5) [35]. In the majority of patients, ctHPVDNA in the serum was detectable up to 6 months prior to diagnosis of recurrent disease with biopsy [35]. Of further interest, in patients that had recurrence and were treated with salvage therapy, many had subsequent response in ctHPVDNA levels [35]. In addition to detecting ctHPVDNA, it is also possible to detect circulating tumor DNA based on mutational analysis of both the primary tumor and plasma ctDNA [39].

More recently, there have been emerging data that HPV integration into the host genome results in worse outcomes. After HPV infection, HPV DNA may remain independent of the host as an episome, or integrate into the host DNA. Genomic studies of HPV-associated HNSCC suggest that the integration site is random, but when integration occurs it can amplify local genes, modify local DNA methylation, and increase mRNA transcripts, including those of oncogenes that promote tumorigenesis [4, 40]. In one study, 71% of HPV-associated HNSCC demonstrated integration, while in The Cancer Genome Atlas (TCGA) study only 28% of HPV-positive patients showed integration [40]. Secondary analysis of TCGA showed that patients with episomal HPV DNA had significantly better 5-year overall survival of 72% compared to 30% in those with

HPV integration into the host genome [41]. Interestingly, HPV-positive HNSCC with integration had similar 5-year overall survival to those with HPV-negative HNSCC [41]. In contrast, a recent study of 35 patients with HPV-associated HNSCC showed 60% of patients exhibited integration which conferred a significant improvement in disease-specific survival [42]. With these conflicting results, it is unclear what role HPV integration truly plays in HNSCC.

From a cancer screening perspective, HPV infection and E6 seropositivity are possible ways to identify at risk populations, but due to the high infection rate in the United States and relatively low incidence of oropharyngeal cancer, the number needed to screen to correctly identify a cancer is approximately 10,500 adults [43]. Hence, HPV and E6 testing are not feasible screening tests at this time and further investigation is ongoing.

EGFR

Epidermal growth factor receptor (EGFR) is a transmembrane receptor tyrosine kinase (RTK) that is in ErbB family of structurally similar RTKs which includes Her2/neu, Her3, and Her4. EGFR activation promotes downstream activation of the PI3K/Akt/mTOR, Jak-STAT, and Ras/Raf/Mek/Erk pathways which promote cell growth and division. In the setting of EGFR overexpression or constitutive activation, there is overactivation of downstream growth signals which promotes tumorigenesis. One of the first studies to analyze EGFR found that 19% of the HNSCC analyzed had amplification of the EGFR gene, and 53% of the tumors had EGFR overexpression by Western blot [44]. A subsequent study found that mRNA levels of EGFR were increased 69-fold in over 90% of the tumors analyzed compared to the mucosa of healthy controls [45]. Furthermore, it was shown that EGFR was also elevated in histologically normal tissues of the same HNSCC patients with a 29-fold increase in EGFR mRNA expression [45].

A retrospective analysis of 91 HNSCC patients treated with surgical resection with curative intent without known metastases showed that when patients were clustered based on EGFR expression via immunohistochemistry, patients in the tertile with high EGFR had higher risk of recurrence (84%) compared to 37% and 17% risk of recurrence in the medium- and low-risk tertiles, respectively [46]. EGFR overexpression also conferred a higher risk of death (71%) compared to 33% and 0% in the medium- and low-risk tertiles, respectively [46]. A retrospective FISH analysis of EGFR gene amplification in a cohort of 86 HNSCC patients found that those with greater than four gene copies in at least 40% tumor cells or greater than 15 copies in 10% of tumor cells had significantly worse progression free survival (median of 18 months versus 25 months in patients without EGFR gene amplification) and overall survival (median time to death of 20 months versus 29 months in patients without EGFR gene amplification) [47]. Interestingly, mutational analysis of these tumors was negative for activating mutations, thus copy number may be a more important underlying factor in patient outcomes rather than mutational burden [47].

EGFR expression was analyzed in a post hoc fashion from the prospective continuous hyperfractionated accelerated radiotherapy (CHART) trial in head and neck cancers from the United Kingdom. In this trial, patients with at least T2N0 disease, and no metastatic disease, were enrolled and randomized to either conventional radiation in 2 Gy daily fractions to 66 Gy or a total of 54 Gy in 1.5 Gy fraction delivered three times daily [48]. EGFR expression levels in HNSCC tissues from the CHART trial were divided based on EGFR expression being either above or below median in expression in trial patients. Patients with EGFR expression below the median level had no benefit from hyperfractionated accelerated radiotherapy with regard to 3-year local control [49]. Conversely, in patients with EGFR expression greater than the group median, patients who underwent hyperfractionated accelerated radiotherapy had significantly improved local control, and thus EGFR expression may be predictive in nature for a subset of HNSCC patients [49]. A similar post hoc analysis of RTOG 9005, which divided HNSCC patients into either high or low EGFR expression relative to median group expression, found that EGFR overexpression was linked to significantly worse overall survival, disease free survival, and higher likelihood of local recurrence, though the authors did not explicitly state the rates of these findings at the 2-year and 5-year marks [50]. Conversely, EGFR overexpression in RTOG 9005 was not associated with distant failure [50].

The prognostic value of EGFR has also been studied in undifferentiated nasopharyngeal carcinoma. At the University of Hong Kong, patients with undifferentiated NPC, that were treated with induction cisplatin and epirubicin, followed by radiation therapy, were retrospectively analyzed for EGFR expression and its relationship to patient outcomes [51]. In this study, EGFR expression was classified based on the number of cells expressing EGFR and grouped as those with <5% expression, >5% but less than 25% expression, and >25% expression. In patients with >25% cells expressing EGFR, there was significantly worse disease-specific survival at 5 years (48%) and 10 years (40%) compared with low EGFR expressing patients with 5-year and 10-year DSS of 86% and 78%, respectively [51]. Relapse free survival was also significantly worse in those with >25% EGFR expression with 5-year and 10-year rates of 36% and 30%, respectively, and 80% and 73% in the EGFR low expressing patients, respectively [51]. Interestingly, distant failure was not significantly different in high or low EGFR expressing tumors [51].

In a meta-analysis of 33 publications for the role of EGFR expression in overall survival, only 14 of 33 studies reported a positive association with EGFR overexpression [52]. When all 33 studies were pooled, EGFR overexpression was significantly linked to worse overall survival outcomes, with a hazard ratio of 1.694 [52]. When analyzing the effect of EGFR on tumor subsites including oral cavity, larynx, and oropharynx, the overall survival was consistently worse in EGFR overexpressers [52]. The link between EGFR overexpression and worse overall survival was independent of geographic location of studies as well [52].

EGFR has consistently been shown to be underexpressed in HPV-associated HNSCC [53–57]. The recently published results of RTOG 1016 further highlight the inverse relationship between HPV-associated HNSCC and EGFR, and why targeting EGFR with cetuximab, an antibody directed against EGFR, does not always improve outcomes. RTOG 1016 was designed as a non-inferiority trial that specifically enrolled patients with HPV-positive OPSCC and randomized 849 patients to radiotherapy with either concurrent cetuximab or cisplatin, with a goal of showing less toxicity in the cetuximab arm [58]. The patients in the cetuximab arm had significantly worse 5-year overall survival, progression free survival, locoregional failure rate compared to the cisplatin arm [58]. Furthermore, there was no significant difference in reported adverse events between the two study arms [58]. What likely explains these results is that EGFR is underexpressed in HPV-associated HNSCC, thus targeting EGFR in this group of patients will likely have minimal benefit, or possibly worse outcomes if cisplatin is omitted from their treatment. Further abrogating the benefit of EGFR inhibition is that approximately 20–30% of HPV-associated HNSCC harbor mutations in phosphoinositide 3-kinase (PI3K) which lies downstream of EGFR, thus making EGFR inhibition moot in those patients [4].

PI3K/PIK3CA

PI3K, whose alpha subunit is encoded by the gene *PIK3CA*, is downstream of EGFR activation. PI3K is a kinase in the PI3K/Akt/mTOR pathway which is involved in upregulation of transcription, translation, cell growth, and proliferation. In early studies in which specific mutations were being studied, PIK3CA mutations were found only in a minority of patients, with one study identifying a rate of <3% of HNSCC [59]. Another study of 37 oral SCC found mutations in PIK3CA in only four cases, while only two of 37 cases in HNSCC from Vietnamese and Indian patients contained PIK3CA mutations [60, 61]. PIK3CA amplification, on the other hand, was present in 32% of the cases in one study [59]. PIK3CA amplification, in general, did not have a significant effect on 2-year overall survival or disease free survival, but analysis of the subset of patients with node-negative disease, PIK3CA amplification correlated with worse 2-year disease free survival of 31% compared to patients with normal copy number that had a 2-year disease free survival of 90% [59]. Thus, PIK3CA gene amplification is a useful prognostic tool in the node-negative HNSCC patient. Curiously, a German study of PIK3CA copy number in HNSCC found that 48% of HNSCC tumors had >2 copies of PIK3CA and this was significantly correlated with likelihood of node positivity [62].

In addition to being downstream of EGFR, PIK3CA is the most commonly mutated gene in HPV-associated OPSCC [63]. As discussed above, early studies of PIK3CA in HNSCC did not identify a large number of mutations, as these studies approached HNSCC as a homogenous group, and did not stratify patients based on HPV status. In addition to the aforementioned studies, a more comprehensive analysis of HNSCC found that PIK3CA mutations were present in approximately 8% of tumors [64]. When testing for common activating mutations at codons 542%, 545%, and 1047, 19.5% of all HNSCC patients had PIK3CA mutations [65]. When patients were then stratified by HPV status, there was enrichment of PIK3CA mutations (28% positive) in the HPV-positive cohort versus 10% in the HPV-negative cohort [65]. An early study with Sanger sequencing in HPV-positive HNSCC found 31% of the tumors harbored PIK3CA mutations [66]. Work by the Cancer Genome Atlas Network to identify common genetic alterations in HNSCC, found that PIK3CA genetic alterations, which includes mutations, amplification, and deletions, occurred in over 50% of HPV-positive HNSCC, though the sample size of HPV tumors was relatively small accounting for only 13% of all HNSCC tumors analyzed in this study (Fig. 9.1) [4]. Conversely, only 34% of HPV-negative HNSCC had PIK3CA genetic alterations [4]. Within all HNSCC, PIK3CA was amplified in 45% of the tumors analyzed [4]. In further exploration of the data from TCGA, 29.5% of HPV-positive tumors contained PIK3CA missense mutations [67]. In a separate study using RNAseq, nearly 30% of HPV-associated oral SCC had PIK3CA mutations, and 87% of these were gain of function, i.e., activation mutations that promoted the PI3K/Akt/mTOR axis [63]. Further bolstering these findings, in a group of prospectively collected OPSCC tissues, 8/33 HPV-associated oropharyngeal SCC were found to have activating PIK3CA mutations while zero of 14 HPV-negative tumors had PIK3CA mutations [68]. Interestingly, these activating mutations did not increase Akt activity, as measured by phosphorylation, but rather increased mTOR activation as measured by increased phosphorylation of mTOR targets S6 and IRS1 [68]. A recent analysis of patients enrolled in two separate phase II trials of deintensified chemoradiation for HPV-positive patients, found that patients with tumors containing PIK3CA mutations resulted in significantly worse 3-year disease free survival of 68%, versus 93.4% in PIK3CA wild-type tumors [34].

EBV

EBV is commonly associated with nasopharyngeal cancer (NPC). EBV circulating tumor DNA is detectable in NPC patients, specifically through the detection of the BamHI-W, a 3 kb repeat sequence in the EBV genome, and EBNA-1, a regulator of EBV replication and genome maintenance [69]. In one of the first studies to look at EBV viral load following treatment of NPC, EBV DNA became undetectable in 7 of 7 patients with clinical disease regression [69]. A relatively small study from Malaysia evaluating cell-free EBV DNA in plasma in patients with untreated NPC had a median of 2043 copies/mL versus a median of zero copies in both healthy controls and patients with treated NPC [70]. Of note, the mean cell-free EBV DNA loads in untreated NPC patients was 11,553 copies/mL, while the healthy controls had a mean EBV load of 133 copies/mL and the treated NPC patients had a mean of 2000 copies/mL [70]. In a larger cohort from China, consisting of 150 NPC patients and

75 healthy controls, the levels of BamHI-W EBV DNA was evaluated in primary tumor tissue, peripheral blood cells (PBCs), and plasma (cell-free DNA). Within plasma, EBV DNA was detected in 92% of patients prior to treatment, whereas only 12% of healthy controls had detectable EBV DNA [71]. Following treatment of NPC, the rate of EBV DNA detection in the serum was significantly lower at 19% [71]. During EBV infection, EBV can enter both B cells and epithelial cells [72, 73]. Interestingly, 24% of NPC patients' PBCs had detectable EBV DNA, and was not significantly different when compared to post-treatment EBV DNA levels or healthy controls, which may indicate that EBV shedding from tumor into the plasma is a relevant biomarker for detection of NPC as well as treatment response [71].

From a prognostic perspective, EBV DNA load in the plasma correlates positively with the historical China 1992 TNM staging system [71, 74]. EBV cell-free DNA level is also associated with patient outcomes. In a study of 139 NPC patients, those with serum EBV DNA level greater than 40,568 copies/mL at enrollment had significantly worse survival [75]. The serum EBV DNA levels were even helpful in prognosticating worse survival outcomes in stage III and IV NPC patients [75]. Furthermore, for every tenfold increase in serum EBV DNA load, there was a relative risk of 1.6 that death would occur [75]. Unfortunately, the authors did not show how the data supported the cutoff of 40,568 copies of EBV DNA/mL. A follow-up study from the same group in Hong Kong further supported the use of serum EBV DNA levels in predicting recurrence in NPC, by incorporating post-treatment serum EBV DNA levels into their model [76]. When placing patients into risk groups, those with serum EBV DNA greater than 500 copies/mL after definitive treatment were considered to be in a higher risk group. The overall survival at 1 year for patients with serum EBV DNA greater than 500 copies/mL was 76%, which was significantly lower than the group with fewer than 500 copies/mL with 97% alive at 1 year [76]. When pre-treatment serum EBV DNA levels were incorporated into patient survival, those with low pre- and post-treatment serum levels of EBV DNA had 98% overall survival at 1 year and those with high pre-treatment level plus low post-treatment level, reflecting a good treatment response, had 95% overall survival at 1 year [76]. In contrast, those with low pre-treatment and high post-treatment serum EBV DNA levels had a 1-year overall survival of 75% and patients with high pre- and post-treatment serum EBV DNA had 1-year overall survival of 61%, reflecting a poor treatment response [76]. Similarly, a study from Taiwan used a cutoff of 1500 copies/mL of EBV DNA prior to treatment and found that those with low pre-treatment EBV DNA levels had 2-year overall survival of 100%, which was significantly higher than those with high pre-treatment levels at 83.4% [77]. The challenge with these studies is that each group has used different EBV DNA cutoffs to identify their high- and low-risk groups, which makes it challenging to apply these results to individual patients when determining prognosis. Post-treatment, it does appear that undetectable serum EBV DNA predicts improved overall survival and freedom from recurrence.

Also comparable to prior studies, in a study of 99 patients with non-metastatic stage III or IV nasopharyngeal carcinoma those with detectable post-treatment plasma EBV DNA had worse 2-year overall survival of 56.3% compared to those with undetectable post-treatment levels with survival of 96.7% [77]. Plasma EBV DNA has also been utilized in long-term monitoring of NPC patients, and elevation of EBV DNA was shown to precede clinically detectable relapse by 6 months [77]. When studying EBV infection in resected or biopsied primary tumor tissue, up to 100% of tissues can be EBV DNA positive [71]. Testing for EBV DNA in this study was 100% sensitive, but only had specificity of 60% for identifying NPC patients, as 40% of nasopharyngitis cases were positive for EBV DNA in the same series [71]. An alternative view of this early data was that the negative predictive value of a negative EBV DNA test of biopsied tissue was 100%. Due the low specificity, alternative biomarkers were required to differentiate tumor from benign tissue. Utilization of in situ hybridization to detect EBV encoded RNA 1 (EBER1), found that 100% of tumors were EBER1 positive, while zero cases of nasopharyngitis were positive [71].

EBV infection is present in up to 90% of the world population, thus screening is a challenge due to the high prevalence of the virus [78]. A study from southern China screened over 20,000 men for NPC with plasma EBV DNA levels. 5.5% of the men had detectable EBV DNA in the serum, and upon repeat serology 4 weeks later 27.8% of those men (309 men) were still positive [79]. Of this group of 309 men, 300 underwent screening for NPC with either MRI or endoscopic exam, and 34 men were found to have NPC [79]. It is unclear what role screening may have in the United States, where NPC is much more rare than East Asia.

Immune Checkpoint Inhibitors

Programmed cell death protein 1 (PD-1) is expressed on activated T cells and its ligands, programmed death ligand 1 and 2 (PD-L1 and PD-L2), are normally expressed on macrophages and dendritic cells to downregulate the immune response by promoting T cell apoptosis. A subset of tumors express PD-L1, which permits tumor evasion of the immune system and subsequent tumor survival. Consequently, PD-1 inhibitors, including pembrolizumab and nivolumab, were developed to limit PD-1 driven immune suppression and promote the host immune response to tumor cells. PD-L1 expression is the clinical measure used to predict if a patient will respond to an immune checkpoint inhibitor. PD-L1 expression is commonly reported as the combined positive score (CPS) which is the number of any cells staining positive for PD-L1 divided by the total number of tumor cells multiplied by 100. Clinically, the KEYNOTE trials studied the role of PD-1 inhibition with pembrolizumab in persistent, recurrent, and metastatic head and neck cancers.

An early trial of pembrolizumab in head and neck cancer was the phase Ib KEYNOTE-012. Sixty five percent of the patients enrolled in the KEYNOTE-012 trial had tumors that expressed PD-L1 [80, 81]. In the initial analysis, there was a significantly better overall response rate of 22% in patients with PD-L1 expression, versus 4% overall response rate in patients with no PD-L1 expression [80]. Next, the phase II trial of pembrolizumab in recurrent and metastatic head and neck cancer, KEYNOTE-055 showed similar results to the KEYNOTE-012 with 16% of patients showing either complete or partial response to therapy [82]. Eighty two percent of patients in KEYNOTE-055 had tumor that had at least a CPS of 1%, and the majority of patients, 77%, had HPV-negative disease [82]. In the KEYNOTE-040 trial, a randomized phase III trial evaluating pembrolizumab versus standard of care methotrexate, docetaxel, or cetuximab in recurrent or metastatic HNSCC, the hazard ratio for death in the intention to treat analysis was significant at 0.80 in the pembrolizumab arm [83]. Though not significant, the median overall survival in the pembrolizumab arm was 8.4 months compared to 6.9 months in the standard of care arm [83]. Similar to KEYNOTE-040, the CheckMate 141 trial studied nivolumab versus methotrexate, docetaxel, or cetuximab in recurrent or metastatic HNSCC [84]. Patients who received nivolumab had a significant decrease in the hazard of death compared to standard of care, with a hazard ratio of 0.70 [84]. The median overall survival was longer in the nivolumab arm at 7.5 months, versus 5.1 months in the standard of care arm [84]. The KEYNOTE-048 trial then studied pembrolizumab with or without chemotherapy versus cetuximab plus chemotherapy in patients with untreated locally recurrent or metastatic HNSCC. Pembrolizumab alone and in combination with chemotherapy significantly improved overall survival in patients with either cutoff of PD-L1 expression at least 20% or at least 1% [85].

Following positive results in the recurrent and metastatic setting with PD-1 inhibition, the PD-L1 inhibitor avelumab was given adjuvantly in the randomized, double-blind, placebo controlled JAVELIN Head and Neck 100 trial in patients with at least stage 3 disease HNSCC or T4, N2c, or N3 disease for HPV-positive HNSCC with unknown PD-L1 status [86]. At the preplanned interim analysis, the primary outcome of superior progression free survival was not met and the trial was discontinued due to futility. Curiously, the avelumab arm had worse median progression free survival of 16.9 months, compared to placebo group of 23.0 months [87]. Of the 697 patients enrolled, 70% had a CPS less than 25%. The generally low expression of PD-L1 in this trial may explain why immune checkpoint inhibitors failed to improve patient outcomes when treating adjuvantly rather than in the setting of recurrent or metastatic disease. The PD-L1 inhibitor durvalumab has been studied in both the first line setting (KESTREL Trial, NCT02551159) and second line setting (EAGLE, NCT02369874) for recurrent or metastatic disease, which were negative trials as durvalumab did not provide any overall survival benefit [88, 89].

PD-1 inhibition plays an important role in recurrent and metastatic HNSCC, and its success in patients correlates positively with high CPS scores. Due the importance of CPS scores with regard to response rate, selection of patients for future checkpoint inhibitor trials may show improved outcomes in highly selected study populations.

Hypoxia

Tumor hypoxia in head and neck cancer has prognostic and predictive value for patients. Eppendorf pO_2 histography of 397 mostly neck node metastases of untreated head and neck tumors at multiple institutions that were planned for radiotherapy as part of their management found that patients with more hypoxic tumors had significantly worse overall survival [90]. The limitation of tumor hypoxia measurements is that it is invasive, and there appears to be significant variability between institutions [90]. An alternative to measuring oxygen values in vivo is measuring carbonic anhydrase IX levels in biopsy specimens with immunohistochemistry, which is upregulated in the setting of hypoxia. Patients with high carbonic anhydrase IX levels, in addition to low tumor infiltrating lymphocytes, had a 2-year local control rate of 14%, versus patients with low carbonic anhydrase IX levels and a high number of tumor infiltrating lymphocytes who had local control rates of 80% [91]. Hypoxia inducible factor 1 alpha (HIF1α) is also upregulated in hypoxic settings and is a surrogate marker for hypoxic tumors. In patients with low tumor HIF1α expression, i.e., normoxic tumors, there were significantly better hazards of progression free survival (0.64), locoregional control (0.61), and overall survival (0.63), relative to patients with high HIF1α expression [92]. Interestingly, elevated HIF1α also predicted improved response to the EGFR inhibitor nimotuzumab in combination with chemoradiation, versus chemoradiation alone [92]. Positron emission tomography (PET) technology has also been utilized to evaluate tumor hypoxia. 18F-Fluoromisonidazole (^{18}FMISO) and ^{18}F-fluoroazomycin arabinoside (^{18}FAZA) are radiotracers that specifically target hypoxic areas of tumors. ^{18}FAZA-PET was utilized in the TROG02.02 trial and found that patients with more hypoxic tumors had worse locoregional control, failure-free survival, and overall survival than normoxic tumors [93]. Similarly, ^{18}FMISO was used in patients enrolled in a phase I trial of HNSCC, which showed that the rate of hypoxic tumors was similar in p16-positive OPSCC (74%) and p16-negative HNSCC tumors (80%) [94].

Since tumor hypoxia is common and patients with hypoxic tumors appear to have resistance to definitive therapy and subsequently worse outcomes, trials were developed utilizing tirapazamine, a small molecule that selectively targets hypoxic cells and results in DNA damage and cell death. A phase I trial of 16 patients with HNSCC treated with 70 Gy radiation with concurrent cisplatin and tirapazamine showed promising results with a 3-year progression free survival of 88% and 3-year overall survival of 69% [95]. Fifteen patients enrolled on this phase I trial had ^{18}FMISO PET performed, 14 of 15 patients had hypoxic tumors prior to initiating treatment, and by the end of treatment only one patient had a hypoxic tumor [95]. With these promising phase I data, a randomized phase II trial, TROG 98.02, enrolled patients with stage III/

IV HNSCC treated with 70 Gy with either concurrent cisplatin plus tirapazamine or cisplatin with infusional fluorouracil boost [96]. Patients in the tirapazamine arm had non-significant improvement in 3-year failure-free survival of 55% and 3-year overall survival of 60% versus 45% and 46% in the fluorouracil arm, respectively [96]. From a toxicity perspective, the fluorouracil arm had more severe skin reactions, whereas the rate of grade 3 and 4 neutropenia were significantly increased in the tirapazamine arm at 37% versus 17% in the fluorouracil arm [96]. Following TROG 98.02, the randomized phase III trial TROG02.02 was developed which studied 861 patients with stage III or IV HNSCC undergoing definitive radiation with either concurrent cisplatin or cisplatin plus tirapazamine [97]. Unfortunately, TROG02.02 was a negative study and did not show significant differences between the two arms with respect to 2-year overall survival rate, failure-free survival, and time to locoregional failure [97]. A post hoc analysis of 41 patients from TROG02.02 with HPV-negative HNSCC who underwent ^{18}FAZA-PET, and were subsequently stratified by hypoxia status, showed that patients with hypoxic tumors who received cisplatin and tirapazamine had significant improvement in the time to locoregional failure and failure-free survival, with no significant difference in overall survival [93]. Despite the early promise of tirapazamine, the results of TROG02.02 limited its adoption clinically. There may be a subset of HPV-negative HNSCC patients with hypoxic tumors that could derive benefit from inclusion of tirapazamine in their systemic therapy regimen, but more trial work will need to be conducted before wider adoption. Further limiting tirapazamine use is that ^{18}FMISO and ^{18}FAZA PET are limited in clinical use at this time.

Biomarkers in Non-squamous Histologies

Although the most common tumors of the aerodigestive tract are squamous cell in origin, there is a diverse array of rarer tumor types that are also present and have unique molecular profiles. Sinonasal undifferentiated carcinoma (SNUC) is a blanket term that previously applied to a group of aggressive sinonasal tumors. With the advent of advanced sequencing techniques, tumors that were previously described as SNUC are now being classified based on genetic findings such as translocation of nuclear protein in testis (NUT) that results in NUT midline carcinomas, HPV-related carcinoma with adenoid cystic-like features, and SMARCB1 deficient sinonasal carcinoma [98–101]. The importance of molecular testing in the sinonasal carcinomas is highlighted in a series of 39 patients with SMARCB1 deficient carcinomas, as these tumors exhibit diverse histological findings that may lead to misdiagnosis [101]. Unfortunately, SMARCB1 deficiency does not confer better outcomes, and two-thirds of the patients in the aforementioned case series were deceased 2 years from diagnosis [101]. HPV-related carcinoma with adenoid cystic-like features is also a recently described entity of the sinonasal tract that appears to be secondary to HPV-33 and HPV-35 infection, unlike OPSCC which is driven by HPV-16 infection [102]. Similar to SMARCB1 deficient carcinomas, histologically and morphologically these sinonasal HPV-related carcinomas have a diverse appearance [102]. Unlike SMARCB1 deficient carcinomas, those with HPV-related carcinomas of the sinonasal tract may have improved outcomes, and a case series of 49 patients did not identify any tumor related deaths at a mean follow-up of 42 months [103].

In salivary gland cancer there is a subset termed mammary analog secretory carcinoma (MASC) which is defined by histopathologic features similar to secretory carcinoma of the breast, and commonly present in childhood to adolescence [104]. Greater than 90% of these tumors exhibit a t(12;15) (p13;q25) ETV6-NTRK3 translocation and specifically stain positively for mammaglobin, while other salivary tumors do not express mammaglobin [105, 106]. Specifically identifying MASC tumors is important because there are two tropomyosin receptor kinase (TRK) inhibitors available, larotrectinib and entrectinib, that are utilized in both locally advanced and metastatic MASC with combined partial and complete response rates ranging from 57% to 79% [107–109].

In non-secretory salivary gland cancers, certain tumors have been found to express HER2 or androgen receptor [110, 111]. Targeted therapy against HER2, trastuzumab, is commonly utilized in HER2 positive breast cancer. A phase II trial of patients with locally advanced, recurrent, or metastatic HER2 positive salivary gland carcinomas treated with docetaxel and trastuzumab demonstrated an overall response rate of 70.2% with a progression free survival of 8.9 months [112]. Androgen receptor expression is present in approximately one-fifth of salivary gland cancers and is associated with worse overall survival and disease free survival [113]. A phase II trial of bicalutamide, an androgen receptor antagonist, and leuprolide, a gonadotropin-releasing hormone agonist, in patients with androgen receptor positive locally advanced or metastatic salivary gland carcinoma showed an overall response rate of 41.7% [114]. Thus, identifying androgen receptor and HER2 expression in salivary tumors is prognostic, in the case of androgen receptor, and predictive for tumor response to targeted therapies.

Differentiated thyroid cancer recurrence can be monitored by measuring serum thyroglobulin. Elevated serum thyroglobulin detected recurrences in 52% of patients in one study, and when combined with ultrasound, 96% of recurrences were identified [115]. Following thyroidectomy and iodine ablation, 80% of patients with thyroglobulin levels above 2.0 ng/mL after recombinant human thyroid stimulation hormone had recurrent disease at 5 years post-ablation [116]. In contrast, only one of 68 patients with thyroglobulin level below 0.5 ng/mL had tumor recurrence at 5 years [116].

In papillary thyroid cancer constitutive RET activation is commonly found in children with radiation exposure, with 77% of pediatric papillary thyroid carcinomas secondary to fallout from the Chernobyl nuclear accident showing RET rearrangements, compared to 65% in sporadic tumors [117]. TRK is another tyrosine kinase that is constitutively activated in papillary thyroid cancer, though at a lower rate than RET, but likely contributes to oncogenesis through activation of PI3K and Ras/MAPK pathways [118, 119]. BRAF is a serine-threonine kinase that is mutated in papillary thyroid cancer, with prevalence of 69% in one published series [120]. The presence of BRAF mutations in papillary thyroid cancer appears to be prognostic of worse mortality, with a retrospective analysis of 1849 patients showing a nearly fivefold increase in mortality in patients with BRAF mutations [121]. BRAF mutations are also associated with larger primary tumor, metastasis, and worse disease free survival compared to tumors with wild-type BRAF [122]. There are some early data showing promising results in metastatic thyroid cancer with known BRAF V600E mutations with the BRAF mutant inhibitor dabrafenib, with 13/14 patients showing some response to treatment [123]. Currently, BRAF mutational status has some prognostic value, but targeting these mutations therapeutically has not been widely assessed. Activating mutations of PIK3CA are found in thyroid cancer as well, though rarely in papillary thyroid cancer, and most commonly in anaplastic thyroid cancer with one series identifying up to 23% of tumors with activating mutations [124]. At this time PIK3CA mutations are associated with more aggressive thyroid cancer, but it is unclear if PIK3CA mutations in more well-differentiated tumors confers a worse prognosis.

Conclusion

Over the past four decades, as we have learned more about the mechanism of tumorigenesis in the head and neck, this has allowed for more opportunity to identify biomarkers for both prognostication and prediction of therapeutic benefit of specific treatments (Table 9.1). The current trend of increasing rates of HPV-associated oropharyngeal SCC, and their overall superior outcomes compared to HPV-negative patients, further stresses the importance of using biomarkers when clinically staging patients. Recently, HPV-positive disease has been classified separately by AJCC as it has become clear that it is its own disease entity [125]. Within the medical oncology field, there has been wide adoption of PD-1 inhibitors in patients with

Table 9.1 Summary of biomarkers in HNSCC

Biomarker	Prognostic	Predictive	Screening	Used regularly in clinical settings?
HNSCC				
P53	Yes	Unclear	No	No
DDR	No	No	No	No
Apoptosis	No	No	No	No
HPV	Yes	Yes	No	Yes
EGFR	Yes	Yes	No	No
PI3K/PIK3CA	Yes	No	No	No
EBV	Yes	Yes	Yes	Yes
PD-1/PD-L1	Yes	Yes	No	Yes
Hypoxia	Yes	Yes	No	No
Non-HNSCC				
SMARCB1	Yes	No	No	Yes
HPV	Yes	No	No	Yes
ETV6-NTRK3 translocation	No	Yes	No	Yes
Mammoglobin	No	Yes	No	Yes
Her2	Yes	No	No	Yes
Androgen receptor	Yes	Yes	No	Yes
Thyroglobulin	Yes	No	No	Yes
RET	No	No	No	No
TRK	No	No	No	No
BRAF	Yes	Unclear	No	No
PI3K/PIK3CA	Unclear	No	No	No

metastatic HNSCC. Looking forward, utilization of blood-based biomarkers will likely become more universal as the cost and availability of these tests make them more readily accessible to monitor for disease recurrence and treatment response.

Conflict of Interest Statement for Bhishamjit Chera and Gaorav Gupta: Coinventor on a patent application regarding a method for measuring tumor-derived viral nucleic acids in blood samples, which is owned by the University of North Carolina at Chapel Hill and licensed to Naveris (Natick, MA). Ownership interest in Naveris.

References

1. Poeta ML, Manola J, Goldwasser MA, Forastiere A, Benoit N, Califano JA, et al. TP53 mutations and survival in squamous-cell carcinoma of the head and neck. N Engl J Med. 2007;357(25):2552–61.
2. Field JK, Spandidos DA, Malliri A, Gosney JR, Yiagnisis M, Stell PM. Elevated P53 expression correlates with a history of heavy smoking in squamous cell carcinoma of the head and neck. Br J Cancer. 1991;64(3):573–7.
3. Koch WM. Clinical implications of biomarkers in head and neck cancer. Curr Oncol Rep. 1999;1(2):129–37.
4. Cancer Genome Atlas Network. Comprehensive genomic characterization of head and neck squamous cell carcinomas. Nature. 2015;517(7536):576–82.
5. Ma L, Ronai A, Riede UN, Köhler G. Clinical implication of screening p53 gene mutations in head and neck squamous cell carcinomas. J Cancer Res Clin Oncol. 1998;124(7):389–96.
6. Koch WM, Brennan JA, Zahurak M, Goodman SN, Westra WH, Schwab D, et al. p53 mutation and locoregional treatment failure in head and neck squamous cell carcinoma. J Natl Cancer Inst. 1996;88(21):1580–6.
7. Masuda M, Shinokuma A, Hirakawa N, Nakashima T, Komiyama S. Expression of bcl-2-, p53, and Ki-67 and outcome of patients with primary nasopharyngeal carcinomas following DNA-damaging treatment. Head Neck. 1998;20(7):640–4.
8. Jin YT, Kayser S, Kemp BL, Ordonez NG, Tucker SL, Clayman GL, et al. The prognostic significance of the biomarkers p21WAF1/CIP1, p53, and bcl-2 in laryngeal squamous cell carcinoma. Cancer. 1998;82(11):2159–65.
9. Nemunaitis J, Clayman G, Agarwala SS, Hrushesky W, Wells JR, Moore C, et al. Biomarkers predict p53 gene therapy efficacy in recurrent squamous cell carcinoma of the head and neck. Clin Cancer Res. 2009;15(24):7719–25.

10. Carrassa L, Damia G. DNA damage response inhibitors: mechanisms and potential applications in cancer therapy. Cancer Treat Rev. 2017;60:139–51.
11. Wang F, Zhu Y, Huang Y, McAvoy S, Johnson WB, Cheung TH, et al. Transcriptional repression of WEE1 by Kruppel-like factor 2 is involved in DNA damage-induced apoptosis. Oncogene. 2005;24(24):3875–85.
12. Sun XS, Tao Y, Le Tourneau C, Pointreau Y, Sire C, Kaminsky MC, et al. Debio 1143 and high-dose cisplatin chemoradiotherapy in high-risk locoregionally advanced squamous cell carcinoma of the head and neck: a double-blind, multicentre, randomised, phase 2 study. Lancet Oncol. 2020;21(9):1173–87.
13. Mantovani F, Banks L. The human papillomavirus E6 protein and its contribution to malignant progression. Oncogene. 2001;20(54):7874–87.
14. Helt AM, Funk JO, Galloway DA. Inactivation of both the retinoblastoma tumor suppressor and p21 by the human papillomavirus type 16 E7 oncoprotein is necessary to inhibit cell cycle arrest in human epithelial cells. J Virol. 2002;76(20):10559–68.
15. Shin MK, Balsitis S, Brake T, Lambert PF. Human papillomavirus E7 oncoprotein overrides the tumor suppressor activity of p21Cip1 in cervical carcinogenesis. Cancer Res. 2009;69(14):5656–63.
16. Zerfass-Thome K, Zwerschke W, Mannhardt B, Tindle R, Botz JW, Jansen-Dürr P. Inactivation of the cdk inhibitor p27KIP1 by the human papillomavirus type 16 E7 oncoprotein. Oncogene. 1996;13(11):2323–30.
17. Rischin D, Young RJ, Fisher R, Fox SB, Le QT, Peters LJ, et al. Prognostic significance of p16INK4A and human papillomavirus in patients with oropharyngeal cancer treated on TROG 02.02 phase III trial. J Clin Oncol. 2010;28(27):4142–8.
18. Syrjänen K, Syrjänen S, Lamberg M, Pyrhönen S, Nuutinen J. Morphological and immunohistochemical evidence suggesting human papillomavirus (HPV) involvement in oral squamous cell carcinogenesis. Int J Oral Surg. 1983;12(6):418–24.
19. Löning T, Ikenberg H, Becker J, Gissmann L, Hoepfer I, zur Hausen H. Analysis of oral papillomas, leukoplakias, and invasive carcinomas for human papillomavirus type related DNA. J Invest Dermatol. 1985;84(5):417–20.
20. Gillison ML, Koch WM, Capone RB, Spafford M, Westra WH, Wu L, et al. Evidence for a causal association between human papillomavirus and a subset of head and neck cancers. J Natl Cancer Inst. 2000;92(9):709–20.
21. Capone RB, Pai SI, Koch WM, Gillison ML, Danish HN, Westra WH, et al. Detection and quantitation of human papillomavirus (HPV) DNA in the sera of patients with HPV-associated head and neck squamous cell carcinoma. Clin Cancer Res. 2000;6(11):4171–5.
22. Kreimer AR, Clifford GM, Boyle P, Franceschi S. Human papillomavirus types in head and neck squamous cell carcinomas worldwide: a systematic review. Cancer Epidemiol Biomark Prev. 2005;14(2):467–75.
23. Mehanna H, Beech T, Nicholson T, El-Hariry I, McConkey C, Paleri V, et al. Prevalence of human papillomavirus in oropharyngeal and nonoropharyngeal head and neck cancer--systematic review and meta-analysis of trends by time and region. Head Neck. 2013;35(5):747–55.
24. Fakhry C, Westra WH, Li S, Cmelak A, Ridge JA, Pinto H, et al. Improved survival of patients with human papillomavirus-positive head and neck squamous cell carcinoma in a prospective clinical trial. J Natl Cancer Inst. 2008;100(4):261–9.
25. Ang KK, Harris J, Wheeler R, Weber R, Rosenthal DI, Nguyen-Tân PF, et al. Human papillomavirus and survival of patients with oropharyngeal cancer. N Engl J Med. 2010;363(1):24–35.
26. Wang H, Sun R, Lin H, Hu WH. P16INK4A as a surrogate biomarker for human papillomavirus-associated oropharyngeal carcinoma: consideration of some aspects. Cancer Sci. 2013;104(12):1553–9.
27. Lindquist D, Romanitan M, Hammarstedt L, Näsman A, Dahlstrand H, Lindholm J, et al. Human papillomavirus is a favourable prognostic factor in tonsillar cancer and its oncogenic role is supported by the expression of E6 and E7. Mol Oncol. 2007;1(3):350–5.
28. Fakhry C, Lacchetti C, Rooper LM, Jordan RC, Rischin D, Sturgis EM, et al. Human papillomavirus testing in head and neck carcinomas: ASCO clinical practice guideline endorsement of the College of American Pathologists guideline. J Clin Oncol. 2018;36(31):3152–61.
29. Fakhry C, Zhang Q, Nguyen-Tan PF, Rosenthal D, El-Naggar A, Garden AS, et al. Human papillomavirus and overall survival after progression of oropharyngeal squamous cell carcinoma. J Clin Oncol. 2014;32(30):3365–73.
30. Chera BS, Amdur RJ, Tepper J, Qaqish B, Green R, Aumer SL, et al. Phase 2 trial of de-intensified chemoradiation therapy for favorable-risk human papillomavirus-associated oropharyngeal squamous cell carcinoma. Int J Radiat Oncol Biol Phys. 2015;93(5):976–85.
31. Chera BS, Amdur RJ, Tepper JE, Tan X, Weiss J, Grilley-Olson JE, et al. Mature results of a prospective study of deintensified chemoradiotherapy for low-risk human papillomavirus-associated oropharyngeal squamous cell carcinoma. Cancer. 2018;124(11):2347–54.
32. Mehanna H, Robinson M, Hartley A, Kong A, Foran B, Fulton-Lieuw T, et al. Radiotherapy plus cisplatin or cetuximab in low-risk human papillomavirus-positive oropharyngeal cancer (De-ESCALaTE HPV): an open-label randomised controlled phase 3 trial. Lancet. 2019;393(10166):51–60.
33. Hanna GJ, Supplee JG, Kuang Y, Mahmood U, Lau CJ, Haddad RI, et al. Plasma HPV cell-free DNA monitoring in advanced HPV-associated oropharyngeal cancer. Ann Oncol. 2018;29(9):1980–6.

34. Chera BS, Kumar S, Beaty BT, Marron D, Jefferys S, Green R, et al. Rapid clearance profile of plasma circulating tumor HPV type 16 DNA during chemoradiotherapy correlates with disease control in HPV-associated oropharyngeal cancer. Clin Cancer Res. 2019;25(15):4682–90.
35. Chera BS, Kumar S, Shen C, Amdur R, Dagan R, Green R, et al. Plasma circulating tumor HPV DNA for the surveillance of cancer recurrence in HPV-associated oropharyngeal cancer. J Clin Oncol. 2020;38(10):1050–8.
36. Cao H, Banh A, Kwok S, Shi X, Wu S, Krakow T, et al. Quantitation of human papillomavirus DNA in plasma of oropharyngeal carcinoma patients. Int J Radiat Oncol Biol Phys. 2012;82(3):e351–8.
37. Dahlstrom KR, Li G, Hussey CS, Vo JT, Wei Q, Zhao C, et al. Circulating human papillomavirus DNA as a marker for disease extent and recurrence among patients with oropharyngeal cancer. Cancer. 2015;121(19):3455–64.
38. Hilke FJ, Muyas F, Admard J, Kootz B, Nann D, Welz S, et al. Dynamics of cell-free tumour DNA correlate with treatment response of head and neck cancer patients receiving radiochemotherapy. Radiother Oncol. 2020;151:182–9.
39. Egyud M, Sridhar P, Devaiah A, Yamada E, Saunders S, Ståhlberg A, et al. Plasma circulating tumor DNA as a potential tool for disease monitoring in head and neck cancer. Head Neck. 2019;41(5):1351–8.
40. Parfenov M, Pedamallu CS, Gehlenborg N, Freeman SS, Danilova L, Bristow CA, et al. Characterization of HPV and host genome interactions in primary head and neck cancers. Proc Natl Acad Sci U S A. 2014;111(43):15544–9.
41. Nulton TJ, Kim NK, DiNardo LJ, Morgan IM, Windle B. Patients with integrated HPV16 in head and neck cancer show poor survival. Oral Oncol. 2018;80:52–5.
42. Pinatti LM, Sinha HN, Brummel CV, Goudsmit CM, Geddes TJ, Wilson GD, et al. Association of human papillomavirus integration with better patient outcomes in oropharyngeal squamous cell carcinoma. Head Neck. 2021;43(2):544–57.
43. Gillison ML, Chaturvedi AK, Anderson WF, Fakhry C. Epidemiology of human papillomavirus-positive head and neck squamous cell carcinoma. J Clin Oncol. 2015;33(29):3235–42.
44. Ishitoya J, Toriyama M, Oguchi N, Kitamura K, Ohshima M, Asano K, et al. Gene amplification and overexpression of EGF receptor in squamous cell carcinomas of the head and neck. Br J Cancer. 1989;59(4):559–62.
45. Grandis JR, Tweardy DJ. Elevated levels of transforming growth factor alpha and epidermal growth factor receptor messenger RNA are early markers of carcinogenesis in head and neck cancer. Cancer Res. 1993;53(15):3579–84.
46. Rubin Grandis J, Melhem MF, Gooding WE, Day R, Holst VA, Wagener MM, et al. Levels of TGF-alpha and EGFR protein in head and neck squamous cell carcinoma and patient survival. J Natl Cancer Inst. 1998;90(11):824–32.
47. Chung CH, Ely K, McGavran L, Varella-Garcia M, Parker J, Parker N, et al. Increased epidermal growth factor receptor gene copy number is associated with poor prognosis in head and neck squamous cell carcinomas. J Clin Oncol. 2006;24(25):4170–6.
48. Dische S, Saunders M, Barrett A, Harvey A, Gibson D, Parmar M. A randomised multicentre trial of CHART versus conventional radiotherapy in head and neck cancer. Radiother Oncol. 1997;44(2):123–36.
49. Bentzen SM, Atasoy BM, Daley FM, Dische S, Richman PI, Saunders MI, et al. Epidermal growth factor receptor expression in pretreatment biopsies from head and neck squamous cell carcinoma as a predictive factor for a benefit from accelerated radiation therapy in a randomized controlled trial. J Clin Oncol. 2005;23(24):5560–7.
50. Ang KK, Berkey BA, Tu X, Zhang HZ, Katz R, Hammond EH, et al. Impact of epidermal growth factor receptor expression on survival and pattern of relapse in patients with advanced head and neck carcinoma. Cancer Res. 2002;62(24):7350–6.
51. Chua DT, Nicholls JM, Sham JS, Au GK. Prognostic value of epidermal growth factor receptor expression in patients with advanced stage nasopharyngeal carcinoma treated with induction chemotherapy and radiotherapy. Int J Radiat Oncol Biol Phys. 2004;59(1):11–20.
52. Keren S, Shoude Z, Lu Z, Beibei Y. Role of EGFR as a prognostic factor for survival in head and neck cancer: a meta-analysis. Tumour Biol. 2014;35(3):2285–95.
53. Won HS, Jung CK, Chun SH, Kang JH, Kim YS, Sun DI, et al. Difference in expression of EGFR, pAkt, and PTEN between oropharyngeal and oral cavity squamous cell carcinoma. Oral Oncol. 2012;48(10):985–90.
54. Kumar B, Cordell KG, Lee JS, Worden FP, Prince ME, Tran HH, et al. EGFR, p16, HPV Titer, Bcl-xL and p53, sex, and smoking as indicators of response to therapy and survival in oropharyngeal cancer. J Clin Oncol. 2008;26(19):3128–37.
55. Hong A, Dobbins T, Lee CS, Jones D, Jackson E, Clark J, et al. Relationships between epidermal growth factor receptor expression and human papillomavirus status as markers of prognosis in oropharyngeal cancer. Eur J Cancer. 2010;46(11):2088–96.
56. Reimers N, Kasper HU, Weissenborn SJ, Stützer H, Preuss SF, Hoffmann TK, et al. Combined analysis of HPV-DNA, p16 and EGFR expression to predict prognosis in oropharyngeal cancer. Int J Cancer. 2007;120(8):1731–8.
57. Kong CS, Narasimhan B, Cao H, Kwok S, Erickson JP, Koong A, et al. The relationship between human papillomavirus status and other molecular prognostic markers in head and neck squamous cell carcinomas. Int J Radiat Oncol Biol Phys. 2009;74(2):553–61.

58. Gillison ML, Trotti AM, Harris J, Eisbruch A, Harari PM, Adelstein DJ, et al. Radiotherapy plus cetuximab or cisplatin in human papillomavirus-positive oropharyngeal cancer (NRG Oncology RTOG 1016): a randomised, multicentre, non-inferiority trial. Lancet. 2019;393(10166):40–50.
59. Suda T, Hama T, Kondo S, Yuza Y, Yoshikawa M, Urashima M, et al. Copy number amplification of the PIK3CA gene is associated with poor prognosis in non-lymph node metastatic head and neck squamous cell carcinoma. BMC Cancer. 2012;12:416.
60. Cohen Y, Goldenberg-Cohen N, Shalmon B, Shani T, Oren S, Amariglio N, et al. Mutational analysis of PTEN/PIK3CA/AKT pathway in oral squamous cell carcinoma. Oral Oncol. 2011;47(10):946–50.
61. Murugan AK, Hong NT, Fukui Y, Munirajan AK, Tsuchida N. Oncogenic mutations of the PIK3CA gene in head and neck squamous cell carcinomas. Int J Oncol. 2008;32(1):101–11.
62. Fenic I, Steger K, Gruber C, Arens C, Woenckhaus J. Analysis of PIK3CA and Akt/protein kinase B in head and neck squamous cell carcinoma. Oncol Rep. 2007;18(1):253–9.
63. Gillison ML, Akagi K, Xiao W, Jiang B, Pickard RKL, Li J, et al. Human papillomavirus and the landscape of secondary genetic alterations in oral cancers. Genome Res. 2019;29(1):1–17.
64. Stransky N, Egloff AM, Tward AD, Kostic AD, Cibulskis K, Sivachenko A, et al. The mutational landscape of head and neck squamous cell carcinoma. Science. 2011;333(6046):1157–60.
65. Nichols AC, Palma DA, Chow W, Tan S, Rajakumar C, Rizzo G, et al. High frequency of activating PIK3CA mutations in human papillomavirus-positive oropharyngeal cancer. JAMA Otolaryngol Head Neck Surg. 2013;139(6):617–22.
66. Chiosea SI, Grandis JR, Lui VW, Diergaarde B, Maxwell JH, Ferris RL, et al. PIK3CA, HRAS and PTEN in human papillomavirus positive oropharyngeal squamous cell carcinoma. BMC Cancer. 2013;13:602.
67. Bratman SV, Bruce JP, O'Sullivan B, Pugh TJ, Xu W, Yip KW, et al. Human papillomavirus genotype association with survival in head and neck squamous cell carcinoma. JAMA Oncol. 2016;2(6):823–6.
68. Sewell A, Brown B, Biktasova A, Mills GB, Lu Y, Tyson DR, et al. Reverse-phase protein array profiling of oropharyngeal cancer and significance of PIK3CA mutations in HPV-associated head and neck cancer. Clin Cancer Res. 2014;20(9):2300–11.
69. Lo YM, Chan LY, Lo KW, Leung SF, Zhang J, Chan AT, et al. Quantitative analysis of cell-free Epstein-Barr virus DNA in plasma of patients with nasopharyngeal carcinoma. Cancer Res. 1999;59(6):1188–91.
70. Tan EL, Looi LM, Sam CK. Evaluation of plasma Epstein-Barr virus DNA load as a prognostic marker for nasopharyngeal carcinoma. Singap Med J. 2006;47(9):803–7.
71. Shao JY, Zhang Y, Li YH, Gao HY, Feng HX, Wu QL, et al. Comparison of Epstein-Barr virus DNA level in plasma, peripheral blood cell and tumor tissue in nasopharyngeal carcinoma. Anticancer Res. 2004;24(6):4059–66.
72. Nemerow GR, Wolfert R, McNaughton ME, Cooper NR. Identification and characterization of the Epstein-Barr virus receptor on human B lymphocytes and its relationship to the C3d complement receptor (CR2). J Virol. 1985;55(2):347–51.
73. Tsao SW, Tsang CM, Pang PS, Zhang G, Chen H, Lo KW. The biology of EBV infection in human epithelial cells. Semin Cancer Biol. 2012;22(2):137–43.
74. Min H, Hong M, Ma J, Zhang E, Zheng Q, Zhang J, et al. A new staging system for nasopharyngeal carcinoma in China. Int J Radiat Oncol Biol Phys. 1994;30(5):1037–42.
75. Lo YM, Chan AT, Chan LY, Leung SF, Lam CW, Huang DP, et al. Molecular prognostication of nasopharyngeal carcinoma by quantitative analysis of circulating Epstein-Barr virus DNA. Cancer Res. 2000;60(24):6878–81.
76. Chan AT, Lo YM, Zee B, Chan LY, Ma BB, Leung SF, et al. Plasma Epstein-Barr virus DNA and residual disease after radiotherapy for undifferentiated nasopharyngeal carcinoma. J Natl Cancer Inst. 2002;94(21):1614–9.
77. Lin JC, Wang WY, Chen KY, Wei YH, Liang WM, Jan JS, et al. Quantification of plasma Epstein-Barr virus DNA in patients with advanced nasopharyngeal carcinoma. N Engl J Med. 2004;350(24):2461–70.
78. Tzellos S, Farrell PJ. Epstein-barr virus sequence variation-biology and disease. Pathogens. 2012;1(2):156–74.
79. Chan KCA, Woo JKS, King A, Zee BCY, Lam WKJ, Chan SL, et al. Analysis of plasma Epstein-Barr virus DNA to screen for nasopharyngeal cancer. N Engl J Med. 2017;377(6):513–22.
80. Chow LQM, Haddad R, Gupta S, Mahipal A, Mehra R, Tahara M, et al. Antitumor activity of pembrolizumab in biomarker-unselected patients with recurrent and/or metastatic head and neck squamous cell carcinoma: results from the phase Ib KEYNOTE-012 expansion cohort. J Clin Oncol. 2016;34(32):3838–45.
81. Mehra R, Seiwert TY, Gupta S, Weiss J, Gluck I, Eder JP, et al. Efficacy and safety of pembrolizumab in recurrent/metastatic head and neck squamous cell carcinoma: pooled analyses after long-term follow-up in KEYNOTE-012. Br J Cancer. 2018;119(2):153–9.
82. Bauml J, Seiwert TY, Pfister DG, Worden F, Liu SV, Gilbert J, et al. Pembrolizumab for platinum- and cetuximab-refractory head and neck cancer: results from a single-arm, phase II study. J Clin Oncol. 2017;35(14):1542–9.
83. Cohen EEW, Soulières D, Le Tourneau C, Dinis J, Licitra L, Ahn MJ, et al. Pembrolizumab versus methotrexate, docetaxel, or cetuximab for recurrent or metastatic head-and-neck squamous cell carci-

noma (KEYNOTE-040): a randomised, open-label, phase 3 study. Lancet. 2019;393(10167):156–67.
84. Ferris RL, Blumenschein G Jr, Fayette J, Guigay J, Colevas AD, Licitra L, et al. Nivolumab for recurrent squamous-cell carcinoma of the head and neck. N Engl J Med. 2016;375(19):1856–67.
85. Burtness B, Harrington KJ, Greil R, Soulières D, Tahara M, de Castro G Jr, et al. Pembrolizumab alone or with chemotherapy versus cetuximab with chemotherapy for recurrent or metastatic squamous cell carcinoma of the head and neck (KEYNOTE-048): a randomised, open-label, phase 3 study. Lancet. 2019;394(10212):1915–28.
86. Cohen EE, Ferris RL, Psyrri A, Haddad R, Tahara M, Bourhis J, et al. 910O Primary results of the phase III JAVELIN head & neck 100 trial: avelumab plus chemoradiotherapy (CRT) followed by avelumab maintenance vs CRT in patients with locally advanced squamous cell carcinoma of the head and neck (LA SCCHN). Ann Oncol. 2020;31:S658.
87. Lee NY, Ferris RL, Psyrri A, Haddad RI, Tahara M, Bourhis J, et al. Avelumab plus standard-of-care chemoradiotherapy versus chemoradiotherapy alone in patients with locally advanced squamous cell carcinoma of the head and neck: a randomised, double-blind, placebo-controlled, multicentre, phase 3 trial. Lancet Oncol. 2021;22(4):450–62.
88. Ferris RL, Haddad R, Even C, Tahara M, Dvorkin M, Ciuleanu TE, et al. Durvalumab with or without tremelimumab in patients with recurrent or metastatic head and neck squamous cell carcinoma: EAGLE, a randomized, open-label phase III study. Ann Oncol. 2020;31(7):942–50.
89. AstraZeneca. Update on KESTREL Phase III trial of Imfinzi with or without tremelimumab in the 1st-line treatment of recurrent or metastatic head and neck cancer [press release]. 2021. https://www.astrazeneca.com/content/astraz/media-centre/press-releases/2021/update-on-kestrel-phase-iii-trial-for-imfinzi.html.
90. Nordsmark M, Bentzen SM, Rudat V, Brizel D, Lartigau E, Stadler P, et al. Prognostic value of tumor oxygenation in 397 head and neck tumors after primary radiation therapy. An international multi-center study. Radiother Oncol. 2005;77(1):18–24.
91. Rühle A, Grosu AL, Wiedenmann N, Stoian R, Haehl E, Zamboglou C, et al. Immunohistochemistry-based hypoxia-immune prognostic classifier for head-and-neck cancer patients undergoing chemoradiation - post-hoc analysis from a prospective imaging trial. Radiother Oncol. 2021;159:75–81.
92. Patel U, Pandey M, Kannan S, Samant TA, Gera P, Mittal N, et al. Prognostic and predictive significance of nuclear HIF1α expression in locally advanced HNSCC patients treated with chemoradiation with or without nimotuzumab. Br J Cancer. 2020;123(12):1757–66.
93. Graves EE, Hicks RJ, Binns D, Bressel M, Le QT, Peters L, et al. Quantitative and qualitative analysis of [(18)F]FDG and [(18)F]FAZA positron emission tomography of head and neck cancers and associations with HPV status and treatment outcome. Eur J Nucl Med Mol Imaging. 2016;43(4):617–25.
94. Trinkaus ME, Hicks RJ, Young RJ, Peters LJ, Solomon B, Bressel M, et al. Correlation of p16 status, hypoxic imaging using [18F]-misonidazole positron emission tomography and outcome in patients with loco-regionally advanced head and neck cancer. J Med Imaging Radiat Oncol. 2014;58(1):89–97.
95. Rischin D, Peters L, Hicks R, Hughes P, Fisher R, Hart R, et al. Phase I trial of concurrent tirapazamine, cisplatin, and radiotherapy in patients with advanced head and neck cancer. J Clin Oncol. 2001;19(2):535–42.
96. Rischin D, Peters L, Fisher R, Macann A, Denham J, Poulsen M, et al. Tirapazamine, cisplatin, and radiation versus fluorouracil, cisplatin, and radiation in patients with locally advanced head and neck cancer: a randomized phase II trial of the Trans-Tasman Radiation Oncology Group (TROG 98.02). J Clin Oncol. 2005;23(1):79–87.
97. Rischin D, Peters LJ, O'Sullivan B, Giralt J, Fisher R, Yuen K, et al. Tirapazamine, cisplatin, and radiation versus cisplatin and radiation for advanced squamous cell carcinoma of the head and neck (TROG 02.02, HeadSTART): a phase III trial of the Trans-Tasman Radiation Oncology Group. J Clin Oncol. 2010;28(18):2989–95.
98. Stelow EB, Bellizzi AM, Taneja K, Mills SE, Legallo RD, Kutok JL, et al. NUT rearrangement in undifferentiated carcinomas of the upper aerodigestive tract. Am J Surg Pathol. 2008;32(6):828–34.
99. Bishop JA, Ogawa T, Stelow EB, Moskaluk CA, Koch WM, Pai SI, et al. Human papillomavirus-related carcinoma with adenoid cystic-like features: a peculiar variant of head and neck cancer restricted to the sinonasal tract. Am J Surg Pathol. 2013;37(6):836–44.
100. Bishop JA, Guo TW, Smith DF, Wang H, Ogawa T, Pai SI, et al. Human papillomavirus-related carcinomas of the sinonasal tract. Am J Surg Pathol. 2013;37(2):185–92.
101. Agaimy A, Hartmann A, Antonescu CR, Chiosea SI, El-Mofty SK, Geddert H, et al. SMARCB1 (INI-1)-deficient sinonasal carcinoma: a series of 39 cases expanding the morphologic and clinicopathologic spectrum of a recently described entity. Am J Surg Pathol. 2017;41(4):458–71.
102. Chen CC, Yang SF. Human papillomavirus-related carcinoma with adenoid cystic-like features of the sinonasal tract (also known as human papillomavirus-related multiphenotypic sinonasal carcinoma). Arch Pathol Lab Med. 2019;143(11):1420–4.
103. Bishop JA, Andreasen S, Hang JF, Bullock MJ, Chen TY, Franchi A, et al. HPV-related multiphenotypic sinonasal carcinoma: an expanded series of 49 cases of the tumor formerly known as HPV-related carcinoma with adenoid cystic carcinoma-like features. Am J Surg Pathol. 2017;41(12):1690–701.

104. Albert CM, Davis JL, Federman N, Casanova M, Laetsch TW. TRK fusion cancers in children: a clinical review and recommendations for screening. J Clin Oncol. 2019;37(6):513–24.
105. Skálová A, Vanecek T, Sima R, Laco J, Weinreb I, Perez-Ordonez B, et al. Mammary analogue secretory carcinoma of salivary glands, containing the ETV6-NTRK3 fusion gene: a hitherto undescribed salivary gland tumor entity. Am J Surg Pathol. 2010;34(5):599–608.
106. Bishop JA, Yonescu R, Batista D, Begum S, Eisele DW, Westra WH. Utility of mammaglobin immunohistochemistry as a proxy marker for the ETV6-NTRK3 translocation in the diagnosis of salivary mammary analogue secretory carcinoma. Hum Pathol. 2013;44(10):1982–8.
107. Drilon A, Laetsch TW, Kummar S, DuBois SG, Lassen UN, Demetri GD, et al. Efficacy of larotrectinib in TRK fusion-positive cancers in adults and children. N Engl J Med. 2018;378(8):731–9.
108. Doebele RC, Drilon A, Paz-Ares L, Siena S, Shaw AT, Farago AF, et al. Entrectinib in patients with advanced or metastatic NTRK fusion-positive solid tumours: integrated analysis of three phase 1-2 trials. Lancet Oncol. 2020;21(2):271–82.
109. Hong DS, DuBois SG, Kummar S, Farago AF, Albert CM, Rohrberg KS, et al. Larotrectinib in patients with TRK fusion-positive solid tumours: a pooled analysis of three phase 1/2 clinical trials. Lancet Oncol. 2020;21(4):531–40.
110. Nabili V, Tan JW, Bhuta S, Sercarz JA, Head CS. Salivary duct carcinoma: a clinical and histologic review with implications for trastuzumab therapy. Head Neck. 2007;29(10):907–12.
111. Fan CY, Melhem MF, Hosal AS, Grandis JR, Barnes EL. Expression of androgen receptor, epidermal growth factor receptor, and transforming growth factor alpha in salivary duct carcinoma. Arch Otolaryngol Head Neck Surg. 2001;127(9):1075–9.
112. Takahashi H, Tada Y, Saotome T, Akazawa K, Ojiri H, Fushimi C, et al. Phase II trial of trastuzumab and docetaxel in patients with human epidermal growth factor receptor 2-positive salivary duct carcinoma. J Clin Oncol. 2019;37(2):125–34.
113. Szewczyk M, Marszałek A, Sygut J, Golusiński P, Golusiński W. Prognostic markers in salivary gland cancer and their impact on survival. Head Neck. 2019;41(9):3338–47.
114. Fushimi C, Tada Y, Takahashi H, Nagao T, Ojiri H, Masubuchi T, et al. A prospective phase II study of combined androgen blockade in patients with androgen receptor-positive metastatic or locally advanced unresectable salivary gland carcinoma. Ann Oncol. 2018;29(4):979–84.
115. Pelttari H, Laitinen K, Schalin-Jäntti C, Välimäki MJ. Long-term outcome of 495 TNM stage I or II patients with differentiated thyroid carcinoma followed up with neck ultrasonography and thyroglobulin measurements on T4 treatment. Clin Endocrinol. 2008;69(2):323–31.
116. Kloos RT, Mazzaferri EL. A single recombinant human thyrotropin-stimulated serum thyroglobulin measurement predicts differentiated thyroid carcinoma metastases three to five years later. J Clin Endocrinol Metab. 2005;90(9):5047–57.
117. Nikiforov YE, Rowland JM, Bove KE, Monforte-Munoz H, Fagin JA. Distinct pattern of ret oncogene rearrangements in morphological variants of radiation-induced and sporadic thyroid papillary carcinomas in children. Cancer Res. 1997;57(9):1690–4.
118. Stoleru B, Popescu AM, Tache DE, Neamtu OM, Emami G, Tataranu LG, et al. Tropomyosin-receptor-kinases signaling in the nervous system. Maedica (Bucur). 2013;8(1):43–8.
119. Bongarzone I, Vigneri P, Mariani L, Collini P, Pilotti S, Pierotti MA. RET/NTRK1 rearrangements in thyroid gland tumors of the papillary carcinoma family: correlation with clinicopathological features. Clin Cancer Res. 1998;4(1):223–8.
120. Cohen Y, Xing M, Mambo E, Guo Z, Wu G, Trink B, et al. BRAF mutation in papillary thyroid carcinoma. J Natl Cancer Inst. 2003;95(8):625–7.
121. Xing M, Alzahrani AS, Carson KA, Viola D, Elisei R, Bendlova B, et al. Association between BRAF V600E mutation and mortality in patients with papillary thyroid cancer. JAMA. 2013;309(14):1493–501.
122. Abubaker J, Jehan Z, Bavi P, Sultana M, Al-Harbi S, Ibrahim M, et al. Clinicopathological analysis of papillary thyroid cancer with PIK3CA alterations in a Middle Eastern population. J Clin Endocrinol Metab. 2008;93(2):611–8.
123. Falchook GS, Millward M, Hong D, Naing A, Piha-Paul S, Waguespack SG, et al. BRAF inhibitor dabrafenib in patients with metastatic BRAF-mutant thyroid cancer. Thyroid. 2015;25(1):71–7.
124. García-Rostán G, Costa AM, Pereira-Castro I, Salvatore G, Hernandez R, Hermsem MJ, et al. Mutation of the PIK3CA gene in anaplastic thyroid cancer. Cancer Res. 2005;65(22):10199–207.
125. Amin MB, Greene FL, Edge SB, Compton CC, Gershenwald JE, Brookland RK, et al. The Eighth Edition AJCC cancer staging manual: continuing to build a bridge from a population-based to a more "personalized" approach to cancer staging. CA Cancer J Clin. 2017;67(2):93–9.

Reirradiation for Head and Neck Cancer

10

Vladimir Avkshtol and David J. Sher

Key Points

- Locoregional failure is reported in 10–50% of head and neck patients, and second primary tumors are the second leading cause of death of head and neck cancer patients. Given the limitations from the first course of treatment, there are no consensus management guidelines for treating locoregional failure or second primary disease.
- Proper patient selection for reirradiation is vital due to the considerable acute and late toxicity that can be associated with the treatment. Generally, patients who are optimal candidates for reirradiation present with low volume recurrence or new primary disease, longer duration from initial radiation treatment, and intact performance status and organ function. Since a limited number of patients meet these criteria, the toxicity of disease progression versus toxicity of re-irradiation must be carefully assessed for each case.
- Patients who are candidates for complete surgical resection should undergo surgery, which is consistently associated with improved outcomes. Post-operative reirradiation of the tumor bed should be considered in cases with high-risk factors, especially in those with positive margins or ENE. Recommended post-operative reirradiation dose is typically 60 Gy in 2 Gy per fraction. Elective nodal coverage is not recommended, even when lymph node metastases are present. Second infield recurrences are still common even with post-operative reirradiation.
- Most patients with recurrent or new primary disease are not surgical candidates and must be considered for definitive reirradiation. Radiation with IMRT to equivalent dose of ≥66 Gy in 2 Gy per fraction (Biologically effective dose [BED] with alpha/beta = 10 of 79.2 Gy) is recommended. Either conventional fractionation or hyperfractionation is appropriate. Elective nodal irradiation is not typically recommended. Concurrent chemotherapy should be utilized when feasible, preferably with an agent with

V. Avkshtol · D. J. Sher (✉)
Department of Radiation Oncology, University of Texas Southwestern Medical Center, Dallas, TX, USA
e-mail: david.sher@utsw.edu

R. A. Chandra, R. J. Li (eds.), *Multidisciplinary Management of Head and Neck Cancer*,
https://doi.org/10.1007/978-3-031-05973-5_10

which the patient has not been treated previously.
- Stereotactic ablative reirradiation therapy is a strong consideration in patients whose disease is less than 25 cm^3 in size, with less than 180° carotid artery tumor encasement, no skin invasion, and non-larynx or hypopharynx primary site. The general dose recommendation is 40 Gy in 5 fractions on non-consecutive days for small tumors, and 44 Gy in 5 fractions for bigger tumors (>25 cm^3). Concurrent systemic therapies appear to be well tolerated but is an area of open investigation.
- Early charged ion therapy reirradiation data with protons or carbon ion are promising, but still too limited to make a firm recommendation. Brachytherapy can be a useful treatment technique in cases where external beam radiation is contraindicated but must be performed in centers with appropriate expertise.

Introduction

Despite advances in treatment of head and neck cancer, local-regional failure (LRF) is reported in 10–50% of patients [1–5]. Survivors are also at risk of developing second primary tumors, which are the second leading cause of death in head and neck squamous cell carcinoma patients, accounting for around 21% of all deaths [6]. To date, no consensus management strategy has been developed for these patients. Recurrent disease or second primary tumors occurring in the previous radiation field present difficult tradeoffs for treating physicians because of the effects of prior treatment on normal tissue, treatment resistance development, and the typical infiltrative nature of recurrent disease [7, 8]. Surgical resection is the treatment strategy of choice, but many patients are not candidates for this option [9]. Optimal additional perioperative treatments are not known. Definitive reirradiation is an alternative but is associated with valid concerns for treatment morbidity and mortality. Contemporary techniques such as intensity-modulated radiation therapy (IMRT) have made reirradiation technically easier, and the growing data on stereotactic ablative radiation therapy (SABR), heavy ions, and targeted systemic therapies have introduced new tools in the arsenal of head and neck radiation and medical oncologists. The decision to pursue reirradiation should be made between an experienced treating physician and the patient, with consideration of prognostic factors, treatment techniques, disease outcomes, and toxicity.

Prognostic Factors and Patient Selection

It is of paramount importance to identify relevant prognostic factors in selecting patients for reirradiation due to the considerable acute and late toxicity that can be associated with the treatment. At the same time, it is important to remember that progression of disease can cause the same or worse morbidity than is reported with reirradiation. In fact, a nomogram that examined the relationship between late reirradiation toxicity and competing risks determined that the risk of disease progression or death is approximately four times the risk of reirradiation-related severe late toxicity [10].

Ward et al. published a risk-stratification model from the Multi-Institution Reirradiation (MIRI) Collaborative using recursive partitioning analysis (RPA) on 412 patients from seven institutions treated with definitive or adjuvant reirradiation using IMRT to a median dose of 60 Gy [11]. They found that three factors were most strongly associated with overall survival (OS): time between radiation treatments (<2 years versus ≥2 years), resection of recurrent disease regardless of margin status, and organ dysfunction (pre-existing feeding tube or tracheostomy with an intact larynx). The RPA had three distinct classes: patients >2 years out from the first course of radiation with resected tumors (class I), patients >2 years out with unresected tumors or ≤2 years out without organ dysfunction (class II), and patients ≤2 years out with organ dysfunction (class III). The 2-year OS was 61.9% (95% CI 51.9–73.9%), 40.0% (95% CI 33.9–47.2%); and 16.8% (95% CI 10.0–28.1%) for class I, class II, and class III, respectively. Of note, while these results were validated by other groups, some studies could not, owing perhaps to a different reirradiation population [12, 13].

The optimal time between radiation treatments for head and neck cancers is not known. As expected, data suggests that severe reirradiation toxicity decreases and local control (LC) increases with increasing time between treatments [14]. In RTOG 9610, the median OS difference was seen for patients who were retreated <1 year versus >1 year from prior radiation (5.8 months vs. 9.8 months; $p = 0.036$) and <3 years versus >3 years (7.7 months vs. 9.8 months; $p = 0.033$) [15]. The most common recommendation is that the patient should be at least 6 months out from their previous radiation treatment to be considered for reirradiation, due to both toxicity and efficacy concerns [16–19]. Recurrent gross disease that is seen less than 6 months from previous radiation therapy most likely represents a residual, radiation-resistant process and these patients should be considered for alternative treatments.

There is a trend of improved OS with patients who present for reirradiation with a second primary rather than recurrent disease [11, 15, 20]. In a retrospective analysis of 100 patients treated with reirradiation alone, the 5-year OS and LRC was 37% and 60% for second primaries versus 17% and 27% for recurrent disease [21]. This favorable outcome is likely tied to the responsiveness of a second primary to reirradiation compared to more infiltrative and characteristically radioresistant recurrent tumors [7, 8].

Tumor size or irradiation volume is also critical information in assessing patients for reirradiation. Patients with small volume disease are most likely to benefit from aggressive local-regional treatment [11–13, 16, 22–24]. Orlandi et al. found that a reirradiation volume of ≤36 cm^3 was associated with OS (5-year OS of 48.8% vs. 37.8%) [13]. Takiar et al. found that clinical target volume at reirradiation ≥50 cm^3 was associated with increased grade ≥3 toxicity (5-year grade ≥3 toxicity of 61% versus 31%) [24].

Location of the recurrence is also significant, as patients with recurrences in the nasopharynx or larynx consistently show improved survival over patients with disease in the oral cavity or hypopharynx [11–13, 24, 25]. There is particular abundant data for reirradiation of locoregionally recurrent nasopharynx cancer with evidence of long-term survivors, but the risk of lethal nasopharyngeal necrosis and hemorrhage must always be taken into account [26–28]. Baseline organ dysfunction is also associated with worse outcomes after reirradiation [11, 13, 14, 16]. In general, patients with serious sequelae from prior treatment, such as osteoradionecrosis, soft tissue necrosis, carotid artery stenosis, pharyngocutaneous fistula, radiation myelopathy, radiation skin/mucosal ulceration, should not be subjected to another definitive radiation course to the region because of risk of severe or fatal toxicity [16, 29, 30].

Patients who have received previous concurrent chemoradiation rather than radiation alone have worse disease outcomes after reirradiation, likely representing a more treatment resistant recurrent disease [9, 31]. In their prospective observational study, Goodwin et al. found that patients who previously received chemotherapy for their disease had a median survival of 8.8 months versus 26.9 months for those who did not receive previous chemotherapy ($p < 0.001$) [9].

It is generally safe to assume that patients with low volume recurrence or new primary disease, ≥2 years out from initial radiation treatment, and who have good performance status and no organ dysfunction are excellent candidates for reirradiation. However, not many patients will fit all these criteria. Each case is unique and must be assessed carefully. Patient selection must consider the patient's willingness to accept the possible toxicity that may come with reirradiation. Assessment should also include close inspection of the initial radiation plan because an early failure may be a marginal miss rather than an aggressive radioresistant disease.

Post-operative Reirradiation

Patients with local-regional recurrence who are candidates for surgery should get resection of their recurrent, radioresistant disease. Several retrospective studies suggest that patient who undergo salvage surgery have better disease outcomes than those treated with reirradiation alone [23, 32–34]. Patients undergoing surgical resection of their recurrent disease can achieve long-term disease control in the 25–40% range [20, 35]. Per a meta-analysis of 32 studies, the 5-year overall survival was approximately 39% in patients undergoing surgery, and it was highest in early larynx cancer, 83% [9]. However, even after a complete salvage resection, the risk of another recurrence is as high as 59% and adjuvant reirradiation often has to be considered [36].

The indications for post-operative radiation are controversial, with multiple factors that need to be considered to minimize toxicity and maximize benefit. The Groupe d'Etude des Tumours Tete et Cou (GETTEC) and the Groupe d'Oncologie Radiotherapie Tete et Cou (GORTEC) conducted a prospective, multicenter, randomized trial to confirm early promising retrospective results of post-operative reirradiation [19]. Patients were randomized to adjuvant chemoradiation (60 Gy in 12 weeks given with concomitant hydroxyurea and continuous infusion fluorouracil [FU]) or observation after an R0 or R1 surgical resection. The three-dimensional conformal radiotherapy was given 1 week on and 1 week off, 2 Gy per fractions, 5 days/week, for six cycles. Extranodal extension (ENE) and positive margins were seen in 26% and 29% of patients, respectively. Adjuvant chemoradiation improved both local-regional control (LRC; HR 2.73, 95% CI 1.66–4.51) and disease-free survival (DFS; 1.68 (95% CI, 1.13–2.50). There was no overall survival difference between the two arms ($p = 0.50$) and there was no subset analysis in patients who had ENE and/or positive margins. Although the radiation arm had less deaths related to local-regional recurrence, the arm had more deaths related to treatment (5 versus 0), distant metastases, and second primary. Acute grade 3/4 chemoradiation toxicity was seen in 28% of patients. The 2-year late grade 3/4 toxicity was seen in 39% of patients in the radiation arm and 10% in the observation arm ($p = 0.06$).

In a meta-analysis of 16 studies assessing patients treated with post-operative reirradiation, variable outcomes in 2-year OS (24–48% with a median of 28%) and 2-year LRC (21–100% with a median of 55%) were seen [37]. Similarly, acute grade 3/4 mucositis and/or dysphagia/pharyngitis varied from 11% to 52% (median 44%). Grade 3/4 late toxicity of fibrosis (2–44%) and feeding-tube dependency or pharynx stenosis (2–70%) also widely varied from study to study. Late treatment reported deaths were between 3% and 5%. These vastly varied outcomes were attributed to the heterogeneity of the post-operative reirradiation population and the inherent differences in their treatments. Although similar disease high-risk factors must be assessed for adjuvant reirradiation as for initial adjuvant radiation treatment, this increased toxicity must be carefully weighed against the benefit of adjuvant reirradiation.

The development of microvascular free flaps has allowed for bigger surgeries in recurrent patients that were previously limited by a lack of reconstruction options in fibrotic tissue from previous treatments. These techniques have expanded the potential patient population that could be offered salvage surgery. In addition, microvascular free flap reconstruction reduces the morbidity of post-operative radiation by introducing well-vascularized and untreated tissue and bone to the reirradiation field [38–40]. One study found that patient who had flap reconstruction experiences similar disease outcomes and acute toxicities than patients with primary closures, but had significantly less late grade 1/2 toxicities (81.4% vs. 54.1%, $p = 0.006$) and late grade 3/4 toxicities (47.5% vs. 21.6%, $p = 0.02$) [40]. When appropriate, free flap reconstruction should be encouraged for patients who will need adjuvant reirradiation.

The role of debulking outside the need of immediate palliation of symptoms is not known. There is limited data to support the idea that surgical cytoreduction reduces adjuvant radiation doses or field sizes. Smaller retrospective studies have suggested a benefit of surgical debulking prior to reirradiation [32, 41, 42]. However, in a non-randomized prospective phase II trial, 35 subjects were treated with induction chemotherapy, there was no benefit found in patients getting surgical resection prior to chemoradiation [43]. In the retrospective MIRI Collaborate analysis of 412 patients, those who had gross residual disease after surgery had similar 2-year LRF when compared to patients treated with definitive radiation (47.4% vs. 46.3%) and had worse 2-year OS than patients with negative margin [11]. Therefore, surgical cytoreduction without intent of getting a complete resection is not recommended outside of unique situations.

Reirradiation of the tumor bed should be considered in cases with high-risk factors, especially in those with positive margins or ENE [19, 44]. Routine reirradiation should not be offered in resected recurrent disease without classical risk factors for recurrence, given the high risk of acute and long-term toxicity and treatment-related death. Recommended post-operative reirradiation dose is usually 60 Gy in 2 Gy fraction, but this may change depending on the presence of high-risk factors, previous radiation dose, and time interval between treatments. Retrospective studies suggest that doses at or above 58–60 Gy are associated with improvement in LRC and even OS [32, 45, 46]. There does not appear to be any benefit in adjuvant hyperfractionation versus once daily fractionation [45]. Routine prophylactic coverage of lymph nodes is not recommended, even when lymph node metastases are present [45]. Even after post-operative reirradiation, patients with recurrent head and neck carcinoma need to be followed closely. Second in-field recurrences are still reported in up to 26–100% of the cases [44, 47, 48].

Definitive Reirradiation

Conventional Radiation

Around 53–62% of patients who present with recurrent or new primary disease are not considered surgical candidates due to poor performance status, anatomical considerations, or patient choice [11–13]. Definitive reirradiation may offer them the only option for a long-term cure. Two randomized trials, GORTEC 98-03 and RTOG 04-21, compared chemotherapy alone to chemoradiation in this patient population, but both were closed early due to poor accrual [49]. However, the RTOG has successfully completed two multi-institutional prospective phase II trials. RTOG 9610 recruited 86 patients from 1996 to 1999 and treated them with 4 weekly cycles of chemoradiation separated by 1 week of rest [15]. Each cycle consisted of 5 days of twice-daily radiation of 1.5 Gy/fraction with concurrent hydroxyurea/5 FU. The 2-year OS was 15.2% and the 5-year OS was 3.8%. A total of 19.4% of patients experience late grade 3/4 toxicities and six patients (7.6%) had a treatment-related death. RTOG 9911 recruited 105 patients from 2000 to 2003 and treated them with a similar chemoradiation regimen except the chemotherapy was switched to cisplatin and paclitaxel [18]. The 2-year OS was improved to 25.9% with a 2-year progression free survival (PFS) of 15.8%. The rate of late grade 3/4 toxicities also increased to 33.8% and there eight patients (8%) who had a treatment-related death. On the one hand, these early RTOG trials and other studies showed the feasibility of definitive reirradiation. On the other hand, they also highlighted the fact that long-term survivors were rare, late grade 3/4 toxicity could be as high as 39%, and treatment-related mortality as high as 10% [19, 50]. Since that time, radiation technology, chemotherapy techniques, patient selection including staging modalities, and toxicity management have evolved. More modern series consistently report 2-year OS above 40% with late grade 3/4 toxicity below 15% and rare treatment-related deaths [11–13, 51, 52]. In a meta-analysis of early and modern studies, pooled 2-year LC, 2-year OS, late grade ≥3 toxicity, and late grade 5 toxicity were 52%, 46%, 26%, and 3.1%, respectively [53].

Dose and Fractionation

The analysis of optimal dose or fractionation is made difficult by the many different radiation schemes used in the reirradiation setting. Radiation dose of at least 50 Gy is needed to produce substantial response rates in HNSCC [22, 32]. Many investigators and groups aim for a reirradiation dose of around 60 Gy [15, 17, 31, 32, 49, 54]. One study showed an improvement in the 2-year OS from 8% to 35% in patients who received <58 Gy versus those who received ≥58 Gy [32]. Larger, more modern analyses demonstrate that doses above 60 Gy are needed to get the full benefit of reirradiation [24, 45, 48, 55, 56]. Theoretically, radiation doses at or above original radiation treatment are needed due to the radioresistant nature that can be seen recurrent tumors [7, 21]. In an analysis of 244 patients from eight institutions who completed definitive reirradiation with IMRT, a dose of ≥66 Gy was associated with improved 2-year OS when compared to dose of 60–65.9 Gy or <60 Gy (2-year OS 49.3% versus 34.2% versus 30.4%, respectively; global: $p = 0.009$), and this association remained significant on a multivariable Cox regression analysis [45]. Improvement in local-regional control was also seen with doses of ≥66 Gy. At the same time, dose escalation was not associated with an increase in grade ≥3 late toxicity. These results may be confounded by the fact that these patients may have been selected to receive the higher dose due to factors that would predict for better disease and toxicity outcomes.

Prospective reirradiation trials have used hyperfractionation with twice-daily fractions of 1.5 Gy and conventionally fractionated radiation with a week on week off regimen [15, 18, 19]. The MIRI Collaborative also assessed fractionation in their retrospective assessment from eight institutions [45]. They found that hyperfractionation (twice-daily treatment) was not associated with improved 2-year LRF (50.3% hyperfractionation versus 46.8% daily, $p = 0.412$) or 2-year OS (42.5% versus 37.9%, $p = 0.302$) in definitive reirradiation patients. Hyperfractionation also didn't impact the incidence of late effects. Nevertheless, hyperfractionation may be preferred on first principles for reirradiation due to improved sparing of late responding issues with equal total dose. Given the patient inconvenience and the lack of benefit of accelerated fractionation with concurrent chemotherapy in the upfront setting, though, many groups have moved to once daily radiation without breaks [57, 58].

Modern IMRT allows for higher conformality and dose distribution which could lead to improved disease and toxicity outcomes. One study noted that reirradiation with IMRT led to improved 2-year local-regional PFS (52% vs. 20%, $p < 0.001$) [33]. Another study noted that late toxicity was more frequent in patients treated with three-dimensional conformal radiotherapy than with IMRT (44% and 7%, respectively; $p < 0.05$) [52]. Utilizing IMRT technique for reirradiation should be considered the gold standard.

Treatment Volume and Patterns of Failure

The need for elective nodal irradiation (ENI) is controversial in the reirradiation setting. Some advocate for ENI especially in cases where the at-risk neck has not been irradiated initially, while others believe the risk of toxicity with ENI in the reirradiation setting is too high. When assessing patterns of failure, LRF is the most common site of failure after reirradiation [59–61]. Distant metastatic failure is also common, but usually occurs with concomitant LRF. In one series, 55% of failures after reirradiation were locoregional only, 28% were locoregional and distant, and only 18% were distant alone [61]. Of the LRFs, 48.6% were in-field, 20% were out-of-field, 14.3% were marginal, 8.6% were in-field and out-of-field, 5.7% were marginal and out-of-field, and 2.9% were unknown. Another larger study echoed these results, with 42% in-field failure, 39% out-of-field failures, and only 18% marginal failures [59]. With >80% failures being either in-field or out-of-field, the data suggest that adverse disease biology rather than inadequate volumes lead to failures. In a large MIRI Collaborative, they assessed neck irradiation in 497 patients and found that ENI did not appear to improve 2-year LRF or OS in node negative or node positive patients [45]. These results are persistent even when stratified for post-operative or definitive intent reirradiation, strongly arguing for treating gross disease without elective treatment.

Systemic Therapy

Like newly diagnosed HNSCC patients, concurrent chemotherapy with radiation is traditionally preferred to radiation alone in the recurrent setting. In general, the reirradiation systemic regimen should avoid agents to which the patient has been previously exposed, but there is no consensus on the optimal systemic therapy. In a recent reirradiation meta-analysis, concurrent chemotherapy was administered to 67% (range 0–100%) of head and neck cancer patients [53]. The most common regimens were platinum based and/or cetuximab. Several retrospective studies have shown improvement in outcomes in those who receive concurrent chemotherapy [24, 56]. In a large single center retrospective analysis, 5-year LRC of 59% with concurrent chemotherapy versus 40% without ($p = 0.05$; MVA HR 0.44, 95% CI 0.24–0.79) [24]. There was no difference in 5-year OS, 24% with chemotherapy versus 18% without ($p = 0.62$). It is important to note that the no chemotherapy group may have had factors that precluded them from getting chemotherapy and these factors led them to have worse outcomes.

Combination of FU and hydroxyurea was a popular regimen in early studies [19, 32, 62]. RTOG 9610 tested this regimen, but with a lower than typical dose of FU, with 2-year and 5-year OS rates of 15% and 4%, respectively [15]. Adding cisplatin and paclitaxel to this regimen have also been reported [63, 64]. Concurrent gemcitabine, paclitaxel, and 5-FU demonstrated promising results with 5-year OS of 34.5% and LRC of 54.5%, but this came at the cost of toxicity, with 21.4% of patients developing late grade 4/5 toxicity [41]. Platinum chemotherapy regimens are also popular in the reirradiation setting. RTOG 9911 used a combination of cisplatin and paclitaxel and showed superior results to RTOG 9610 [18]. However, this benefit also appeared to come at the cost of increased toxicity despite using hematopoietic growth factors. Reports on cisplatin/FU [65], docetaxel/cisplatin [66], cisplatin alone [67], carboplatin/paclitaxel [68], carboplatin/pemetrexed [43] have also been published.

Targeted agents have also been added to help overcome the radioresistance of recurrent disease. Prospective studies looking at concurrent cetuximab with conventional reirradiation showed acceptable toxicity and tumor outcomes, with one study having 1-year OS and PFS of 44% and 33%, respectively [69–72]. Combination of cisplatin and cetuximab has also been reported with 1-year OS and recurrence free survival (RFS) of 60.4% and 34.1% [45]. In a retrospective comparison of concurrent cetuximab versus cisplatin-based chemotherapy from a single center in Germany, the disease outcomes appeared comparable, with 1-year OS of 44.4% for cetuximab versus 45.5% for cisplatin-based ($p = 0.352$) and 1-year LC 46.4% for cetuximab versus 54.2% for cisplatin based ($p = 0.625$) [73]. Cetuximab group had more grade ≥3 pain and the cisplatin-based group had more grade ≥3 hematological toxicity. Reirradiation with concurrent Bortezomib added to FU and hydroxyurea showed too much toxicity to be a viable option [74]. Trials on reirradiation with checkpoint inhibitors, such as nivolumab (NCT03317327) and pembrolizumab (NCT02289209), are ongoing.

The role of induction therapy in recurrent HNSCC setting is unknown, but it is falling out of favor due to the negative trials in the first line setting [75–77]. In a phase II trial, two cycles of induction pemetrexed and gemcitabine were followed by surgical resection, if appropriate, and/or conventional chemoradiation [43]. The 1-year OS and PFS were 43% and 20%, respectively. Patients who had any response to induction chemotherapy had improved survival ($p = 0.019$). In a large single center retrospective analysis, the utilization of induction chemotherapy was not associated with LRC, in contrast to a benefit seen with concurrent chemotherapy [24]. A response to induction chemotherapy, however, was associated with LRC and OS. The role of induction chemotherapy is thus best reserved to help predict patient outcomes and to help select patients for more appropriate therapy based on their response.

Organs at Risk (OAR)

Due to the paucity of data on HNSCC reirradiation and quite poor long-term survival, acute and late tolerance data for various head and neck organs to reirradiation is limited. Neurological structures have the most data due to the extrapolation from other disease sites [78–83]. In general, there are many variables that influence OAR tolerance: the interval between two courses, volume undergoing reirradiation, fractionation regimen, radiation technique, addition of systemic therapy, underlying organ dysfunction, and patient-related factors [24, 30, 52, 84]. Dose constraints need to be individualized on a case-by-case basis. The spinal cord should take precedence as one of the most critical OAR. Nieder et al. found that the risk of radiation myelopathy is low with BED_2 135.5 Gy when the interval is ≥6 months and the dose of each course is ≤98 Gy [82]. Wang et al. recommended brainstem reirradiation constraints of D_{max} of 39 Gy after receiving an initial D_{max} of 40 Gy in equivalent dose in 2 Gy fractions with more than 12 months between treatments [80]. Chan et al. reported no temporal lobe necrosis with a cumulative $BED_{2.5} < 150$ Gy [85]. They postulated that after a course of 70 Gy, 45.9 Gy in 30 fraction could be reasonably delivered to the temporal lobes. Chen at al. recommended a cumulative D_{max} of 95 Gy to the brachial plexus. They found that the 1-year freedom from brachial plexus-related neuropathy was 86% and 67% for those with $D_{max} < 95$ Gy and >95 Gy ($p = 0.05$), respectively. There are no established dose constraints for the carotids, although one study placed the median lifetime doses for patients who died from fatal bleeding was 126 Gy [30]. Reirradiation of the nasopharynx is particularly difficult due to concern for lethal nasopharyngeal necrosis, which was reported to be as high as 33% in a recent meta-analysis [27, 28, 86]. Yu et al. showed that female sex, necrosis before reirradiation, cumulative gross tumor volume dose ≥145.5 Gy, and recurrent tumor volume ≥25.38 cm^3 were independent risk factors for lethal nasopharyngeal necrosis and developed a model to predict lethal nasopharyngeal necrosis [26]. In general, dose to critical OARs must be kept as low as possible, but the risk of side effects must be weighed against the morbidity of disease progression [10].

Conventional Radiation Conclusion

In the definitive setting, reirradiation with IMRT to equivalent total doses in 2-Gy fractions of ≥66 Gy (Biologically effective dose [BED] with alpha/beta = 10 of 79.2 Gy) is recommended when feasible. Either conventional fractionation or hyperfractionation is appropriate. ENI is typically not recommended in the reirradiation setting. Concurrent chemotherapy should be utilized when feasible, preferably with an agent with which the patient has not been treated previously.

Stereotactic Ablative Radiation Therapy (SABR)

SABR is a highly conformal radiation technique that *may* provide a radiobiological advantage compared to conventional radiation, although it is an open question whether stereotactic doses in head and neck reirradiation are biologically more potent than conventional dosing. SABR also utilizes a steep dose gradient to reduce the treatment volume. Equally important for a compromised population is the fact that the typical 5 fraction or less SABR regimen is much more feasible given the poor prognosis in the cohort. Table 10.1 outlines selected studies on SABR reirradiation that contain at least 20 patients and are screened to avoid duplication of data. In addition, there is evidence that ablative fractions of radiation can induce an immunogenic response in the patient which may lead to improved tumor control and work synergistically with immunotherapy [99, 100]. The ongoing Phase II trial "SBRT +/− Pembrolizumab in Patients with Local-Regionally Recurrent or Second Primary Head and Neck Carcinoma (KEYSTROKE)" (NCT03546582) is evaluating this interaction by randomizing to patients receiving reirradiation with SABR to either receive pembrolizumab or not.

Table 10.1 Published reports on stereotactic ablative radiotherapy reirradiation

Study	Institution	Patients	Median FU (months)	Dose and fractions	LC	OS	Acute toxicity	Late toxicity
Voynov et al. (2006) [87]	University of Pittsburgh Cancer Institute, USA	22	19 for survivors	10–36 Gy 1–8 fx	2-years 26%	1-year 47% 2-years 22%	≥ G3: 4.5%	≥ G3: 0%
Roh et al. (2009) [88]	The Catholic University of Korea, Korea	36	17.3	18–40 Gy 3–5 fx	2-years 52.2%	1-year 52% 2-years 30.9%	≥ G3: 29.5%	≥ G3: 8.3%
Siddiqui et al. (2009) [89]	Henry Ford Health System, USA	21	7	14–48 Gy 1–8 fx	1-year 60.6%	2-years 14.3%	N/A	≥ G3: 23.8%
Kawaguchi et al. (2010) [90]	Tsurumi University, Japan	22	24	20–42 Gy 2–5 fx	2-years 45.5%	2-years 78.0% (LN negative) and 12.5% (LN positive)	≥ G3: 22.7%	≥ G3: 0%
Rwigema et al. (2011) [91]	University of Pittsburgh Cancer Institute, USA	96	14	20–42 Gy 2–5 fx	2-years LRC (40–50 Gy): 57.8% 2-years LRC (15–36Gy): 31.7%	1-year 58.9% 2-years 28.4%	≥ G3: 5.2%	≥ G3: 3.1%
Lartigau et al. (2013) [92]	Multi-institution, France	56	11.4	15–50 Gy 1–5 fx	N/A	1-year 47.5% 2-years 31%	N/A	≥ G3: 32.1%
Yazici et al. (2013) [93]	Hacettepe University, Turkey	75	N/A	30 Gy 5 fx	77.4%	1-year 42% (daily RT) and 84% (qad RT) 2-years 23% (daily RT) and 36% (qad RT)	N/A	N/A
Kress et al. (2015) [94]	Georgetown University, USA	85	17 for survivors	16–41 Gy 3–5 fx	1-year LRC 57.8 2-years LRC 28%	1-year 51.1% 2-years 24%	≥ G3: 2.4%	≥ G3: 5.9%
Vargo et al. (2015) [95]	University of Pittsburgh Cancer Institute, USA	48	18 for survivors	40–44 Gy 5 fx	1-year LRPF 37%	1-year 40%	≥ G3: 6.3%	≥ G3: 6.3%
Yamazaki et al. (2016) [96]	Multi-institutions, Japan	107	15	15–39 Gy 3–8 fx	2-years 64%	1-year 55% 2-years 35%	N/A	≥ G3: 20.6% (G5: 8.4%)
Vargo et al. (2018) [97]	Multi-institutional, USA	197	7.1	16–50 Gy 1–8 fx	LRC 43%	2-years 16.3%	≥ G3: 11.7%	≥ G3: 11.6%
Gogineni et al. (2019) [98]	Northwell Health, USA	60	9.3	35–45 Gy 5fx	1-year 79% 2-years 79%	1-year 59% 2-years 45%	N/A	≥ G3: 5%
Orlandi et al. (2019) [13]	Fondazione IRCCS Istituto Nazionale dei Tumori, Italy	38	N/A	29–30 Gy 5 fx	N/A	2-years 64.1% 5-years 23.3%	N/A	N/A

FU follow up, *fx* fractions, *G* grade, *LC* local control, *LRC* local-regional control, *LRPF* local-regional progression free survival, *N/A* not available, *OS* overall survival

Dose and Fractionation

Retrospective data from the University of Pittsburgh Cancer Institute demonstrated the feasibility of reirradiation with SABR in the head and neck setting [87]. Since that time several studies have examined the use of SABR in the treatment of recurrent or second primary head and neck cancers. These non-randomized trials administered doses from 15 to 48 Gy in 1–8 fractions, with 5 fraction regimens being the most common [88, 89, 94, 96, 101–103]. In a phase I dose-escalation study of SABR reirradiation in HNSCC, Heron et al. were able to reach their highest dose tier of 44 Gy in 5 fractions (8.8 Gy per fraction) without any acute grade 3/4 or dose-limiting toxicity [104]. An extended retrospective analysis from the University of Pittsburgh revealed that doses as high as 50 Gy in 5 fractions are possible [91]. They also established a clear SABR dose-volume response, with improved local control seen on the steepest gradient between 35 and 40 Gy. For tumor size ≤25 cm^3, radiation doses above 44 Gy in 5 fractions did not increase the rate of complete responses than those getting 40 Gy in 5 fractions. For tumors >25 cm^3, no optimum dose was identified as patients in the 44–50 Gy group continued to see an increase in complete response when compared to the 40 Gy group. A steep dose–response relationship for doses 35–45 Gy (in 5 fractions) was also found by an American Association of Physicists in Medicine working group when they retrospectively evaluated the optimal dose in 300 patients from eight studies [105]. Although volume was not incorporated in their model, their recommendation was 45 Gy in 5 fraction for smaller tumors, given the flat dose-response curve above this level. Doses between 45 and 50 Gy in 5 fractions were recommended for larger tumors, but the lack of clinical data at this equivalent dose does limit the extent to which it can and should be used at this juncture.

Systemic Therapy

One method of further treatment escalation with SABR is adding concurrent systemic therapies. Initial SABR reirradiation trials did not include systemic therapies due to concern of toxicity. Since that time, multiple studies have showed the safety and efficacy of cetuximab. A retrospective-matched cohort study found that adding concurrent cetuximab to SABR reirradiation significantly increased LC (2-year LC 33.6% versus 49.2%) and OS (2-year OS 21.1% versus 53.3%) [106]. It is important to note that the no systemic therapy group may have had unaccounted factors that precluded them from getting cetuximab and these factors led them to have worse disease outcomes. There were no significant differences in the incidence of grade 1–3 toxicity events between the two cohorts and there were no grade 4/5 toxicities. Concurrent administration of cetuximab also didn't appear to show influence on quality of life [107]. Experience with other systemic therapy agents, such as cisplatin, 5-FU, carboplatin/paclitaxel, have also been published [90, 98, 108].

SABR Versus IMRT

A third publication from the MIRI Collaborative group assessed the efficacy of SABR versus IMRT reirradiation [97]. The authors first validated the RPA classification published by Ward et al. [11]. Compared to the IMRT cohort, the SABR cohort had shorter interval between radiation courses (1.2 years versus 3.1 years), was more likely to have received prior systemic therapy (64% versus 35%), and had less pretreatment organ dysfunction (19% versus 35%) [97]. In the unadjusted analysis, the 2-year OS was better in the IMRT group (35.4% versus 16.3%; $p < 0.001$). However, in an adjusted analysis the OS superiority was no longer statistically significant (HR 0.877; 95% CI 0.702–1.097). Similarly, on unadjusted analysis there was superiority in LRF with IMRT (45.5% versus 57%; $p = 0.014$) that was no longer significant on adjusted analysis (HR 1.154; 95% CI 0.886–1.505). Additional subgroup analysis by RPA class was done to further minimize bias in the retrospective study. No significant difference in 2-year OS was detected between IMRT and SABR in the RPA class III ($n = 61$; 16.2% for IMRT versus 3.6% for SABR; $p = 0.42$). In the RPA class II ($n = 353$), 2-year OS was significantly higher in the IMRT group (39.1% versus 18.6%; $p < 0.001$). The difference in OS in the RPA class II could mostly be accounted for by tumor size, 2-year OS in tumors ≤25 cm^3 or rT0-2 was 50% with IMRT and 38.5% with SABR to ≥35 Gy ($p = 0.42$). For tumors >25 cm^3 or rT3-4, 2-year OS was 28.2% with IMRT and 8.8% with SABR to ≥35 Gy ($p < 0.001$).

Post-operative SABR

There is considerably less post-operative SABR reirradiation data, although it continues to offer the same advantage of conveniently short regimens [98, 109]. Vargo et al. published a retrospective series on 28 patients who had either positive surgical margin or ENE [109]. The median SABR dose was 40 Gy in 5 fractions to a median treatment volume of 23.1 cm^3. Follow-up was short, but they did report acceptable 1-year LRC of 51%, DFS of 90%, and OS of 64%. The rate of acute and late grade ≥3 toxicities was low at 0% and 8%, respectively. Additional data are lacking to further justify this approach.

Organs at Risk (OARs)

Although the reported rates of toxicity with SABR are similar if not lower than with conventional radiation, there are concerns of severe toxicity given the high dose per fraction [97]. Carotid blowout syndrome (CBS) is of particular concern as early reports appeared to show a higher rate than in conventional reirradiation [101, 103, 110, 111]. An early study noted that 17.3% of SABR patients developed CBS, although it only occurred in patients who received the full prescribed dose to the carotid artery due to the tumor surrounding the artery [101]. A larger earlier study reported CBS rates of 8.4% with a median of 5 months after reirradiation [110]. Only skin invasion, and not dose to the carotid artery, was identified as a statistically significant prognostic factor for CBS on multivariate analysis. More modern series show low rates of CBS, even irrespective of dose to the carotid arteries [92, 95, 112]. This difference may be due to fractionation schedule, e.g., daily fractions versus non-consecutive days [95, 113]. Better patient selection may also explain the decrease in CBS after SABR reirradiation, such as avoiding patients that have tumors that entrap >180° of the carotid artery or large treatment volumes [93, 114]. In a dose-response model study on patients receiving 5 fraction SABR reirradiation, the risk of carotid bleeding was 0.8% and 5% with $D_{0.1cc}$ of 20 Gy and $D_{0.1cc}$ of 50 Gy to the carotid, respectively [115]. No patients had carotid bleeding with a SABR $D_{0.1cc} < 39.4$ Gy to the carotid and CBS with a SABR $D_{0.1cc} < 47.6$ Gy to the carotid. It is important to remember that factors that increase risk of CBS also often preclude patients for getting surgical resection and further progression of disease could put the patient at even higher risk of life-threatening carotid bleeding. Patients with recurrent disease in the larynx or hypopharynx are also associated with increase late toxicity, likely due to the high risk of organ dysfunction after SABR reirradiation [112, 116]. Although there is a paucity of long-term data, careful follow-up after SABR reirradiation is required as multiple reports show continued long-term toxicity and patient-reported quality-of-life decline, whether due to treatment or tumor recurrence [112, 117].

SABR Conclusion

Reirradiation with SABR is a viable and convenient treatment option with ideal patients having recurrent disease <25 cm^3 in size, <180° carotid artery tumor encasement, no skin invasion, and non-larynx or hypopharynx recurrent site. Although no consensus on dose exists, the general recommendation is 40 Gy in 5 fractions on non-consecutive days for small tumors, and 44 Gy in 5 fractions for bigger tumors (>25 cm^3). Concurrent cetuximab and other systemic therapies appear to be well tolerated, with concurrent cetuximab showing an improvement in survival in a retrospective analysis.

Charged Particle Radiation

The finite range of heavy ions (protons, carbon ions) due to the Bragg peak make them an attractive tool for reirradiation given the concern of toxicity with reirradiation. The ROCOCO in silico trial compared intensity-modulated proton therapy (IMPT), intensity-modulated carbon-ion therapy (IMCIT), and photon therapy volumetric modulated arc therapy (VMAT) reirradiation plans to a second dose of 70 Gy [118]. IMCIT and IMPT reduced the D_{mean} in 100% and 68% ($p < 0.02$), respectively, of the 22 OARs compared to VMAT. The dose to 2% of the volume for the brainstem and spinal cord were also statistically significantly reduced compared to VMAT. The primary benefit was seen in contralateral organs, spinal cord, and brainstem. IMCIT generated an 84% lower OAR dosage compared to VMAT versus 60% with IMPT.

Proton

Published reports on IMPT reirradiation show 1-year OS from around 55–81.3% and 2-year LRF from 23% to 50% (Table 10.2) [119–123]. Acute and late grade ≥3 toxicities range from 9.9% to 31.4% and 18.6–24.6%. In a study comparing SBRT, IMRT, and protons in reirradiation of skull base tumors less than 60 cm^3,

Table 10.2 Published reports on intensity-modulated proton therapy reirradiation

Study	Patients	Median FU (months)	LRF	OS	Acute toxicity	Late toxicity
Lin et al. (1999) [119]	16 (NPX only)	23.7	2-year 50%	1-year 50% 2-year 50%	n/a	n/a
McDonald et al. (2016) [120]	61	15.2	2-year 23%	1-year ~55% 2-year 32.7%	G2 47.5%; G3 13.1%; G5 1.6%	G2 22.6%; G3 15.1%; G4 5.7%, G5 3.8%
Phan et al. (2016) [121]	60	13.6	1-year 19.2% 2-year 27.2%	1-year 81.3% 2-year 69.0%	G3 30%	G3 20%; G5 3.3%
Romesser et al. (2016) [122]	92	10.4	1-year 25.1% (2-year ~48%)	1-year 65.2% 2-year ~42%	G3 31.4%	G3 8.5%; G4 7.2%; G5 2.9%
Dionisi et al. (2019) [123]	17 (NPX only)	10	18-month 59.3%	18-month LC 72.9%	G≥3 0%	G3 23.5%; G5 5.9%

Note: Inter-study comparison should not be made due to different patient populations, treatment techniques, and reporting methods for disease and toxicity outcomes
FU follow up, *G* grade, *LRF* local regional failure, *NPX* Nasopharyngeal carcinoma, *OS* overall survival

there was difference in OS or LC between the three modalities [124]. Toxicity appeared less in the SBRT cohort likely due to smaller treatment volume. In another small comparative study looking at IMRT and IMPT reirradiation, IMPT was associated with decrease in physician-reported overall grade 3 acute toxicities, 31% IMPT versus 73% IMRT ($p = 0.01$) [125]. The 2-year OS, LR, and DFS were the same between the two treatment groups. Preliminary results for IMPT reirradiation are promising and, with a growing number of centers offering protons therapy, more data on reirradiation is expected. However, it is too early to make any conclusion on IMPT benefit over the more widely available and cheaper photon reirradiation. Current data is limited by selection bias, small numbers with short follow-up, and continued refinement and improvement in IMPT techniques over time. Moreover, the majority of serious toxicities in reirradiation are due to in-field dose rather than the lower dose bath, raising the question of whether high-quality IMRT will really be inferior to IMPT in most cases.

Carbon

The high relative biological effectiveness of carbon ion compared to photon and proton therapy may improve tumor control in the more radioresistant recurrent tumors [126]. At the time of this writing, there are only 12 centers in five countries treating patients with carbon ions [127]. Due to the scarcity of IMCIT resources, published series may be biased toward more unusual and difficult cases, thus skewing results. Published reports on IMCIT show 1-year OS from 72% to 95.9% and 1-year LC from about 70–84.9% (Table 10.3) [128, 129, 132]. Acute and late grade ≥3 toxicities range from 0.7% to 10.4% and 10.4–37.5%. A subset analysis of nasopharynx patients treated with IMCIT were also published [131]. The series report 2-year OS and LC rates of 83.7% and 58%, respectively. Nasopharyngeal necrosis was seen in 16% of patients, and 4.85% of patients died due to massive hemorrhage. It is important to note that there are no guidelines for dose tolerance of OARs for heavy ion reirradiation, and thorough risk-benefit evaluation is required for each individual patient [130].

Table 10.3 Published Reports on Intensity-Modulated Carbon-Ion Therapy Reirradiation

Study	Patients	Median FU (months)	LC	OS	Acute toxicity	Late toxicity
Gao et al. (2019) [128]	141	14.7	1-year 84.9% (regional 97.7%)	1-year 95.9%	G≥3 0.7%	G≥3 10.6%
Hayashi et al.(2019) [129]	48	27.1	1-year ~70% 2-year 40.5%	1-year ~80% 2-year 59.6%	G≥3 10.4%	G≥3 37.5%
Held et al. (2019) [130]	229	28.5	1-year 60%	1-year 72%	G≥3 3.1%	G≥3 14.5%
Hu et al. (2020)*[131][a]	206 (NPX only)	22.8	1-year ~89% 2-year 58%	1-year ~92% 2-year 83.7%	G≥3 0%	G≥3 19.4%

Note: Inter-study comparison should not be made due to different patient populations, treatment techniques, and reporting methods for disease and toxicity outcomes
FU follow-up, *G* grade, *LC* local control, *NPX* Nasopharyngeal carcinoma, *OS* overall survival
*The study cohort in Hu et al. has major overlap with the study cohort in Gao et al., which consised of 78% NPX patients

IMCIT also shows promise in limited publications, but it will likely take some time before it is available to more than a tiny minority of patients.

Brachytherapy

Like heavy ions, brachytherapy can be an advantageous tool in reirradiation by protecting adjacent normal tissue through delivering a focused radiation dose from inside the target. Brachytherapy has been utilized for years for recurrent head and neck carcinoma, but its use has declined over time due to development of more conformal external beam techniques, limited available expertise in performing the procedures, and limitations in the patient population that are best suited for this technique [133]. Brachytherapy can be delivered via permanent implants (using low-dose-rate brachytherapy [LDR] or high-dose-rate brachytherapy [HDR]) or via removable catheters through which typically high activity radioisotopes are inserted (using HDR or pulsed-dose-rate brachytherapy [PDR]).

Definitive brachytherapy reirradiation studies report 2-year local control and OS rates from 27.5% to 92.5% and 18.2–55.6%, respectively [134–141]. Similar to external beam radiation, there is evidence that patients who get surgical resection along with brachytherapy have improved disease outcomes [133, 142, 143]. Rudzianskas et al. found that patients who had surgical resection followed by HDR-brachytherapy had 2-year OS of 62% versus 35% in those who had HDR-brachytherapy alone ($p = 0.035$). There is also some evidence that concurrent chemotherapy with brachytherapy results in disease control benefit [144]. One study showed 10-year LC rates of 76% with concomitant chemotherapy versus 39% without ($p = 0.014$). At centers with appropriate expertise, brachytherapy can be a useful tool in the treatment of recurrent or secondary primary head and neck cancers, especially in situations where previous radiation precludes further external beam treatment.

Summary

Head and neck radiotherapy techniques have improved dramatically over the past two decades, and this innovation has translated into progressively more favorable outcomes in the curative management of locoregional recurrence. The introduction of immunotherapy has markedly improved survival in patients with metastatic HNSCC, but the data suggest that the new systemic therapy regimens have only minimal impact on prognosis in patients with locoregional recurrence [145]. Therefore, improved survival in this condition will be dependent on continued optimization of the benefits versus risks of local therapy. The continued evolution and implementation of IMRT, adaptive therapy, SABR, and particle therapy may be expected to provide progressively superior dosimetry, but a dramatic sea change in prognosis may still require novel radiosensitizers or protectants that meaningfully improve the therapeutic ratio. There is a delicate balance in this space between oncologic efficacy versus treatment-associated morbidity and mortality, and the decision to pursue aggressive curative-intent treatment must incorporate patient-specific data on prognosis, expected toxicity and, of course, patient preferences.

Case Study

Patient is a 73-year-old female with a history of squamous cell carcinoma of the right tonsil, AJCC Seventh Edition cStage IVA (T1 N2a M0), p16 unknown. She completed definitive chemoradiation (7000 cGy in 35 fractions with weekly cisplatin) in February 2015 with a complete metabolic response. In January 2020, she noted mild discomfort in the back of her throat, which progressed to left otalgia and throat pain. Endoscopic biopsy in November 2020 of a 2.5 cm ulcerated left base of tongue mass demonstrated poorly differentiated, non-keratinizing squamous cell carcinoma. Neck CT with contrast demonstrated an ulcerated enhancing mass along the mucosa of the left base of the tongue extending to the glossotonsillar sulcus and deeply invasive into the intrinsic tongue musculature and focally involving the hyoglossus muscle. PET/CT did not show any evidence of FDG avid locoregional or distant metastatic disease. The GTV and PTV volumes were 5.3 cm^3 and 12.2 cm^3, respectively. She was treated with definitive SABR to the left of base of tongue mass with pembrolizumab on a clinical trial. Radiation prescription was 4000 cGy in 5 fractions to cover 95% of the tumor volume, delivered every other day, and with a CBCT daily. PTV was made by expanding the GTV radially in all directions by 3 mm. Patient tolerated the treatment well with only Grade 1 mucositis. She remains disease free on last follow-up.

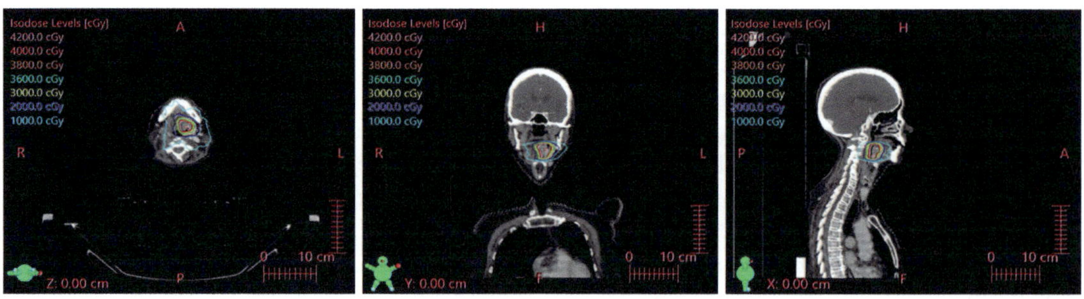

DVH	Structure	Structure Status	Volume (cm3)	Min Dose (cGy)	Max Dose (cGy)	Mean Dose (cGy)
	Eye_L	Approved	4.0	6.3	17.7	11.5
	Esophagus_S	Approved	5.1	2.5	11.9	5.7
	Epiglottis	Approved	0.6	258.9	3331.6	1098.0
	Cavity_Oral_UTSW	Approved	159.8	27.4	4734.3	1033.7
	Cavity_Oral	Approved	204.4	19.7	4791.4	734.1
	BrainStem_03	Approved	42.4	5.0	55.4	12.8
	BrainStem	Approved	25.6	5.5	43.7	12.2
	Bone_Mandible	Approved	56.3	10.3	3486.5	757.0
	Eye_R	Approved	3.9	3.5	15.8	10.2
	Glnd_Submand_L	Approved	5.8	40.1	4231.8	896.3
	Larynx	Approved	19.9	16.3	4594.6	242.6
	Lips	Approved	18.9	0.0	1195.4	324.5
	PACS	Approved	3.9	12.2	26.0	18.1
	Parotid_L	Approved	21.9	23.4	1957.3	545.4
	Parotid_R	Approved	23.3	14.6	921.1	301.6
	Pharynx	Approved	46.2	23.0	4783.6	1142.7
	SoftPalate	Approved	0.4	317.4	2473.4	1042.2
	SpinalCord	Approved	19.7	0.0	669.9	107.9
	PTV+2-Body	Approved	16946.0	0.0	1929.1	19.1
	GTV_4000	Approved	5.3	4055.7	4815.0	4422.4

References

1. Gillison ML, et al. Radiotherapy plus cetuximab or cisplatin in human papillomavirus-positive oropharyngeal cancer (NRG Oncology RTOG 1016): a randomised, multicentre, non-inferiority trial. Lancet. 2019;393(10166):40–50.
2. Cooper JS, et al. Long-term follow-up of the RTOG 9501/intergroup phase III trial: postoperative concurrent radiation therapy and chemotherapy in high-risk squamous cell carcinoma of the head and neck. Int J Radiat Oncol Biol Phys. 2012;84(5):1198–205.
3. Argiris A, et al. Head and neck cancer. Lancet. 2008;371(9625):1695–709.
4. Brockstein B, et al. Patterns of failure, prognostic factors and survival in locoregionally advanced head and neck cancer treated with concomitant chemoradiotherapy: a 9-year, 337-patient, multi-institutional experience. Ann Oncol. 2004;15(8):1179–86.
5. Pignon JP, et al. Meta-analysis of chemotherapy in head and neck cancer (MACH-NC): an update on 93 randomised trials and 17,346 patients. Radiother Oncol. 2009;92(1):4–14.
6. Baxi SS, et al. Causes of death in long-term survivors of head and neck cancer. Cancer. 2014;120(10):1507–13.
7. Weichselbaum RR, et al. Radioresistant tumor cells are present in head and neck carcinomas that recur after radiotherapy. Int J Radiat Oncol Biol Phys. 1988;15(3):575–9.
8. Ho AS, et al. Decision making in the management of recurrent head and neck cancer. Head Neck. 2014;36(1):144–51.
9. Goodwin WJ Jr. Salvage surgery for patients with recurrent squamous cell carcinoma of the upper aerodigestive tract: when do the ends justify the means? Laryngoscope. 2000;110(3 Pt 2 Suppl 93):1–18.
10. Ward MC, et al. A competing risk nomogram to predict severe late toxicity after modern re-irradiation for squamous carcinoma of the head and neck. Oral Oncol. 2019;90:80–6.
11. Ward MC, et al. Refining patient selection for reirradiation of head and neck squamous carcinoma in the IMRT era: a multi-institution cohort study by the MIRI collaborative. Int J Radiat Oncol Biol Phys. 2018;100(3):586–94.
12. Lee J, et al. Intensity-modulated radiotherapy-based reirradiation for head and neck cancer: a multi-institutional Study by Korean Radiation Oncology Group (KROG 1707). Cancer Res Treat. 2020;52(4):1031–40.
13. Orlandi E, et al. Long-term outcome of re-irradiation for recurrent or second primary head and neck cancer: a multi-institutional study of AIRO-Head and Neck working group. Head Neck. 2019;41(10):3684–92.
14. Tanvetyanon T, et al. Prognostic factors for survival after salvage reirradiation of head and neck cancer. J Clin Oncol. 2009;27(12):1983–91.
15. Spencer SA, et al. Final report of RTOG 9610, a multi-institutional trial of reirradiation and chemotherapy for unresectable recurrent squamous

cell carcinoma of the head and neck. Head Neck. 2008;30(3):281–8.
16. Chen AM, Phillips TL, Lee NY. Practical considerations in the re-irradiation of recurrent and second primary head-and-neck cancer: who, why, how, and how much? Int J Radiat Oncol Biol Phys. 2011;81(5):1211–9.
17. Buglione M, et al. Reirradiation in head and neck recurrent or second primary tumor: efficacy, safety, and prognostic factors. Tumori. 2015;101(5):585–92.
18. Langer CJ, et al. Phase II study of low-dose paclitaxel and cisplatin in combination with split-course concomitant twice-daily reirradiation in recurrent squamous cell carcinoma of the head and neck: results of Radiation Therapy Oncology Group Protocol 9911. J Clin Oncol. 2007;25(30):4800–5.
19. Janot F, et al. Randomized trial of postoperative reirradiation combined with chemotherapy after salvage surgery compared with salvage surgery alone in head and neck carcinoma. J Clin Oncol. 2008;26(34):5518–23.
20. Bachar GY, et al. Long-term outcome analysis after surgical salvage for recurrent tonsil carcinoma following radical radiotherapy. Eur Arch Otorhinolaryngol. 2010;267(2):295–301.
21. Stevens KR Jr, Britsch A, Moss WT. High-dose reirradiation of head and neck cancer with curative intent. Int J Radiat Oncol Biol Phys. 1994;29(4):687–98.
22. Riaz N, et al. A nomogram to predict loco-regional control after re-irradiation for head and neck cancer. Radiother Oncol. 2014;111(3):382–7.
23. Embring A, et al. Overlapping volumes in re-irradiation for head and neck cancer - an important factor for patient selection. Radiat Oncol. 2020;15(1):147.
24. Takiar V, et al. Reirradiation of head and neck cancers with intensity modulated radiation therapy: outcomes and analyses. Int J Radiat Oncol Biol Phys. 2016;95(4):1117–31.
25. Ohizumi Y, et al. Prognostic factors of reirradiation for recurrent head and neck cancer. Am J Clin Oncol. 2002;25(4):408–13.
26. Yu YH, et al. A model to predict the risk of lethal nasopharyngeal necrosis after re-irradiation with intensity-modulated radiotherapy in nasopharyngeal carcinoma patients. Chin J Cancer. 2016;35(1):59.
27. Leong YH, et al. Long-term outcomes after reirradiation in nasopharyngeal carcinoma with intensity-modulated radiotherapy: a meta-analysis. Head Neck. 2018;40(3):622–31.
28. Kong F, et al. Long-term survival and late complications of intensity-modulated radiotherapy for recurrent nasopharyngeal carcinoma. BMC Cancer. 2018;18(1):1139.
29. McDonald MW, et al. ACR appropriateness criteria retreatment of recurrent head and neck cancer after prior definitive radiation expert panel on radiation oncology-head and neck cancer. Int J Radiat Oncol Biol Phys. 2011;80(5):1292–8.
30. Dionisi F, et al. Organs at risk's tolerance and dose limits for head and neck cancer re-irradiation: a literature review. Oral Oncol. 2019;98:35–47.
31. Choe KS, et al. Prior chemoradiotherapy adversely impacts outcomes of recurrent and second primary head and neck cancer treated with concurrent chemotherapy and reirradiation. Cancer. 2011;117(20):4671–8.
32. Salama JK, et al. Long-term outcome of concurrent chemotherapy and reirradiation for recurrent and second primary head-and-neck squamous cell carcinoma. Int J Radiat Oncol Biol Phys. 2006;64(2):382–91.
33. Lee N, et al. Salvage re-irradiation for recurrent head and neck cancer. Int J Radiat Oncol Biol Phys. 2007;68(3):731–40.
34. Ahlawat P, et al. Reirradiation with IMRT for recurrent head and neck cancer: a single-institutional report on disease control, survival, and toxicity. Rep Pract Oncol Radiother. 2017;22(4):331–9.
35. Parsons JT, et al. Salvage surgery following radiation failure in squamous cell carcinoma of the supraglottic larynx. Int J Radiat Oncol Biol Phys. 1995;32(3):605–9.
36. Duprez F, et al. Intensity-modulated radiotherapy for recurrent and second primary head and neck cancer in previously irradiated territory. Radiother Oncol. 2009;93(3):563–9.
37. Merlotti A, et al. What is the role of postoperative re-irradiation in recurrent and second primary squamous cell cancer of head and neck? A literature review according to PICO criteria. Crit Rev Oncol Hematol. 2017;111:20–30.
38. Kim AJ, et al. Salvage surgery with free flap reconstruction: factors affecting outcome after treatment of recurrent head and neck squamous carcinoma. Laryngoscope. 2007;117(6):1019–23.
39. Suh JD, et al. Reirradiation after salvage surgery and microvascular free flap reconstruction for recurrent head and neck carcinoma. Otolaryngol Head Neck Surg. 2008;139(6):781–6.
40. Ho AS, et al. Impact of flap reconstruction on radiotoxicity after salvage surgery and reirradiation for recurrent head and neck cancer. Ann Surg Oncol. 2016;23(Suppl 5):850–7.
41. Milano MT, et al. Twice-daily reirradiation for recurrent and second primary head-and-neck cancer with gemcitabine, paclitaxel, and 5-fluorouracil chemotherapy. Int J Radiat Oncol Biol Phys. 2005;61(4):1096–106.
42. Biagioli MC, et al. Intensity-modulated radiotherapy with concurrent chemotherapy for previously irradiated, recurrent head and neck cancer. Int J Radiat Oncol Biol Phys. 2007;69(4):1067–73.
43. Villaflor VM, et al. Phase II trial of pemetrexed-based induction chemotherapy followed by concomitant chemoradiotherapy in previously irradiated patients with squamous cell carcinoma of the head and neck. Ann Oncol. 2011;22(11):2501–7.

44. Kasperts N, et al. Results of postoperative reirradiation for recurrent or second primary head and neck carcinoma. Cancer. 2006;106(7):1536–47.
45. Caudell JJ, et al. Volume, dose, and fractionation considerations for IMRT-based reirradiation in head and neck cancer: a multi-institution analysis. Int J Radiat Oncol Biol Phys. 2018;100(3):606–17.
46. Ruhle A, et al. The value of moderate dose escalation for re-irradiation of recurrent or second primary head-and-neck cancer. Radiat Oncol. 2020;15(1):81.
47. Benchalal M, et al. Hyperfractionation in the reirradiation of head and neck cancers. Result of a pilot study. Radiother Oncol. 1995;36(3):203–10.
48. Curtis KK, et al. Outcomes of patients with locoregionally recurrent or new primary squamous cell carcinomas of the head and neck treated with curative intent reirradiation at Mayo Clinic. Radiat Oncol. 2016;11:55.
49. Tortochaux J, et al. Randomized phase III trial (GORTEC 98-03) comparing re-irradiation plus chemotherapy versus methotrexate in patients with recurrent or a second primary head and neck squamous cell carcinoma, treated with a palliative intent. Radiother Oncol. 2011;100(1):70–5.
50. Strojan P, et al. Recurrent and second primary squamous cell carcinoma of the head and neck: when and how to reirradiate. Head Neck. 2015;37(1):134–50.
51. Awan MJ, et al. Final results of a multi-institutional phase II trial of reirradiation with concurrent weekly cisplatin and cetuximab for recurrent or second primary squamous cell carcinoma of the head and neck. Ann Oncol. 2018;29(4):998–1003.
52. Kharofa J, et al. Continuous-course reirradiation with concurrent carboplatin and paclitaxel for locally recurrent, nonmetastatic squamous cell carcinoma of the head-and-neck. Int J Radiat Oncol Biol Phys. 2012;83(2):690–5.
53. Lee J, et al. Reirradiation with intensity-modulated radiation therapy for recurrent or secondary head and neck cancer: meta-analysis and systematic review. Head Neck. 2020;42(9):2473–85.
54. Sher DJ, et al. Efficacy and toxicity of reirradiation using intensity-modulated radiotherapy for recurrent or second primary head and neck cancer. Cancer. 2010;116(20):4761–8.
55. Duprez F, et al. High-dose reirradiation with intensity-modulated radiotherapy for recurrent head-and-neck cancer: disease control, survival and toxicity. Radiother Oncol. 2014;111(3):388–92.
56. Choi SH, et al. Re-irradiation using intensity-modulated radiotherapy for recurrent and second primary head and neck cancer. Anticancer Res. 2018;38(5):3165–73.
57. Nguyen-Tan PF, et al. Randomized phase III trial to test accelerated versus standard fractionation in combination with concurrent cisplatin for head and neck carcinomas in the Radiation Therapy Oncology Group 0129 trial: long-term report of efficacy and toxicity. J Clin Oncol. 2014;32(34):3858–66.
58. Bourhis J, et al. Concomitant chemoradiotherapy versus acceleration of radiotherapy with or without concomitant chemotherapy in locally advanced head and neck carcinoma (GORTEC 99-02): an open-label phase 3 randomised trial. Lancet Oncol. 2012;13(2):145–53.
59. Ng SP, et al. Patterns of failure after definitive reirradiation for patients with head and neck cancer. Int J Radiat Oncol Biol Phys. 2019;105:E380–1.
60. Popovtzer A, et al. The pattern of failure after reirradiation of recurrent squamous cell head and neck cancer: implications for defining the targets. Int J Radiat Oncol Biol Phys. 2009;74(5):1342–7.
61. Margalit DN, et al. Patterns of failure after reirradiation with intensity-modulated radiation therapy and the competing risk of out-of-field recurrences. Oral Oncol. 2016;61:19–26.
62. Spencer SA, et al. Concomitant chemotherapy and reirradiation as management for recurrent cancer of the head and neck. Am J Clin Oncol. 1999;22(1):1–5.
63. Haraf DJ, Weichselbaum RR, Vokes EE. Re-irradiation with concomitant chemotherapy of unresectable recurrent head and neck cancer: a potentially curable disease. Ann Oncol. 1996;7(9):913–8.
64. Chmura SJ, et al. Long-term outcome of re-irradiation with concomitant chemotherapy for locally recurrent squamous cell carcinoma of the head and neck. Int J Radiat Oncol Biol Phys. 2003;57(2):S409–10.
65. Machtay M, et al. Pilot study of postoperative reirradiation, chemotherapy, and amifostine after surgical salvage for recurrent head-and-neck cancer. Int J Radiat Oncol Biol Phys. 2004;59(1):72–7.
66. Hehr T, et al. Reirradiation alternating with docetaxel and cisplatin in inoperable recurrence of head-and-neck cancer: a prospective phase I/II trial. Int J Radiat Oncol Biol Phys. 2005;61(5):1423–31.
67. Mabanta SR, et al. Salvage treatment for neck recurrence after irradiation alone for head and neck squamous cell carcinoma with clinically positive neck nodes. Head Neck. 1999;21(7):591–4.
68. Marwaha A, et al. Clinical efficacy and tolerability of continuous course reirradiation with concurrent weekly carboplatin-paclitaxel for locally recurrent, nonmetastatic squamous cell carcinoma of the head and neck (SCCHN). J Clin Oncol. 2010;28(15 Suppl):5518.
69. Balermpas P, et al. Reirradiation with cetuximab in locoregional recurrent and inoperable squamous cell carcinoma of the head and neck: feasibility and first efficacy results. Int J Radiat Oncol Biol Phys. 2012;83(3):e377–83.
70. Tao Y, et al. Randomized trial comparing two methods of re-irradiation after salvage surgery in head and neck squamous cell carcinoma: once daily split-course radiotherapy with concomitant chemotherapy or twice daily radiotherapy with cetuximab. Radiother Oncol. 2018;128(3):467–71.

71. Zwicker F, et al. Reirradiation with concurrent cetuximab immunotherapy in recurrent head and neck cancer. Strahlenther Onkol. 2011;187:32–8.
72. Milanović D, et al. Reirradiation plus EGFR inhibition in locally recurrent and unresectable head and neck cancer. Strahlenther Onkol. 2013;189:842–8.
73. Dornoff N, et al. Re-irradiation with cetuximab or cisplatin-based chemotherapy for recurrent squamous cell carcinoma of the head and neck. Strahlenther Onkol. 2015;191(8):656–64.
74. Van Waes C, et al. Inhibition of nuclear factor-kappaB and target genes during combined therapy with proteasome inhibitor bortezomib and reirradiation in patients with recurrent head-and-neck squamous cell carcinoma. Int J Radiat Oncol Biol Phys. 2005;63(5):1400–12.
75. Hitt R, et al. A randomized phase III trial comparing induction chemotherapy followed by chemoradiotherapy versus chemoradiotherapy alone as treatment of unresectable head and neck cancer. Ann Oncol. 2014;25(1):216–25.
76. Cohen EE, et al. Phase III randomized trial of induction chemotherapy in patients with N2 or N3 locally advanced head and neck cancer. J Clin Oncol. 2014;32(25):2735–43.
77. Geoffrois L, et al. Induction chemotherapy followed by cetuximab radiotherapy is not superior to concurrent chemoradiotherapy for head and neck carcinomas: results of the GORTEC 2007-02 phase III randomized trial. J Clin Oncol. 2018;36(31):3077–83.
78. Maranzano E, et al. Re-irradiation of brain metastases and metastatic spinal cord compression: clinical practice suggestions. Tumori. 2005;91(4):325–30.
79. Mayer R, Sminia P. Reirradiation tolerance of the human brain. Int J Radiat Oncol Biol Phys. 2008;70(5):1350–60.
80. Wang HZ, et al. The tolerance of brainstem in reirradiation with intensity modulated radiation therapy in recurrent nasopharyngeal carcinoma. Int J Radiat Oncol Biol Phys. 2016;96(2):E340.
81. Grosu AL, et al. Retreatment of the spinal cord with palliative radiotherapy. Int J Radiat Oncol Biol Phys. 2002;52(5):1288–92.
82. Nieder C, et al. Update of human spinal cord reirradiation tolerance based on additional data from 38 patients. Int J Radiat Oncol Biol Phys. 2006;66(5):1446–9.
83. Sminia P, et al. Re-irradiation of the human spinal cord. Strahlenther Onkol. 2002;178(8):453–6.
84. Lee JY, et al. Predictors of severe long-term toxicity after re-irradiation for head and neck cancer. Oral Oncol. 2016;60:32–40.
85. Chan OS, et al. Reirradiation with intensity-modulated radiotherapy for locally recurrent T3 to T4 nasopharyngeal carcinoma. Head Neck. 2017;39(3):533–40.
86. Agas RAF, et al. Reirradiation for recurrent nasopharyngeal carcinomas: experience from an academic tertiary center in a low- to middle-income country. J Glob Oncol. 2019;5:1–14.
87. Voynov G, et al. Frameless stereotactic radiosurgery for recurrent head and neck carcinoma. Technol Cancer Res Treat. 2006;5(5):529–35.
88. Roh KW, et al. Fractionated stereotactic radiotherapy as reirradiation for locally recurrent head and neck cancer. Int J Radiat Oncol Biol Phys. 2009;74(5):1348–55.
89. Siddiqui F, et al. Stereotactic body radiation therapy for primary, recurrent, and metastatic tumors in the head-and-neck region. Int J Radiat Oncol Biol Phys. 2009;74(4):1047–53.
90. Kawaguchi K, et al. Stereotactic radiosurgery may contribute to overall survival for patients with recurrent head and neck carcinoma. Radiat Oncol. 2010;5:51.
91. Rwigema JC, et al. The impact of tumor volume and radiotherapy dose on outcome in previously irradiated recurrent squamous cell carcinoma of the head and neck treated with stereotactic body radiation therapy. Am J Clin Oncol. 2011;34(4):372–9.
92. Lartigau EF, et al. Multi institutional phase II study of concomitant stereotactic reirradiation and cetuximab for recurrent head and neck cancer. Radiother Oncol. 2013;109(2):281–5.
93. Yazici G, et al. A simple strategy to decrease fatal carotid blowout syndrome after stereotactic body reirradiaton for recurrent head and neck cancers. Radiat Oncol. 2013;8:242.
94. Kress MA, et al. Safety and efficacy of hypofractionated stereotactic body reirradiation in head and neck cancer: long-term follow-up of a large series. Head Neck. 2015;37(10):1403–9.
95. Vargo JA, et al. A prospective phase 2 trial of reirradiation with stereotactic body radiation therapy plus cetuximab in patients with previously irradiated recurrent squamous cell carcinoma of the head and neck. Int J Radiat Oncol Biol Phys. 2015;91(3):480–8.
96. Yamazaki H, et al. Reirradiation using robotic image-guided stereotactic radiotherapy of recurrent head and neck cancer. J Radiat Res. 2016;57(3):288–93.
97. Vargo JA, et al. A multi-institutional comparison of SBRT and IMRT for definitive reirradiation of recurrent or second primary head and neck cancer. Int J Radiat Oncol Biol Phys. 2018;100(3):595–605.
98. Gogineni E, et al. Quality of life outcomes following organ-sparing SBRT in previously irradiated recurrent head and neck cancer. Front Oncol. 2019;9:836.
99. Walle T, et al. Radiation effects on antitumor immune responses: current perspectives and challenges. Ther Adv Med Oncol. 2018;10:1758834017742575.
100. Ho AY, et al. Optimizing radiation therapy to boost systemic immune responses in breast cancer: a critical review for breast radiation oncologists. Int J Radiat Oncol Biol Phys. 2020;108(1):227–41.
101. Cengiz M, et al. Salvage reirradiaton with stereotactic body radiotherapy for locally recurrent head-

and-neck tumors. Int J Radiat Oncol Biol Phys. 2011;81(1):104–9.
102. Rwigema JC, et al. Fractionated stereotactic body radiation therapy in the treatment of previously-irradiated recurrent head and neck carcinoma: updated report of the University of Pittsburgh experience. Am J Clin Oncol. 2010;33(3):286–93.
103. Kodani N, et al. Stereotactic body radiation therapy for head and neck tumor: disease control and morbidity outcomes. J Radiat Res. 2011;52(1):24–31.
104. Heron DE, et al. Stereotactic body radiotherapy for recurrent squamous cell carcinoma of the head and neck: results of a phase I dose-escalation trial. Int J Radiat Oncol Biol Phys. 2009;75(5):1493–500.
105. Vargo JA, et al. Head and neck tumor control probability: radiation dose-volume effects in stereotactic body radiation therapy for locally recurrent previously-irradiated head and neck cancer: report of the AAPM working group. Int J Radiat Oncol Biol Phys. 2021;110:137.
106. Heron DE, et al. Concurrent cetuximab with stereotactic body radiotherapy for recurrent squamous cell carcinoma of the head and neck: a single institution matched case-control study. Am J Clin Oncol. 2011;34(2):165–72.
107. Vargo JA, et al. Prospective evaluation of patient-reported quality-of-life outcomes following SBRT +/- cetuximab for locally-recurrent, previously-irradiated head and neck cancer. Radiother Oncol. 2012;104(1):91–5.
108. Echevarria M, et al. Phase I dose escalation of stereotactic body radiation therapy and concurrent cisplatin for re-irradiation of unresectable, recurrent head and neck squamous cell carcinoma. J Clin Oncol. 2020;38(15 Suppl):6543.
109. Vargo JA, et al. Adjuvant stereotactic body radiotherapy+/-cetuximab following salvage surgery in previously irradiated head and neck cancer. Laryngoscope. 2014;124(7):1579–84.
110. Yamazaki H, et al. Frequency, outcome and prognostic factors of carotid blowout syndrome after hypofractionated re-irradiation of head and neck cancer using CyberKnife: a multi-institutional study. Radiother Oncol. 2013;107(3):305–9.
111. McDonald MW, Moore MG, Johnstone PA. Risk of carotid blowout after reirradiation of the head and neck: a systematic review. Int J Radiat Oncol Biol Phys. 2012;82(3):1083–9.
112. Ling DC, et al. Risk of severe toxicity according to site of recurrence in patients treated with stereotactic body radiation therapy for recurrent head and neck cancer. Int J Radiat Oncol Biol Phys. 2016;95(3):973–80.
113. Vargo JA, et al. Examining tumor control and toxicity after stereotactic body radiotherapy in locally recurrent previously irradiated head and neck cancers: implications of treatment duration and tumor volume. Head Neck. 2014;36(9):1349–55.
114. Yamazaki H, et al. Carotid blowout syndrome in pharyngeal cancer patients treated by hypofractionated stereotactic re-irradiation using CyberKnife: a multi-institutional matched-cohort analysis. Radiother Oncol. 2015;115(1):67–71.
115. Ling DC, et al. Dose-response modeling the risk of carotid bleeding events after stereotactic body radiation therapy for previously irradiated head and neck cancer. J Radiosurg SBRT. 2019;6(2):83–9.
116. Ling DC, et al. Dose-response model for severe late laryngeal toxicity after stereotactic body radiation therapy for previously-irradiated head and neck cancer. J Radiosurg SBRT. 2020;7(2):89–94.
117. Thomas J, et al. Long-term patient-reported quality of life after stereotactic body radiation therapy for recurrent, previously-irradiated head and neck cancer. Front Oncol. 2020;10:83.
118. Eekers DBP, et al. Benefit of particle therapy in re-irradiation of head and neck patients. Results of a multicentric in silico ROCOCO trial. Radiother Oncol. 2016;121(3):387–94.
119. Lin R, et al. Nasopharyngeal carcinoma: repeat treatment with conformal proton therapy--dose-volume histogram analysis. Radiology. 1999;213(2):489–94.
120. McDonald MW, et al. Reirradiation of recurrent and second primary head and neck cancer with proton therapy. Int J Radiat Oncol Biol Phys. 2016;96(4):808–19.
121. Phan J, et al. Reirradiation of head and neck cancers with proton therapy: outcomes and analyses. Int J Radiat Oncol Biol Phys. 2016;96(1):30–41.
122. Romesser PB, et al. Proton beam reirradiation for recurrent head and neck cancer: multi-institutional report on feasibility and early outcomes. Int J Radiat Oncol Biol Phys. 2016;95(1):386–95.
123. Dionisi F, et al. Clinical results of proton therapy reirradiation for recurrent nasopharyngeal carcinoma. Acta Oncol. 2019;58(9):1238–45.
124. Ng SP, et al. Patient outcomes after reirradiation of small skull base tumors using stereotactic body radiotherapy, intensity modulated radiotherapy, or proton therapy. J Neurol Surg B Skull Base. 2020;81(6):638–44.
125. Shuja M, et al. Clinical outcomes after re-irradiation in recurrent head and neck cancers treated with intensity modulated proton and photon therapies. Int J Radiat Oncol Biol Phys. 2019;105(1):E382–3.
126. Jones B. A simpler energy transfer efficiency model to predict relative biological effect for protons and heavier ions. Front Oncol. 2015;5:184.
127. Particle Therapy Co-Operative Group. Particle therapy facilities in clinical operation. Villigen: Particle Therapy Co-Operative Group; 2020. https://www.ptcog.ch/index.php/facilities-in-operation. Accessed 11 Feb 2021.
128. Gao J, et al. Salvage carbon-ion radiation therapy for locoregionally recurrent head and neck malignancies. Sci Rep. 2019;9(1):4259.
129. Hayashi K, et al. Feasibility of re-irradiation using carbon ions for recurrent head and neck malignancies after carbon-ion radiotherapy. Radiother Oncol. 2019;136:148–53.

130. Held T, et al. Dose-limiting organs at risk in carbon ion re-irradiation of head and neck malignancies: an individual risk-benefit tradeoff. Cancers (Basel). 2019;11(12):2016.
131. Hu J, et al. Clinical outcomes of carbon-ion radiotherapy for patients with locoregionally recurrent nasopharyngeal carcinoma. Cancer. 2020;126(23):5173–83.
132. Held T, et al. Carbon ion reirradiation for recurrent head and neck cancer: a single-institutional experience. Int J Radiat Oncol Biol Phys. 2019;105(4):803–11.
133. Rodin J, et al. A systematic review of treating recurrent head and neck cancer: a reintroduction of brachytherapy with or without surgery. J Contemp Brachytherapy. 2018;10(5):454–62.
134. Hepel JT, et al. Salvage high-dose-rate (HDR) brachytherapy for recurrent head-and-neck cancer. Int J Radiat Oncol Biol Phys. 2005;62(5):1444–50.
135. Grimard L, et al. Interstitial low-dose-rate brachytherapy in the treatment of recurrent head and neck malignancies. Head Neck. 2006;28(10):888–95.
136. Meng N, et al. Permanent implantation of iodine-125 seeds as a salvage therapy for recurrent head and neck carcinoma after radiotherapy. Cancer Investig. 2012;30(3):236–42.
137. Wiegand S, et al. Interstitial HDR brachytherapy for advanced recurrent squamous cell carcinoma of the head and neck. Anticancer Res. 2013;33(1):249–52.
138. Zhu L, et al. An investigation of 125I seed permanent implantation for recurrent carcinoma in the head and neck after surgery and external beam radiotherapy. World J Surg Oncol. 2013;11:60.
139. Strnad V, et al. Re-irradiation with interstitial pulsed-dose-rate brachytherapy for unresectable recurrent head and neck carcinoma. Brachytherapy. 2014;13(2):187–95.
140. Hegde JV, et al. Head and neck cancer reirradiation with interstitial high-dose-rate brachytherapy. Head Neck. 2018;40(7):1524–33.
141. Jiang P, et al. Five-year outcome of ultrasound-guided interstitial permanent (125)I seeds implantation for local head and neck recurrent tumors: a single center retrospective study. J Contemp Brachytherapy. 2019;11(1):28–34.
142. Narayana A, et al. High-dose-rate interstitial brachytherapy in recurrent and previously irradiated head and neck cancers--preliminary results. Brachytherapy. 2007;6(2):157–63.
143. Rudzianskas V, et al. Reirradiation of recurrent head and neck cancer using high-dose-rate brachytherapy. Acta Otorhinolaryngol Ital. 2012;32(5):297–303.
144. Strnad V, et al. Reirradiation for recurrent head and neck cancer with salvage interstitial pulsed-dose-rate brachytherapy: long-term results. Strahlenther Onkol. 2015;191(6):495–500.
145. Burtness B, et al. Pembrolizumab alone or with chemotherapy versus cetuximab with chemotherapy for recurrent or metastatic squamous cell carcinoma of the head and neck (KEYNOTE-048): a randomised, open-label, phase 3 study. Lancet. 2019;394(10212):1915–28.

Medical Dentistry 11

Michael J. Dienberg

Key Points

- Medical dentistry is the integration and practice of dentistry in a medical/hospital setting, functions as an ancillary hospital service, removes barriers to oral care and is ideal for team-based multidisciplinary management of head and neck cancer.
- Refer to experienced dentist with oncologic head and neck management experience for timely oral optimization prior to head and neck cancer treatment.
- A hospital dentist operates in a hospital-based clinic and is well suited to coordinate timely oral optimization and multidisciplinary management of head and neck cancer team-based care.
- Mitigate risk of oral complications and poor outcomes by close evaluation and management of oral infections, posterior mandible molars, mandibular tori, and periodontal disease.
- Overcome barriers to the care of head and neck cancer patients: sustainable clinical practices, support staff, ease of access, and maximization of longitudinal oral care.

Introduction

Hospital dentistry is not available in all health systems, but is a well-known concept [1–3]. Integration of dental practitioners into medical centers, hospitals, and multidisciplinary medical teams is essential to achieve optimal clinical outcomes for many complex patients [4]. Cancer patients are surviving longer and require longer duration of post-treatment care; in the case of head and neck cancer, with diligent medical and dental follow-up [5, 6]. Some community-based local primary care dentists (PCD) may not have training or experience in managing head and neck cancer (HNC) patients. Despite limited training, much of long-term oral care and management during cancer treatment is provided by local PCDs [7–12]. Referral of HNC patients to an experienced dental practitioner for oral optimization may be necessary in select cases [8, 13]. Delays in cancer care are a risk when access to oncology-focused dental care is limited [4, 14].

Medical dentistry is the integration and practice of dentistry in a medical/hospital setting. Medical dentistry is a member of many multidisciplinary teams caring for complex patients. Areas of focus include patients with special needs, cancer, immune compromise, incurable end-of-life disease, solid organ transplantation, cleft lip and/or palate, craniofacial anomalies, hemophilia, and other bleeding disorders. Many of these patients are unable to obtain oral care in

M. J. Dienberg (✉)
Oregon Health & Science University, Portland, OR, USA
e-mail: dienberg@ohsu.edu

R. A. Chandra, R. J. Li (eds.), *Multidisciplinary Management of Head and Neck Cancer*,
https://doi.org/10.1007/978-3-031-05973-5_11

a traditional dental care setting. In some institutions, hospital dentistry has integrated well into the multidisciplinary management of HNC. Often individual HNC patients require more tailored dental care than that which is outlined in existing guidelines, with the goal to optimize oral status prior to cancer treatment [6, 8, 15, 16]. Most HNC oral optimization and maintenance guidelines are structured in a similar format: pre-, intra-, and post-HNC treatment, addressing specific oral complications throughout HNC care [7, 8, 16–29]. Each guideline tends to have similar components and recommendations, but still requires experienced clinical judgment for optimal dental care. This chapter will identify barriers to oral care and highlight practical principles to reduce or eliminate the barriers. This chapter will distill the current guidelines for oral evaluation and management of head and neck cancers into a simplified guideline, and also highlight clinical pearls and applications [6, 15]. Lastly, this chapter will highlight medical dental practitioner's role in the multidisciplinary management of HNC care.

Multidisciplinary Management of Head and Neck Cancer

It is estimated that in 2021 there will be 54,010 new cases of oral cavity and pharyngeal cancer, which is 2.8% of all new cancer cases, and 10,850 deaths [30]. This is an increase in incidence and number of deaths compared to 2010 [7]. As HNC incidence and survival rates are increasing, so is the need for longer-term oral care [18]. The National Comprehensive Cancer Network (NCCN) offers comprehensive guidelines and recommendations for the management of HNC [27]. The guidelines include principles on dental evaluation and management of HNC patients. In 2015, the Joint Task Force of the Multinational Association of Supportive Care in Cancer/International Society of Oral Oncology (MASCC/ISOO) produced Basic Oral Care (BOC) guidelines [8, 16]. Another robust guideline for oral/dental management of HNC patients was considers the full scope of HNC cancer treatment continues to be referenced by the NCCN [18, 27].

The guidelines usually reduce the management of HNC into pre-HNC treatment oral care, intra-cancer treatment oral care, post-HNC treatment oral care, and HNC complications of treatment.

These guidelines for the dental management of HNC patients may not fully address current barriers to care. Unmet oral care needs pose a risk to patient outcomes in head and neck cancer [4, 31]. One proposal has been to invite community dentists to multidisciplinary head and neck cancer tumor boards for coordination of patient care, although the practicality of this may be limited [4]. Where possible, referral to an experienced HNC dental practitioner should be made [8, 13].

The care of head and neck cancer patients in individual PCD practices may be infrequent [7–11]. Some oral/dental practitioners have raised the level of awareness of dental issues specific to head and neck cancer in an effort to optimize the quality of care [6]. The need for dental surgeons to staff regional radiation cancer centers has been identified [32, 33]. The National Comprehensive Cancer Network (NCCN) 2021 Virtual Annual Conference highlighted the oral care gap contributing to dental complications negatively affecting quality-of-life [34]. Some organizations developed formal guidelines for the composition of multidisciplinary care teams [5]. In Ontario, Canada, the minimum core team includes a surgical oncologist, radiation oncologist, medical oncologist, pathologist, dentist, clinical nurse specialist, medical imaging physician, speech-language pathologist, registered dietitian, social worker, and primary care physician (Fig. 11.1). The dentist with expertise in dental oncology is required to have completed a 1-year hospital residency that includes training in management of head and neck cancer, or an oral pathology or oral medicine residency training program. The guidelines recommend a multidisciplinary ambulatory care clinic with an orthopantomograph and cone-beam computed tomography (CBCT) imaging machine, instruments, supplies, and staffing to provide efficient oral care to HNC patients [35].

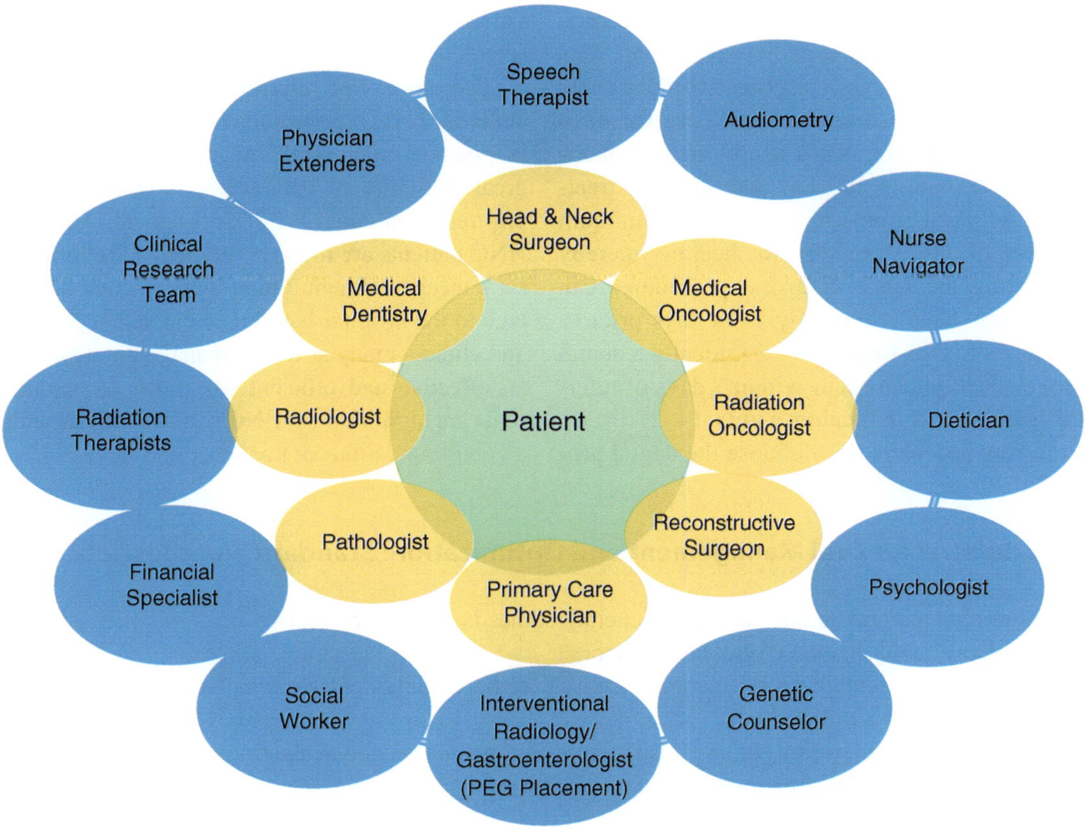

Fig. 11.1 Modified version of schema of a multidisciplinary head and neck cancer care team. *PEG* percutaneous endoscopic gastrostomy

Medical dentistry is the integration and practice of dentistry in a medical/hospital setting. Medical dentistry encompasses multidisciplinary management in a team-based care model with a hospital-based dental practice and care for patients in the same setting as many other medical and surgical specialties and integrated professionals. An analogous example of oral/dental integration into a multidisciplinary patient care is seen in the American Cleft and Craniofacial Anomalies (ACPA) organization. The ACPA team certifying body states that dentists and dental specialties (orthodontics, pediatric dentist) are required to be certified as an approved ACPA team based on the Standards for Approval of Cleft Palate and Craniofacial Teams, based on the Parameters of Care in the USA and Canada [36]. According to the ACPA Commission on Approval of Teams, ACPA team approval is non-exclusionary and voluntary. ACPA does not currently use a team certification or accreditation [36]. Aspects of this model of dental practitioner integration can be used as a structured approach to multidisciplinary team-based HNC care.

Oral Optimization: Coordinated Timely Care

Timely oral care is difficult to obtain if the patient is medically complex. The onset of coordinated timely oral care begins with an urgent referral in the case of HNC patients. This is facilitated by an established referral network, practitioner to practitioner relationships, and standardized protocols for outgoing referral. The receiving team for dental referrals must recognize the urgency and expedite scheduling. The standardized systematic protocol should include levels of urgency (routine, urgent, emergent) to aid in triage of referrals. A useful method of triage is ascertaining the start date of HNC treatment, followed by determination of the minimum number of days necessary for healing, thereby estimating the latest date oral procedures can take place. This is typically 7–14 days prior to the start of HNC treatment [27]. Ideally a dental oral evaluation can occur within 7 days of referral, with oral optimization within 14–21 [4, 8]. Referring providers must choose the dental provider service that aligns with the HNC treatment schedule.

The process of referring a HNC patient to a dentist for oral optimization may seem simple, but in practice, many HNC providers struggle with great difficulty to complete this component of pre-treatment care [4, 37]. In Medical Dentistry, HNC patients are high priority given the timeline for cancer treatment. Often there is great advantage to the hospital-based medical dental practice and clinic—ready at will to manage the patient in an effective and efficient workflow. A standard oral care algorithm for HNC patients may provide a good foundation for this process.

HNC Referral for Oral Management and Optimization: Standard Workflow [38]

1. Determine the oral health status and generalized needs of HNC patient
   1.1. Oral Cavity Status: Good, fair, or poor
   1.2. Teeth: present or not (Full complement of teeth, partially dentate, complete edentulism)
      1.2.1. Signs of Infection
2. HNC patient is referred to Medical Dentistry for consultation and evaluation:
   2.1. Determine HNC treatment modality (if known)
      2.1.1. Surgical therapy
      2.1.2. Nonsurgical therapy
      2.1.3. Combination surgical and nonsurgical therapy
   2.2. HNC Treatment Start, Date: ____________________
      2.2.1. HNC Oral Optimization, Date: ____________________ (allow 7–14 days s/p dental surgery prior to HNC radiation)
   2.3. HNC Referral Contact Information: ____
      2.3.1. Determine RT completion, Date: ____________________
3. Intake referral is processed and referred to Medical Dentistry
4. Intake referral is processed and officially referred to Medical Dentistry service for OR consult and treatment
   4.1. If indicated, contact the team to determine urgency and planned timeline of care
5. Objective for HNC patients is consult in 1 week, sometimes same day
   5.1. Consultation
      5.1.1. Treatment plan and timeline of care is established
         5.1.1.1. Oral Optimization: Determine Setting for Care
            5.1.1.1.1. Ambulatory Clinic under local (1–4 routine dental extractions)
            5.1.1.1.2. Ambulatory Clinic under nitrous oxide and/or local anesthesia (1–6 routine dental extractions)
            5.1.1.1.3. Operating Room (OR) under General Anesthesia: Day Surgery Center or High Acuity OR (difficult extractions)
            5.1.1.1.4. Operating Room (OR) under General Anesthesia: Coordinated with HNC Surgeon (difficult extractions, possible HNC surgery during same encounter)

        - 5.1.1.2. Consent obtained
        - 5.1.1.3. Surgery packet provided to patient
            - 5.1.1.3.1. Preoperative Medicine Clinic (PMC) History and Physical
        - 5.1.1.4. See PMC Visit Type education documentation for guidance
    - 5.1.2. Estimated date range
    - 5.1.3. Consent scanned and uploaded into EHR
    - 5.1.4. Surgical Case Request
        - 5.1.4.1. Surgery/urgency/location/coding/PMC visit

6. If case managed in ambulatory clinic, then schedule within 1–10 days or at appropriate timely intervals
7. If case managed in OR, then schedule within 1–10 days (Coordinate with HNC surgeon if possible):
    - 7.1. Block scheduling (first priority) or standby case request (second Priority)
    - 7.2. Prior authorization of case
        - 7.2.1. Review codes with coding specialist
        - 7.2.2. Send medical codes to medical insurer; send dental codes to dental insurer
    - 7.3. Patient Communication:
        - 7.3.1. Estimate if any portion of the case is not covered by insurance
        - 7.3.2. Non-covered form—determine how and when this is signed
        - 7.3.3. Date of surgery, Date: ____________________
        - 7.3.4. H&P/PMC reminder
        - 7.3.5. Post-op appointment scheduling
    - 7.4. Case Review—readiness
        - 7.4.1. Review chart for H&P, consent (1 week out)
        - 7.4.2. Patient phone call (OR scheduling team—night before)
        - 7.4.3. Cancellation
            - 7.4.3.1. If surgery cancels, put in a telephone encounter for faculty/resident to alert practitioner
    - 7.5. Urgent, Add-on Cases
        - 7.5.1. Check with hospital dentist
8. Communicate with referring HNC team
    - 8.1. Schedule
        - 8.1.1. Re-evaluation and reinforcement of preventative protocol (6–12 weeks s/p RT), Date: ____________________
        - 8.1.2. Determine individualized oral hygiene recall/follow-up interval (q3,6,9,12 months), Interval: ____________________
    - 8.2. Plan:
        - 8.2.1. Fluoride tray (if indicated)
            - 8.2.1.1. Diagnostic dental impressions
            - 8.2.1.2. Fabricate fluoride trays and insertion
        - 8.2.2. Oral positioning devices (if indicated)
            - 8.2.2.1. Diagnostic dental impressions
            - 8.2.2.2. Fabricate oral positioning devices and insertion
        - 8.2.3. Obturator (if indicated)
            - 8.2.3.1. Determine type, fabricate as indicated
        - 8.2.4. Follow-up (oral hygiene recall): q3,6,9,12 months, Date: ____________________

There may be some cases where the timeline and/or circumstances are such that an oral workup, examination, and intervention may be most efficiently coordinated with the HNC surgeon. This is treated in the same manner as if the patient was seen in clinic for a routine consult, but the consult and treatment occur same day as HNC resection surgery in the OR setting. On the front end, this process usually requires a more active effort by the HNC surgeon to determine oral health needs based on experience and rapport with the hospital dental team. The workflow is similar to the standard workflow with several modifications:

HNC Referral for Oral Management and Optimization: Urgent Workflow [38]

1. Determine the oral health status and generalized needs of HNC patients
   - 1.1. Oral Cavity Status: Good, fair, or poor
   - 1.2. Teeth: present or not (Full complement of teeth, partially dentate, complete edentulism)
     - 1.2.1. Signs of infection
     - 1.2.2. Difficult extractions
2. HNC Surgeon refers patient for oral examination and optimization prior to HNC radiation
   - 2.1. HNC Surgeon requests OR coordination under special circumstances
     - 2.1.1. Severe delay in HNC therapy occurred, oral optimization warranted
     - 2.1.2. Transfer of HNC care from outside hospital (OSH) resulted in HNC therapy significant delay
     - 2.1.3. Obvious non-restorable dentition, difficult serial exodontia, and alveoloplasty warranted
     - 2.1.4. PCD cannot meet HNC therapy deadline, often due to serial restorations and/or extractions
     - 2.1.5. Patients informed tooth or teeth indicated for extractions and/or restorations, unable to expedite care locally
     - 2.1.6. Patient proximity to regional hospital greater than 2–3 h, coordinated same day care is optimal to reduce burden of transportation and risk of delays
3. HNC Surgeon contacts hospital dentist directly (email, phone, pager) to discuss case and determine feasibility of coordinated OR case
   - 3.1. Determine possible dates
   - 3.2. Schedule surgery, Date: ___________________
     - 3.2.1. Determine procedure order, communicate with anesthesia and surgery teams
       - 3.2.1.1. Most of the time this depends on surgery team schedules
4. Intake referral is processed and officially referred to Medical Dentistry service for OR consult and treatment
   - 4.1. Communication to medical dental office staff team is advisable
   - 4.2. Communication to surgery scheduler is advisable

5. Hospital Dentist objectives for HNC patients in OR setting
   5.1. Consultation
      5.1.1. Consent obtained in pre-operative setting
      5.1.2. Patient is transferred from pre-operative room to medical dentistry clinic for radiographs
         5.1.2.1. Orthopantomograph with four (4) bitewings radiographs (posterior teeth, interproximal)
            5.1.2.1.1. Radiographs are read and interpreted
         5.1.2.2. Complete intraoral series: eighteen (18) radiographs:
            5.1.2.2.1. Most often the intraoral radiographs are obtained within the OR with patient under general anesthesia, this is case dependent
            5.1.2.2.2. Radiographs are read and interpreted
         5.1.2.3. Photographs obtained:
            5.1.2.3.1. Extraoral
            5.1.2.3.2. Intraoral
      5.1.3. Comprehensive oral examination is completed under general anesthesia in OR
         5.1.3.1. Oral cavity charted (odontogram updated) and treatment planned
            5.1.3.1.1. Dental prophylaxis or debridement is completed
            5.1.3.1.2. Dental restorations are completed
            5.1.3.1.3. Dental exodontia is completed with alveoloplasty (if warranted)
      5.1.4. Treatment plan and timeline of care is established
         5.1.4.1. Oral Optimization: Determine Setting for Care
            5.1.4.1.1. Operating Room (OR) under General Anesthesia: Coordinated with HNC Surgeon (difficult extractions)
6. Communicate with referring HNC team
   6.1. Oral optimization completed, Date: ____________________
   6.2. Schedule
      6.2.1. Re-evaluation and reinforcement of preventative protocol (6–12 weeks s/p RT), Date: ____________________
      6.2.2. Determine individualized oral hygiene recall/follow-up interval (q3,6,9,12 months), Interval: ____________________
   6.3. Plan:
      6.3.1. Fluoride tray (if indicated)
         6.3.1.1. Diagnostic dental impressions
         6.3.1.2. Fabricate fluoride trays and insertion
      6.3.2. Oral positioning devices (if indicated)
         6.3.2.1. Diagnostic dental impressions
         6.3.2.2. Fabricate oral positioning devices and insertion
      6.3.3. Obturator (if indicated)
         6.3.3.1. Determine type, fabricate as indicated
      6.3.4. Follow-up (oral hygiene recall): q3,6,9,12 months, Date: ____________________

Hospital dentistry completes daily inpatient rounds on patients until discharged. Several clinical pearls to maximizing success of expedited and coordinated oral care through urgent workflow is great team-based communication; frequent communication with the patient, significant others and/or family members prior to day of surgery, during inpatient stay and post-operatively. Coordinating treatment and care builds strong rapport with patient, patient's support cohort and HNC team. This process is typically greatly appreciated by all parties involved and lends itself very well to integrated care with radiation medicine and medical oncology. Patients are often unaware of the existence of hospital dentists and find medical dentistry to be a valuable resource in all phases of HNC care.

Head and Neck Cancer: Phases of Oral Care Management

Pre-HNC Therapy—Oral Care and Optimization

A comprehensive oral examination consists of a full mouth series of radiographs, an orthopantomograph, and a complete hard and soft tissue examination. Dental prophylaxis, dental restorations, exodontia, dental impressions, and fabrication of fluoride trays should be completed or obtained prior to HNC radiation therapy. Proactive pre-HNC therapy measures to mitigate post-treatment complications can be accomplished by identification and elimination of acute and chronic oral infections of both odontogenic and non-odontogenic origin, protecting the posterior mandibular bone (body of mandible) by eliminating non-restorable molars and/or teeth with poor prognosis, identify and eliminate periodontal disease, identify signs of potential compromising risk factors such as mandibular tori, impacted third molars (M3), and preserve salivary gland vitality and function. Seize moments to maximize pre-prosthetic surgery opportunities (i.e., alveoloplasty, mandibular tori removal, palatal torus palatinus removal, maxillary and mandibular exostoses reduction) in order to reduce post-HNC therapy risks and complications. Exodontia (dental extraction) must be completed on retained tooth components (retained roots, crowns), teeth with mobility, impacted teeth with communication to the oral cavity, partially erupted teeth, teeth with periodontal involvement (>5 mm attachment/bone loss and/or periodontal pocket) must be completed prior to HNC radiation therapy (RT) [8]. Local oral soft tissue irritants (sharp, rough enamel, tongue rings, lip rings, dental calculus) can cause significant traumatic lesions and must also be completely removed. After the oral cavity and dentition are stabilized, oral home care and maintenance is essential in combating future disease development.

The periodontium contains the structures that hold the tooth to the alveolar crest of the maxilla and mandible. It is comprised of alveolar bone, gingiva and gingival attachments to tooth structure, cementum and periodontal ligament (PDL) [39]. Periodontal disease is defined as gingivitis or periodontitis. Periodontitis is the inflammation of the periodontal tissues producing pocketing of periodontium, loss of tooth attachments, and alveolar bone loss [8]. It can lead to tooth mobility, risk of infection, dental caries, root caries, and loss of periodontium and teeth [39].

Osteoradionecrosis (ORN) is bone death resulting from exposure to therapeutic radiation unrelated to tumor recurrence. ORN is progressive, often ill-defined, with lytic bone changes in the mandible. Signs and symptoms of ORN include bone sclerosis and sequestra (dead bone). ORN is due to radiation-induced hypovascularity and innate compact bone structure [40, 41]. The posterior mandible (body of mandible) is most commonly affected at a frequency ratio of mandible:maxilla 24:1 [42].

Mandibular unilateral torus or bilateral tori (torus mandibularis) are benign exophytic expansile bony lesions of trabecular and cortical bone located near the premolar sites on the lingual aspect of the mandible. Mandibular tori are found in various sizes, shapes, and contours. They are usually asymptomatic but can be easily traumatized due to their exophytic nature and coverage only by a thin and taut mucosa. Presence of torus mandibularis (mandibular tori) is a local factor that may increase the risk of osteoradionecrosis following post-RT tooth/teeth extraction in the posterior mandible [42]. Tori are routinely removed during pre-prosthetic surgery in preparation for an oral prosthesis (removable maxillary or mandibular complete or partial denture).

The effects HNC therapy has on the oral cavity is complex and dynamic. The oral cavity can be described as an aperture with maxillary bones fixed to the skull and the mandible held in place with a sling of muscles articulated at a point of rotation and translation of the temporal mandibular joint (TMJ). The maxilla has a robust blood supply while the mandible has a poor blood supply. Many HNC oral complications can be attributed to the compromised mandibular blood supply and corticated dense alveolar bone necessary to maintain integrity of dentition during the constant brutal physiologic forces applied in normal activities such as mastication, physical fitness, and articulation between maxilla and mandible. The posterior mandible is required to bare a complex and dynamic load favorable to withstand the most punishing and greatest forces generated by the oral cavity. Dense bone is required and deposited to shore up and maintain the integrity of mandibular molar occlusal position, plane of occlusion, allow minor movement, combat the forces generated during mastication and be fracture resistant. The most susceptible and relatively hypovascular portion of the mandible is the posterior mandible. After radiation, if the hypovascular bone becomes injured or greater compromised (i.e., tooth extraction), healing is compromised. Presence of mandibular tori typically indicates more dense and hypovascular bone, further increasing chance of poor healing following post-HNC radiation tooth/teeth extractions.

The impact of HNC treatment is harsh on the oral cavity and surrounding radiated structures. Damage to the salivary glands can be irreversible and compromise the oral environment and microflora [43]. The distribution of saliva is critical to the protection of enamel—the hardest structure in the human body—and tends to be extremely durable and withstand the severe physiologic forces generated on a daily basis. Enamel can become compromised by decrease in saliva production.

Intra-HNC Therapy: Oral Care Maintenance

Throughout Active HNC therapy, the risk for oral mucositis is elevated and oral and dental care is relatively contraindicated; the dental team remains on standby [8]. In case of dental emergency, immediate treatment may be necessary [44]. The role is fairly simple: manage and reinforce oral hygiene, manage symptoms, maintain communication with the multidisciplinary HNC team, and continue management of oral complications in a non-surgical fashion [8].

Post-HNC Therapy: Oral Care—Risks, Impact, and Complications of HNC

Despite the best efforts of hospital dentists to lower the risk level of post-HNC treatment complications, some patients will return with complication status post dental extractions. The keys to managing complications are experience in HNC oral care, and raising awareness to the HNC team as early as possible to minimize adverse outcomes. Raising the level of awareness and involving the oncology team is critical in order to optimize patients for infection management. Most often oral surgical site infections can be managed with debridement, cleaning, and antibiotics.

Complications of head and neck cancer treatment are many—dysphagia, aspiration, trismus, esophageal stricture, stomatitis, neutropenia, anorexia and cachexia, lymphedema, radiation caries, periodontal complications, osteoradionecrosis, xerostomia, candida infection, and mucositis [9, 45].

When attempting to mitigate risk for osteoradionecrosis, prevention is key:

- Dental extraction(s) of indicated teeth pre-HNC therapy; and conversely avoidance of post-radiation extractions when possible
- Complete pre-prosthetic surgery prior to head and neck cancer therapy
- Exostoses removal
- Torus Palatinus (Maxillary palatal torus) removal, if necessary (Bilateral)
- Mandibular tori removal
- Complete necessary alveoloplasty

Home oral care rituals are disrupted during and after head and neck cancer therapy. Even if the patient's pre-treatment home oral care was good, our experience indicates this declines to varying degrees. Key considerations are location of primary tumor, quantity of soft tissue sclerosis, mucus and salivary gland function, and the volume and dose of radiation to the maxilla and/or mandible. Anecdotally, if the tumor is located at or below the lower border of the mandible without tongue involvement, then the patients oral home hygiene remains around the level it was at prior to cancer therapy. If the cancerous tumor is at or above the lower border of the mandible, then the patient's oral care regimen tends to suffer more. This is usually modified by the degree of bony and soft tissue resection and post-operative sclerosis.

Overcoming Barriers to Multidisciplinary Management of Head and Neck Cancer: Oral Care

There are multiple barriers to obtaining reliable, efficient and steadfast oral care and optimization in HNC. Some of the most notable barriers are insufficient amount of practitioners, support staff, access to oral care, adequate and sustainable funding, and lack of oral care continuity. Care coordination is essential to providing and coordinating multidisciplinary team-based care [46]. This role increases the efficacy and efficiency of care delivery and integration with other services through status tracking dynamics and human interface. A care coordinator reduces the practitioner's administrative and logistical burden and allows for the practitioner to focus on patient clinical care. Management of urgent head and neck cancer intake referrals is taxing but necessary for the expedited care model. In our practice, the referral intake for patient consult may range from 1 to 7 days. This type of care is a coordinated effort between our medical dental cohort including clinical care coordinator, surgery scheduler, hospital dental assistants, and hospital dentists.

An experienced OR surgery scheduler is essential in the hand-off of surgical case requests, scheduling the patient for surgery, obtaining pre-authorization, coordinating care with the PMC, and with HNC surgeons. If this role is done well, then I can have a profound impact on scheduled and urgent work queues. Experienced hospital dentists can train dental assistants for their team on-site. Dental assistants can also cross-train in many areas of dentistry and the dental specialties. An OMFS-trained dental assistant brings great value to a hospital dental program as well. Hospital dental assistants work in the operating room, manage preparation of instruments and assist in the management of orofacial trauma, HNC surgery care, hospital-based pediatric dental surgery, and team-based craniofacial anomalies and cleft team care.

A sustainably, adequately funded medical dentistry program and well-equipped hospital dental surgeon faculty are paramount to success of medical dental integration into the multidisciplinary management of head and neck cancer care. Academic medical and teaching hospitals are more likely to support such program, which may require significant funding support. Hospital leadership must be apprised of the importance of oral care in the HNC patient context, as well as application to other seriously ill cohorts of patients [37]. If the patient has dental insurance with partial coverage of services, or no dental benefits, some medical centers have committed funding to support dental evaluation, radiographic imaging, dental restorations, dental extractions, alveoloplasty, and tori removal prior to their necessary medical treatment. This is a main driving force to retain hospital dental staff, and to rapidly accept, evaluate, and treat patients prior to HNC treatment.

Continuity of Oral Care: Many patients report they have not seen a dentist in many years, if ever. A medical dental clinic located within the hospital is extremely advantageous in the care of HNC patients. Hospital dentists maintain hospital privileges and routine care for patients in the hospital setting, including inpatient, emergency department and ambulatory sites. There are three primary settings by which we accomplish our care: (1) Hospital dental clinic/surgery suite, (2) Hospital dental clinic with anxiolytics and/or IV sedation, and (3) Operating rooms in an ambulatory day surgery center or inpatient surgery operating room. Hospital dental services maintain 24/7 on-call services. The scope of our medical dental practice includes management of the oral cavity, in particular the dentoalveolar complex. Hospital dental programs can provide care for community oncologic practices as well [13].

Conclusion

HNC is best managed by a multidisciplinary team. Experienced hospital dentists ideally manage efficient programs in academic or high-volume community hospitals. They are committed to managing medically complex patients in team-based care, which frequently includes HNC patients. The integration of medical dentistry into the multidisciplinary management of HNC is a critical contribution to optimizing post-treatment quality of life.

References

1. Turner JA. The Dentist in the Hospital. J Natl Med Assoc. 1930;22(3):157–8. PMID: 20892402; PMCID: PMC2625241.
2. Kunkel GH, Roberts GK. Hospital dentistry: the Swiss-Army Knife of Dental Medicine. J Dent Educ. 2021;85:112–5. https://doi.org/10.1002/jdd.12476.
3. Bellissimo-Rodrigues WT, et al. Is it necessary to have a dentist within an intensive care unit team? Report of a randomized clinical trial. Int Dent J. 2018;68(6):420–7.
4. Beeram M, Kennedy A, Hales N. Barriers to comprehensive multidisciplinary head and neck care in a community oncology practice. Am Soc Clin Oncol Educ Book. 2021;41:e236–45.
5. Selby P, Popescu R, Lawler M, Butcher H, Costa A. The value and future developments of multidisciplinary team cancer care. Am Soc Clin Oncol Educ Book. 2019;39:332–40.
6. Samim F, Epstein JB, Zumsteg ZS, et al. Oral and dental health in head and neck cancer survivors. Cancers Head Neck. 2016;1:14. https://doi.org/10.1186/s41199-016-0015-8.
7. Murdoch-Kinch CA, Zwetchkenbaoum S. Dental management of the head and neck cancer patient treated with radiation therapy. J Michigan Dent Assoc. 2011;93:28–37.
8. Epstein JB, Barasch A. Oral and dental health in head and neck cancer patients. In: Maghami E, Ho A, editors. Multidisciplinary care of the head and neck cancer patient. Cancer treatment and research, vol. 174. Cham: Springer; 2018. https://doi.org/10.1007/978-3-319-65421-8_4.
9. Dixon HG, Thomson WM, Ting GS. Dentists' knowledge and experiences of treating patients with head and neck cancer. N Z Dent J. 2021;117(1):15–21. 7 p.
10. Barker GJ, Epstein JB, Williams KB, Gorsky M, Rober-Dulacher JE. Current practice and knowledge of oral care for cancer patients: a survey of supportive health care providers. Support Care Cancer. 2005;13:32–41.
11. Epstein JB, Parker IR, Epstein MS, Gupta A, Kutis S, Witkowski DM. A survey of National Cancer Institute-designated comprehensive cancer centers' oral health supportive care practices and resources in the USA. Support Care Concern. 2007;15:357–62.
12. Epstein JB, Parker IR, Epstein MS, Stevenson-Moore P. Cancer-related oral health care services and resources: a survey of oral and dental care in Canadian cancer centers. J Can Dent Assoc. 2004;70:302–4.
13. Maghami E, Ho AS. Multidisciplinary care of head and neck cancer patient. New York, NY: Springer; 2018. https://doi.org/10.1007/978-3-319-65421-8.
14. Neal RD, Tharmanathan P, France B, et al. Is increased time to diagnosis and treatment in symptomatic cancer associated with poorer outcomes? Systematic review. Br J Cancer. 2015;112(suppl 1):S92–S107.
15. National Cancer Institute. Oral complications of chemotherapy and head/neck radiation. Bethesda, MD: National Cancer Institute; 2021. https://www.cancer.gov/about-cancer/treatment/side-effects/mouth-throat/oral-complications-hp-pdq. Accessed 3 Sept 2021.
16. Elad S, Raber-Durlacher JE, Brennan MT. Basic oral care for hematology-oncology patients and hematopoietic stem cell transplantation recipients: a position paper from the joint task force of the Multinational Association of Supportive Care in Cancer/

International Society of Oral Oncology (MASCC/ISOO) and European Society for Blood and Marrow Transplantation (EBMT). Support Care Cancer. 2015;23(1):223–36.
17. Studer G, Glanzmann C, Studer SP, et al. Risk-adapted dental care prior to intensity-modulated radiotherapy (IMRT). Schweiz Monatsschr Zahnmed. 2011;121:216–29.
18. Epstein JB, Thariat J, Bensadoun RJ, et al. Oral complications of cancer and cancer therapy: from cancer treatment to survivorship. CA Cancer J Clin. 2012;62:400–22.
19. National Comprehensive Cancer Network. Principles of dental evaluation and management. National Comprehensive Cancer Network guidelines version 3.2021. Plymouth Meeting, PA: National Comprehensive Cancer Network; 2021.
20. Farah CS, Balasubramaniam R, McCullough MJ. Contemporary oral medicine. Cham: Springer; 2019.
21. Kuriakose MA. Contemporary oral oncology. Cham: Springer; 2017. ISBN 978-3-319-43854-2.
22. Nowak AJ, Caramassimo PS. The handbook of pediatric dentistry, vol. 18. 4th ed. Chicago, IL: American Academy of Pediatric Dentistry; 2011. p. 225–31.
23. Little JW, Falace DA, Miller CS, Rhodus NL. Dental management of the medically compromised patient, vol. 26. 7th ed. St. Louis, MO: Elsevier; 2008. p. 433–60.
24. Hupp JR, Ellis E III, Tucker MR. Comtempory oral and maxillofacial surgery, vol. 18. 5th ed. Amsterdam: Elsevier; 2008. p. 363–79.
25. Jawad H, Hodson N, Nixon P. A review of dental treatment of head and neck cancer patients, before, during and after radiotherapy: Part 1. Br Dent J. 2015;218:65–8. https://doi.org/10.1038/sj.bdj.2015.28.
26. American Association of Oral and Maxillofacial Surgeons. Head and neck cancer screening and prevention. Rosemont, IL: American Association of Oral and Maxillofacial Surgeons; 2020. p. 1–2.
27. National Comprehensive Cancer Network (NCCN). Clinical practice guidelines in oncology head and neck cancers version 3.2021. Plymouth Meeting, PA: National Comprehensive Cancer Network; 2021. https://www.nccn.org/professionals/physician_gls/pdf/head-and-neck_blocks.pdf. Accessed 29 Aug 2021.
28. Hancock PJ, Epstein JB, Sadler GR. Oral and dental management related to radiation therapy for head and neck cancer. J Can Dent Assoc. 2003;69(9):585–90.
29. Sroussi HY, Jessri M, Epstein J. Oral assessment and management of the patient with head and neck cancer. Oral Maxillofac Surg Clin North Am. 2018;30(4):445–58. https://doi.org/10.1016/j.coms.2018.06.006. PMID: 30173900.
30. National Cancer Institute. Surveillance, epidemiology, and end results program. Cancer stat facts: cancer of the oral cavity and pharynx. Bethesda, MD: National Cancer Institute; n.d.. https://seer.cancer.gov/statfacts/html/oralcav.html. Accessed 3 Sept 2021.
31. Prades J, Remue E, van Hoof E, et al. Is it worth reorganising cancer services on the basis of multidisciplinary teams (MDTs)? A systematic review of the objectives and organisation of MDTs and their impact on patient outcomes. Health Policy. 2015;119:464–74.
32. Lizi E. A case for a dental surgeon at regional radiotherapy centres. Br Dent J. 1992;173(1):24–6.
33. De Felice F. Multidisciplinary team in head and neck cancer: a management model. Med Oncol. 2018;36:2.
34. Caudell JJ, Pfister DG, Weber RS. Supportive care in head and neck cancers: multidisciplinary management. J Natl Compr Cancer Netw. 2021;19:625–8. https://doi.org/10.6004/jnccn.2021.5010.
35. Irish J, Kim J, Waldron J, Wei AC, Winquist E, Yoo J, Boasie A, Brouwers M, Meertens E, McNair S, Walker-Dilks C, Expert Panel on Organizational Guidance for the Care of Patients with Head and Neck Cancer in Ontario. Organizational guidance for the care of patients with head-and-neck cancer in Ontario. Curr Oncol. 2020;27(2):e115–22. https://doi.org/10.3747/co.27.5873. PMID: 32489261; PMCID: PMC7253750.
36. The American Cleft Palate-Craniofacial Association (ACPA). Standards for approval of cleft palate and craniofacial teams. Chapel Hill, NC: The American Cleft Palate-Craniofacial Association (ACPA); n.d.. https://acpa-cpf.org/wp-content/uploads/2019/04/Standards-2019-Update.pdf. Accessed 3 Sept 2021.
37. Ward M, Carpenter MD, et al. Oncologists' perspective on dental care around the treatment of head and neck cancer: a pattern of practice survey. J Oncol Pract. 2021;18:e28. https://doi.org/10.1200/OP.20.00913.
38. Long K, Dienberg MJ. Surgery scheduling workflow - HDS standard work. Portland, OR: OHSU Hospital Dental Services; 2019.
39. Rose LF, Mealey BL, Genco RJ, Cohen DW. Periodontics: medicine, surgery, and implants. Amsterdam: Elsevier; 2004.
40. Gomez DR, Estilo CL, Wolden SL, et al. Correlation of osteoradionecrosis and dental events with dosimetric parameters in intensity-modulated radiation therapy for head-and-neck cancer. Int J Radiat Oncol Biol Phys. 2011;81:3207–e213.
41. Lee IJ, Koom WS, Lee CG, et al. Risk factors and dose-effect relationship for mandibular osteoradionecrosis in oral and oropharyngeal cancer patients. Int J Radiat Oncol Biol Phys. 2009;75:1084–91.
42. Koenig LJ, Tamimi D, Petrikowski CG, Perschbacher SE. Diagnostic imaging oral and maxillofacial, vol. Section III. 2nd ed; 2017. p. 342–7.
43. Jensen SB, Pedersen AM, Vissink A, et al. A systematic review of salivary gland hypofunction and xerostomia induced by cancer therapies: management strategies and economic impact. Support Care Cancer. 2010;18:1061–79.

44. Chandra RA. Instructions for dentists evaluating patients undergoing head and neck cancer therapy. Dental practitioner handout. Boston, MA: Division of Oral Medicine and Dentistry; n.d.
45. The Oral Cancer Foundation. n.d.. https://oralcancerfoundation.org/complications/. Accessed 3 Sep 2021.
46. Patio C, et al. Physician assistants and nurse practitioners in head and neck surgery. In: Maghami E, Ho A, editors. Multidisciplinary care of the head and neck cancer patient. Cancer treatment and research, vol. 174. Cham: Springer; 2018. https://doi.org/10.1007/978-3-319-65421-8_2.

12 Supportive Care, Comorbid Conditions, and Survivorship

Z. A. Kohutek and B. A. Murphy

Key Points

- Supportive care impacts patient outcomes by improving patient quality of life, increasing patient treatment compliance, and reducing late toxicities.
- Head and neck cancer patients have a high prevalence of comorbid medical conditions, assessment of which is important in accurately determining a patient's prognosis and selecting most appropriate treatments.
- Survivorship is a complex period of transition during which patients are likely to experience a wide range of challenges that are best met by a multidisciplinary approach.
- Health care providers should involve patient caregivers both during and after treatment in order to maximize patient compliance and outcomes.

Supportive Care in Head and Neck Cancer

Therapies for head and neck cancer, including surgery, chemotherapy, and radiotherapy are associated with significant acute and late toxicities. These toxicities can lead to increased morbidity and mortality, as well as impaired quality of life. Efforts to mitigate such toxicities is of critical importance in order to improve patient outcomes. Supportive care encompasses a wide range of interventions designed to prevent, reduce, and manage toxicities related to the disease and cancer-directed therapies [1].

The benefits of high-quality supportive care in the management of head and neck cancer are often underappreciated by patients and medical professionals alike [2, 3]. In addition to improving quality of life, symptom management can result in increased patient compliance and treatment completion, resulting in improved cancer outcomes. Supportive care can also reduce late effects and as such can improve overall survival. On a larger scale, supportive care can reduce health care costs by preventing hospital admissions during treatment and reduce risk for costly late complications.

Z. A. Kohutek (✉)
Department of Radiation Oncology, Vanderbilt University Medical Center, Nashville, TN, USA
e-mail: zachary.a.kohutek@vumc.org

B. A. Murphy
Ingram Cancer Center, Vanderbilt University Medical Center, Nashville, TN, USA

R. A. Chandra, R. J. Li (eds.), *Multidisciplinary Management of Head and Neck Cancer*,
https://doi.org/10.1007/978-3-031-05973-5_12

Comorbidity Assessment in Head and Neck Cancer

Head and neck cancer patients incur an increased risk for morbidity and mortality from comorbid conditions unrelated to their underlying cancer. Between 36% and 75% of head and neck cancer patients have comorbidities at baseline, and as many as half of these have moderate or severe comorbid medical conditions [4]. This patient population also has high rates of tobacco and alcohol abuse, which are themselves associated with higher rates of cardiovascular and respiratory illness. Furthermore, advanced age is a common risk factor for both head and neck cancer and many comorbid illnesses.

There are multiple definitions and measures of comorbidity status. Tools such as the Adult Comorbidity Evaluation 27 (ACE-27) and Charlson Index (CI) are used in a wide range of clinical settings and both have been validated in head and neck cancer, but numerous head and neck cancer specific comorbidity indices have also been developed, including the Head and Neck Cancer Index (HNCA) and Washington University Head and Neck Comorbidity Index (WUHNCI), which may be more predictive in certain settings [5, 6]. Regardless of how comorbidity is measured, it has been repeatedly demonstrated to have prognostic and predictive value, as patients with increased comorbidity at baseline have worse outcomes including increased severe acute toxicities [7] and worse survival [8, 9].

In addition to considering baseline medical comorbidity, patients should be assessed for socioeconomic comorbidities such as low education and income that can also adversely affect treatment outcomes [10]. Furthermore, the disease itself can secondarily lead to malnutrition, compromised immune status, chronic aspiration, and other conditions that may lead to increased patient frailty, impaired ability to complete treatment, and a significantly higher risk for mortality. Age is frequently used to assess a patient's fitness for treatment, but comorbidity remains a more informative measure of a patient's ability to tolerate a given therapeutic intervention [11]. Assessment of comorbidity can also be helpful in determining an individual patient's supportive care needs during and after treatment.

Pre-treatment Supportive Care

In addition to assessing a patient's comorbidity status and ability to tolerate a particular treatment, clinicians must take steps to maximize a patient's functional status prior to initiation of cancer-directed therapy. If patients are to undergo radiation therapy as part of their treatment, pre-treatment dental evaluation is strongly recommended [12]. Patients should complete all required dental procedures including extraction of diseased teeth prior to initiation of radiation, as post-radiation extractions are associated with an increased risk for late complications including osteoradionecrosis. If patients require pre-treatment extractions, they should ideally be fully healed and have had at least 2 weeks elapse before starting radiotherapy. Patient education in proper oral hygiene is also important. Recommendations include using a soft toothbrush, fluoridated toothpaste, and daily topical fluoride mouthwashes or fluoride gel application using fluoride trays.

In preparation for radiation therapy, specialized nutritional screening, assessment, and counseling are essential to identify patients with baseline malnutrition, maximize nutritional status, and help maintain caloric intake during treatment [13]. Randomized studies have shown that individualized nutritional counseling and use of oral supplements can minimize weight loss, maximize nutritional status, and improve patient quality of life [14, 15]. In the surgical setting, patients with severe malnutrition should be provided with preoperative nutritional support in order to help reduce perioperative mortality.

Patients are also typically referred for evaluation by a speech and swallow specialist prior to initiating radiotherapy for head and neck cancer. Swallow assessment at baseline provides a reference point for future studies, and counseling regarding swallowing including pre-treatment swallowing exercises has been demonstrated to improve swallowing function and dysphagia related quality of life after radiation [16, 17].

In patients undergoing surgery, preoperative counseling is necessary to help prepare patients and their families for anticipated physical, mental, and psychological challenges posed by the procedure. If speech or voice is to be affected, particularly in the case of laryngectomy or glossectomy, it is important to set realistic expectations regarding short-term and long-term functional outcomes. It is also strongly recommended to engage caregivers in these discussions, as this can help maximize patient compliance with therapeutic and supportive interventions.

Acute Symptom Control During Cancer Treatment

Patients undergoing head and neck cancer treatment develop a range of predictable toxicities that each require specific management strategies. Proper management of acute treatment-related toxicities is of paramount importance to maximize patient outcomes. Several of the most common acute toxicities associated with chemoradiation treatment are discussed below.

Mucositis and Odynophagia

Oral mucositis is a nearly universal side effect of radiation to the head and neck region [18]. Patients typically develop mucosal erythema during the second week of radiation, followed by the development of mucosal ulceration and pain. As patients progress through to approximately the fifth week of radiation, they develop confluent ulceration that limits oral intake. Severe mucositis then persists until approximately 2–4 weeks following the completion of radiotherapy, during which time most patients require increasing doses of opiate pain medications and drastic dietary modification that eliminates solid or even soft foods in favor of liquids. In many cases, patients with severe mucositis develop the inability to maintain adequate nutrition orally and require placement of a percutaneous endoscopic gastrostomy tube for enteral nutrition. This is particularly common in patients requiring concomitant chemotherapy, which increases the severity of oral mucositis. Complications of oral mucositis are many, and include pain, malnutrition, dehydration, infection, and psychosocial distress.

Management of mucositis is focused first on preventative measures designed to reduce the severity of mucositis or associated pain. While most preventative interventions lack high level evidence, recent guidelines have supported the use of several agents for the purpose of mucositis prevention, including benzydamine oral rinses, oral glutamine, and topical honey [19]. There is also a growing body of evidence for low-level energy light therapy termed photobiomodulation [20]. Although this therapy depends on very specific protocols using specialized equipment, and long-term risks including potential carcinogenic effects have not been well characterized, photobiomodulation remains an intriguing and possibly highly effective means for prophylactic management of oral mucositis.

In addition to these prophylactic measures, the addition of neuromodulatory drugs can reduce mucositis-related pain. Gabapentin has been demonstrated in multiple single institution randomized controlled trials to reduce mucositis-related pain when administered during radiation treatment and is now considered standard of care [21]. Probiotics also show promise in reducing the severity of oral mucositis and, while data on their effectiveness remains somewhat limited, can be employed with little risk for toxicity [22].

Dermatitis

Radiation-induced dermatitis is common during head and neck treatment [23]. Radiation produces a sunburn-like reaction that can range from mild to severe depending on radiation dose to skin, use of concurrent chemotherapy, and a patient's inherent radiosensitivity. Patients are typically advised to keep the skin clean and dry by daily washing, to avoid direct sun exposure during treatment, and to utilize water-based moisturizers on a daily basis during treatment. Topical corticosteroids have traditionally been used to manage symptoms of pruritus during radiation treatment, but their use in preventing severe radiation dermatitis in head and neck cancer has also recently been demonstrated [24]. As patients progress through their radiation treatments, a subset of patients will develop moist desquamation. In these cases, additional topic agents such as silver sulfadiazine or silver containing dressings may be employed to promote healing.

In recent years, the EGFR inhibitor cetuximab has also been utilized in the treatment of head and neck cancers. This agent can cause an inflammatory dermatitis that is exacerbated by concomitant radiotherapy, and management of this dermatitis varies from radiation dermatitis, often necessitating agents such as doxycycline, minocycline, or isotretinoin [25].

Salivary Dysfunction

Head and neck cancer patients frequently develop thick secretions, excess mucous, or hyposalivation resulting in xerostomia [26]. Head and neck radiotherapy can cause an acute increase in the viscosity of saliva. This can be managed with frequent hydration, sodium bicarbonate oral rinses, and use of mucolytic agents such as guaifenesin. For patients with excess mucous production, treatment with anticholinergic agents such as glycopyrrolate or scopolamine may be indicated, although it should be noted that these therapies can cause excessive dryness which may be unpalatable to some patients. For patients that have early xerostomia during treatment, frequent hydration, use of a cool-mist humidifier, and oral rinses with sodium bicarbonate can be helpful.

Cancer Surveillance

Careful screening for cancer recurrence should be performed regularly by clinicians, and patients should be educated regarding signs and symptoms of cancer recurrence. Although the frequency of routine follow-up should be individualized, in general patients should be evaluated by their physicians every 1–3 months for the first year, every 2–6 months for the second year, every 4–8 months for years 3–5, and annually thereafter. Patients who have a current or prior history of smoking should be assessed to determine if they are candidates for annual lung cancer screening with low-dose CT scans.

Health Promotion in Survivorship

In addition to routine cancer surveillance, patients should be counseled regarding the need for health maintenance with emphasis on key issues relevant to head and neck cancer survivors. Even mild residual taste alterations, hyposalivation, and dysphagia result in dietary adaptations with associated dietary deficiencies. Thus, patients must be aware of the need to maintain a balanced diet with attention to both macro- and micronutrients. Patient frequently experiences significant deconditioning and muscle mass loss associated with treatment. This coupled with fatigue can lead to a marked decline in activity. Patients should be educated about the importance of a systematic approach to physical activity including strength training and aerobic exercise. All head and neck cancer survivors must be educated regarding the importance of thorough dental hygiene practices and the need for frequent dental visits. If survivors are continuing to use tobacco products, they should be referred for counseling or cessation therapy as necessary. During cancer treatment, general medical care is often placed on hold. Upon completion of therapy, patients must also be encouraged to follow-up with their primary care physician for management of pre-existing medical conditions.

Survivorship and Management of Late Effects

As head and neck cancer treatments improve, and with changing demographics related to increased incidence of HPV-related cancers, long-term survival after head and neck cancer treatment is becoming more common. As such, survivorship care has evolved to become a critical component of cancer care, encompassing efforts to mitigate and manage late treatment toxicities. The American Cancer Society has published survivorship care guidelines detailing recommendations for care of head and neck cancer patients [27]. Treatment-related toxicities evolve over time. Some acute effects such as taste alterations may slowly decrease in intensity over time. Other late effects may progress slowly, manifesting weeks to years after completion of therapy. For example, patients may have mild stiffness in the neck and shoulders at the completion of therapy which may progress over time to frank fibrosis with contracture. Thus, patients should be made aware that supportive care regimens must be assessed and adjusted as the clinical picture evolves. Thus, head and neck cancer patients should be thoroughly evaluated for toxicities from treatment as well as specific needs including informational, psychosocial, or health concerns at intervals that are appropriate to their symptom burden and functional deficits. Providers should also be certain to encourage inclusion of caregivers in conversations about survivorship, as this may aid in patient care and support. Some of the most common toxicities of head and neck cancer therapy are described below, along with suggested general management strategies.

Lymphedema

The head and neck region contains an extensive lymphatic network. This is frequently disrupted in cancer patients, either by tumor or by surgery and radiation [28]. Lymphatic dysfunction results in accumulation of lymph fluid within tissues of the head and neck, which itself can promote inflammation and sclerosis. Patients may develop external lymphedema involving facial and submental tissues, which can lead to suboptimal cosmesis or impaired range of motion. Patients can also develop internal lymphedema affecting structures such as the epiglottis, which can potentiate functional impairment such as dysphagia and dysphonia [29]. Clinicians should monitor for lymphedema by use of visual inspection, which may include endoscopic assessment of internal edema. Patients can then be referred to a specialist for consideration of manual lymphatic drainage and compressive bandaging [30].

Dysphagia

Dysphagia after head and neck cancer treatment correlates strongly with patient quality of life [31]. Both surgery and radiation can lead to acute changes in swallowing function. The degree to which therapeutic interventions alter swallowing is in part determined by their extent of impact on dysphagia-aspiration related structures (DARS), including the pharyngeal constrictors and supraglottic larynx. Head and neck cancer survivors should be monitored closely for dysphagia and symptoms of aspiration including cough, dyspnea, or pneumonia, and if appropriate should be referred to speech-language pathologists for swallow evaluation. Swallow therapy is aimed at improving swallowing through exercises to improve strength and mobility, swallowing techniques, and sensory response. While evidence supporting improvement in swallowing function with such interventions is limited [32], these efforts may at least help to limit progression of late effects caused by treatment-induced fibrosis and lymphedema that can worsen over time. Even with optimal therapy, long-term dysphagia can persist as a significant cause of late morbidity and impaired quality of life. Novel and personalized therapeutic interventions will be important to help improve swallowing outcomes in this population.

Dysphagia can also result from esophageal stricture, which can frequently develop after laryngectomy or radiation to the hypopharynx and cervical esophagus [33, 34]. In patients with suspected esophageal stricture, upper endoscopy should be performed for evaluation, and esophageal dilation should be considered [35]. While head and neck cancer patients are at higher risk for complications from this procedure, dilation has been demonstrated to be effective in improving symptoms of dysphagia [36]. Unfortunately, the effects of esophageal dilation are often transient, and patients often require multiple repeat procedures over the course of months to years.

Dysgeusia

Altered taste is a common toxicity that occurs during and persists after head and neck radiotherapy [37]. Severity of dysgeusia is thought to relate to radiation dose to the tongue and oral cavity, as higher doses lead to atrophy of taste buds. Impaired salivary function can also contribute to altered taste. Disturbances in taste typically improve over the course of months to years, but this improvement can be slow and incomplete. With no available pharmacological therapy, dysgeusia can lead to long-term negative effects on quality of life and nutritional status. Patients experiencing persistent or severe alterations in taste may be referred for specialized dietary counseling to assist in developing a nutrition plan that maximizes intake of nourishing foods while minimizing unpleasant foods and seasonings.

Ototoxicity

Sensorineural hearing loss can be caused by exposure to ototoxic drugs such as cisplatin or by high dose radiation to the cochlea [38]. Hearing loss may initially affect higher frequency ranges, but symptoms can be progressive over time. In addition, radiation can cause eustachian tube dysfunction which can lead to muffled hearing, tinnitus, imbalance, and a feeling of fullness in the ear [39]. If this persists, it may require insertion of a tube to ventilate the middle ear. Any patient suffering from hearing loss after head and neck cancer therapy should be referred to an audiologist for further testing including tympanometry, pure tone testing, and speech reception threshold testing. Aural rehabilitation can be considered, which involves use of hearing aids or, for patients with profound sensorineural hearing loss, consideration of cochlear implants.

Dysphonia

Dysphonia refers to the impaired ability to produce sounds of speech, caused either by impaired production of sound as air passes through the larynx (voice) or an inability to shape sounds into words that are understandable (speech). Treatment for head and neck cancer can negatively affect both voice and speech, typically either by surgical extirpation of or damage to tissue in the larynx, oral cavity, or oropharynx. While dysphonia is common in these patients, new or worsening changes in voice in the survivorship setting should be evaluated thoroughly to rule out new or recurrent malignancy.

Restoration of voice after total laryngectomy can be achieved by use of an electrolarynx, an electronic transducer that induces vibration of the vocal tract, or by tracheoesophageal puncture, a surgical procedure that involves placing a one-way valve between the tracheostoma and the cervical esophagus that allows expired air to pass into and produce vibration in the neopharynx [40]. Even after less extensive laryngeal surgery, glottic insufficiency can develop leading to dysphonia. In such cases, surgical techniques such as laryngoplasty can be performed to narrow the glottic gap and improve voice [41]. Patients may also benefit from indirect voice therapy which includes techniques that encourage proper vocal hygiene, including maintaining appropriate hydration and discouraging overuse. These techniques can also be useful in improving voice quality after radiotherapy. In summary, patients with voice or speech disturbance after therapy should be referred to an experienced speech-language pathologist as numerous interventions can be employed to potentially improve function.

Musculoskeletal Impairment

Head and neck cancer treatment can cause impaired function of musculature in the jaw, neck, and shoulders, resulting in functional limitation and decreased quality of life [42]. Trismus, or the inability to fully open the mouth, is a common treatment-related complication that can lead to difficulty eating and speaking [43]. While modern radiotherapy techniques such as intensity modulated radiotherapy (IMRT) may decrease the prevalence of trismus, it is frequently seen in the survivorship setting, particularly in patients with advanced tumors at baseline. Physical therapy including jaw stretching exercises, with or without the use of a dental appliance for stretching or splinting, is recommended.

Neck and shoulder dysfunction are also quite prevalent after head and neck cancer therapy [42]. Spinal accessory nerve palsy is now less common with the advent of modern surgical techniques, but even when the spinal accessory nerve is preserved patients can develop fibrosis, contracture, tissue atrophy, and minor nerve damage from surgery and radiation. This can lead to dystonic spasms, pain, and motor weakness. Neck and shoulder stretching exercises are routinely recommended to help maintain mobility and minimize the development of fibrosis. In patients who develop neuropathic pain and/or muscle spasms, agents such as pregabalin, gabapentin, duloxetine, or botulinum toxin type A injections can be utilized [44]. In addition to fibrosis and dystonia, combined modality therapies can result in postural weakness, which itself increases risk for pain, aspiration, and impaired lung function. Patients with any symptoms of musculoskeletal impairment may benefit from physical and occupational therapy, and if refractory can be referred to a physical medicine and rehabilitation physician.

Oral and Dental Complications

Xerostomia, or oral dryness, is a common sequela of head and neck radiotherapy related to hyposalivation. Even with salivary gland sparing using IMRT, almost 90% of patients develop some degree of xerostomia, with many reporting decreased quality of life due to resultant impairments in swallowing, sleep, and speech [45]. In addition, xerostomia leads to a lower oral pH and decreased intraoral calcium and phosphate levels, causing a shift toward demineralization of dental enamel and development of dental caries [46]. Head and neck cancer survivors should follow closely with a dental professional in order to continue intensive preventive care. Patients should also practice a daily dental hygiene regimen including frequent brushing with remineralizing toothpaste, daily flossing, use of fluoride, and dietary modification to a low-sucrose diet. Xerostomia can often be managed symptomatically with over-the-counter salivary substitutes. Some patients with residual salivary function may benefit from use of salivary stimulants such as sugar-free candy or gum. Sialogogues such as pilocarpine, a muscarinic agonist, can also be prescribed in an effort to increase salivary flow [47].

Osteoradionecrosis is a late sequela of radiotherapy in which an area of exposed bone, most typically in the mandible, does not heal [48]. Clinically this process can manifest as pain, facial deformity, fracture, and orocutaneous fistula formation. Osteoradionecrosis typically develops between a few months and several years after radiation treatment. Management includes the use of broad-spectrum antibiotics and daily use of saline or chlorhexidine rinses in early-stage disease. In later stages, patients may be referred for hyperbaric oxygen therapy or referred to a head and neck surgeon for debridement of necrotic tissue. In severe cases, patients may require extensive bone resection with free flap reconstruction.

Body Image

Body image dissatisfaction is present in up to 1 in 5 head and neck cancer patients and is most common in younger patients and those who undergo extensive surgery or have wound healing issues [49]. Distress related to body image increases risk for depressive symptoms and can significantly impact patients' coping, anxiety, and daily functioning [50]. Patients with altered body image may feel a "loss of self" that impacts their ability to regain their pre-cancer identity. Furthermore, patients may seek social avoidance as a strategy to reduce embarrassment associated with any perceived changes in body image, thus further exacerbating feelings of social isolation that are already prevalent in head and neck cancer population [51]. Patients should be screened for body image distress and should be referred for professional counseling when appropriate.

Chronic Systemic Symptoms

Bodily stressors such as head and neck cancer and its treatments lead to neurohumoral activation and a robust inflammatory response. While the physiologic purpose of these processes is to achieve recovery and restore homeostasis, when exaggerated or persistent they may cause maladaptive permanent physiologic changes that are detrimental in the long term, producing chronic symptoms such as fatigue, pain, sleep disturbance, neurocognitive dysfunction, mood disorders, and metabolic alterations resulting in cachexia and sarcopenia. These symptoms are thought to be mediated in part by peripheral proinflammatory cytokines including interleukin (IL)-1, IL-6, and tumor necrosis factor alpha (TNF-α), which are elevated during and after head and neck cancer treatment [52]. This proinflammatory environment can cause neuroinflammation that, together with psychological stressors, can lead to long-term alterations in neural function. While each of these symptoms will be discussed independently, it should be noted that they often present in combination as symptom clusters [53].

Fatigue

Chronic fatigue is the result of complex physiologic processes that lead to decreased energy and physical exhaustion. Fatigue is extremely prevalent in head and neck cancer survivors and has been shown to cause significant disability and decreased quality of life [53]. Fatigue in these patients has been linked to younger age, radiation treatment, closer temporal proximity of treatment, and depressive symptoms [54]. Within the broad categorization of fatigue there are multiple distinct clinical syndromes which manifest with different symptoms [55]. Central fatigue, for instance, is characterized by behavioral changes including depression and lack of motivation, while peripheral fatigue is typified by muscle fatigue and exhaustion [56]. In addition, fatigue can be caused by hormonal or hematologic abnormalities related to cancer treatments.

Patients with fatigue should undergo laboratory studies to assess for other potentially modifiable causes of fatigue including hypothyroidism and anemia. Given the high propensity for hypothyroidism to occur after head and neck radiotherapy, survivors should undergo regular screening for hypothyroidism with TSH testing every 6–12 months. Medications used to manage other conditions could also potentiate fatigue, and careful medication reconciliation should be performed. Patients should also be evaluated for possible mood or sleep disorders which can contribute to fatigue. Physical causes of fatigue such as obstructive sleep apnea should be considered and treated if identified.

Chronic Pain

Nearly half of head and neck cancer survivors suffer from chronic pain [57, 58], either due to focal musculoskeletal impairment, peripheral neuropathy, mucosal sensitivity, or central pain syndromes that result in more widespread musculoskeletal pain [59]. Pain in these patients has been linked to mood disorders including depression and anxiety and has been shown to cause impaired quality of life [58]. Patients often require long-term treatment with opiate pain medications or neuromodulatory agents such as gabapentin, pregabalin, or duloxetine. It is important to treat underlying mood disorders, as these could contribute to the aberrant processing of pain signals.

Sleep and Neurocognitive Disturbance

Neurocognitive function including attention, thinking, and memory are frequently impaired at baseline in head and neck cancer patients [60]. These domains can become further compromised after cancer therapy. Much study has been done in patients with nasopharyngeal and skull base tumors, in whom the brain often receives significant radiation doses. Such patients have been found to display baseline and posttreatment deficits in a wide range of neurocognitive domains including concentration, verbal recall, visual recall, memory, and visuospatial abilities [61]. Even in patients without directed brain radiation, similar neurocognitive deficits may develop [60]. Neuropsychiatric symptoms including irritability, distractibility, and restlessness may also occur in a large proportion of head and neck cancer patients and can persist long term [53].

Many head and neck cancer survivors suffer from sleep disturbances including insomnia, hypersomnolence, and sleep-related breathing disturbances. Patients with such sleep issues should be referred to a sleep specialist and undergo a polysomnogram if there is evidence of sleep apnea. Other sleep hygiene measures can be recommended depending on a particular patient's specific constellation of symptoms. Examples include use of a cool-mist humidifier to help with airway dryness, nasal strips and more upright positioning to help reduce snoring, and removal of dentures to avoid nighttime irritation.

Mood Disorders

Mood disorders, in particular depression and anxiety, are exceedingly common in the head and neck cancer population [62]. Risk factors for head and neck cancer include nicotine and alcohol dependence, which themselves are associated with a higher prevalence of mood disorders at baseline. There is also evidence that genetic polymorphisms may selectively predispose head and neck patients to depression [63]. In patients that are already at high risk for mood disorders, the emotional stress of cancer treatment may itself trigger anxiety and depression. There is also emerging evidence that the same proinflammatory cytokines that are elevated in head and neck cancer are involved with the pathogenesis of depression [64]. Emotional distress can also be related to inability to cope with long-term sequelae of treatment, financial toxicities, or caregiver strain. Patients should be assessed at least annually for emotional distress, depression, and anxiety using a validated screening tool, and referral for appropriate mental health resources including psychiatric counseling or medications should be considered.

Metabolic Alterations

Proinflammatory cytokines associated with head and neck cancer and cancer treatments lead to a state of increased muscle catabolism and reduced anabolism [65]. These processes can lead to both cachexia, a condition manifested by unintentional weight loss, and sarcopenia, defined as a reduction in skeletal muscle mass. Cachexia is present in many head and neck cancer patients prior to treatment [66], and a proinflammatory state resulting from treatment can exacerbate weight and muscle loss. Sarcopenia is also present in a substantial proportion of patients with head and neck cancer at baseline and up to 80% of patients have sarcopenia after completion of radiation treatment. Sarcopenic patients tend to have worse overall survival [67], indicating the importance of muscle wasting in patient long-term outcomes. Patients should be counseled on proper nutrition and dietary modifications in order to achieve and maintain a healthy weight.

References

1. Bonomo P, et al. Quality assessment in supportive care in head and neck cancer. Front Oncol. 2019;9:926.
2. Berman R, et al. Supportive care: an indispensable component of modern oncology. Clin Oncol (R Coll Radiol). 2020;32(11):781–8.
3. Monnery D, et al. Multi-professional-delivered enhanced supportive care improves quality of life for patients with incurable cancer. Int J Palliat Nurs. 2018;24(10):510–4.
4. Paleri V, et al. Comorbidity in head and neck cancer: a critical appraisal and recommendations for practice. Oral Oncol. 2010;46(10):712–9.

5. Piccirillo JF, et al. Development of a new head and neck cancer-specific comorbidity index. Arch Otolaryngol Head Neck Surg. 2002;128(10):1172–9.
6. Piccirillo JF, et al. Comparison of comorbidity indices for patients with head and neck cancer. Med Care. 2004;42(5):482–6.
7. Monteiro AR, et al. ACE-27 as a prognostic tool of severe acute toxicities in patients with head and neck cancer treated with chemoradiotherapy: a real-world, prospective, observational study. Support Care Cancer. 2021;29(4):1863–71.
8. Gollnitz I, et al. Role of comorbidity on outcome of head and neck cancer: a population-based study in Thuringia, Germany. Cancer Med. 2016;5(11):3260–71.
9. Stordeur S, et al. Comorbidity in head and neck cancer: is it associated with therapeutic delay, post-treatment mortality and survival in a population-based study? Oral Oncol. 2020;102:104561.
10. Choi SH, et al. Socioeconomic and other demographic disparities predicting survival among head and neck cancer patients. PLoS One. 2016;11(3):e0149886.
11. Derks W, de Leeuw RJ, Hordijk GJ. Elderly patients with head and neck cancer: the influence of comorbidity on choice of therapy, complication rate, and survival. Curr Opin Otolaryngol Head Neck Surg. 2005;13(2):92–6.
12. Buglione M, et al. Oral toxicity management in head and neck cancer patients treated with chemotherapy and radiation: dental pathologies and osteoradionecrosis (Part 1) literature review and consensus statement. Crit Rev Oncol Hematol. 2016;97:131–42.
13. Talwar B, et al. Nutritional management in head and neck cancer: United Kingdom National multidisciplinary guidelines. J Laryngol Otol. 2016;130(S2):S32–40.
14. Isenring EA, Bauer JD, Capra S. Nutrition support using the American Dietetic Association medical nutrition therapy protocol for radiation oncology patients improves dietary intake compared with standard practice. J Am Diet Assoc. 2007;107(3):404–12.
15. Ravasco P, et al. Impact of nutrition on outcome: a prospective randomized controlled trial in patients with head and neck cancer undergoing radiotherapy. Head Neck. 2005;27(8):659–68.
16. Carroll WR, et al. Pretreatment swallowing exercises improve swallow function after chemoradiation. Laryngoscope. 2008;118(1):39–43.
17. Kulbersh BD, et al. Pretreatment, preoperative swallowing exercises may improve dysphagia quality of life. Laryngoscope. 2006;116(6):883–6.
18. Maria OM, Eliopoulos N, Muanza T. Radiation-induced oral mucositis. Front Oncol. 2017;7:89.
19. Elad S, et al. MASCC/ISOO clinical practice guidelines for the management of mucositis secondary to cancer therapy. Cancer. 2020;126(19):4423–31.
20. Campos TM, et al. Photobiomodulation in oral mucositis in patients with head and neck cancer: a systematic review and meta-analysis followed by a cost-effectiveness analysis. Support Care Cancer. 2020;28(12):5649–59.
21. Smith DK, et al. Preventive use of gabapentin to decrease pain and systemic symptoms in patients with head and neck cancer undergoing chemoradiation. Head Neck. 2020;42(12):3497–505.
22. Shu Z, et al. The effectiveness of probiotics in prevention and treatment of cancer therapy-induced oral mucositis: a systematic review and meta-analysis. Oral Oncol. 2020;102:104559.
23. Rosenthal A, Israilevich R, Moy R. Management of acute radiation dermatitis: a review of the literature and proposal for treatment algorithm. J Am Acad Dermatol. 2019;81(2):558–67.
24. Yokota T, et al. Phase 3 randomized trial of topical steroid versus placebo for prevention of radiation dermatitis in patients with head and neck cancer receiving chemoradiation. Int J Radiat Oncol Biol Phys. 2021;111:794.
25. Pinto C, et al. Management of skin reactions during cetuximab treatment in association with chemotherapy or radiotherapy: update of the Italian expert recommendations. Am J Clin Oncol. 2016;39(4):407–15.
26. Bomeli SR, et al. Management of salivary flow in head and neck cancer patients--a systematic review. Oral Oncol. 2008;44(11):1000–8.
27. Nekhlyudov L, et al. Head and neck cancer survivorship care guideline: American Society of Clinical Oncology clinical practice guideline endorsement of the American Cancer Society Guideline. J Clin Oncol. 2017;35(14):1606–21.
28. Deng J, et al. Prevalence of secondary lymphedema in patients with head and neck cancer. J Pain Symptom Manag. 2012;43(2):244–52.
29. Deng J, Ridner SH, Murphy BA. Lymphedema in patients with head and neck cancer. Oncol Nurs Forum. 2011;38(1):E1–E10.
30. Deng J, Sinard RJ, Murphy B. Patient experience of head and neck lymphedema therapy: a qualitative study. Support Care Cancer. 2019;27(5):1811–23.
31. Hunter KU, et al. Toxicities affecting quality of life after chemo-IMRT of oropharyngeal cancer: prospective study of patient-reported, observer-rated, and objective outcomes. Int J Radiat Oncol Biol Phys. 2013;85(4):935–40.
32. Perry A, et al. Therapeutic exercises for affecting post-treatment swallowing in people treated for advanced-stage head and neck cancers. Cochrane Database Syst Rev. 2016;8:CD011112.
33. Laurell G, et al. Stricture of the proximal esophagus in head and neck carcinoma patients after radiotherapy. Cancer. 2003;97(7):1693–700.
34. Sweeny L, et al. Incidence and outcomes of stricture formation postlaryngectomy. Otolaryngol Head Neck Surg. 2012;146(3):395–402.
35. Prisman E, Miles BA, Genden EM. Prevention and management of treatment-induced

pharyngo-oesophageal stricture. Lancet Oncol. 2013;14(9):e380–6.
36. Moss WJ, et al. Esophageal dilation in head and neck cancer patients: a systematic review and meta-analysis. Laryngoscope. 2018;128(1):111–7.
37. Deshpande TS, et al. Radiation-related alterations of taste function in patients with head and neck cancer: a systematic review. Curr Treat Options in Oncol. 2018;19(12):72.
38. Theunissen EA, et al. Sensorineural hearing loss in patients with head and neck cancer after chemoradiotherapy and radiotherapy: a systematic review of the literature. Head Neck. 2015;37(2):281–92.
39. Christensen JG, et al. Otitis media with effusion after radiotherapy of the head and neck: a systematic review. Acta Oncol. 2018;57(8):1011–6.
40. Tang CG, Sinclair CF. Voice restoration after total laryngectomy. Otolaryngol Clin N Am. 2015;48(4):687–702.
41. Bertelsen C, Reder L. Efficacy of type I thyroplasty after endoscopic cordectomy for early-stage glottic cancer: literature review. Laryngoscope. 2018;128(3):690–6.
42. Ghiam MK, et al. Assessment of musculoskeletal impairment in head and neck cancer patients. Support Care Cancer. 2017;25(7):2085–92.
43. Bensadoun RJ, et al. A systematic review of trismus induced by cancer therapies in head and neck cancer patients. Support Care Cancer. 2010;18(8):1033–8.
44. Cohen EE, et al. American Cancer Society head and neck cancer survivorship care guideline. CA Cancer J Clin. 2016;66(3):203–39.
45. Cooperstein E, et al. Vanderbilt Head and Neck Symptom Survey version 2.0: report of the development and initial testing of a subscale for assessment of oral health. Head Neck. 2012;34(6):797–804.
46. Pinna R, et al. Xerostomia induced by radiotherapy: an overview of the physiopathology, clinical evidence, and management of the oral damage. Ther Clin Risk Manag. 2015;11:171–88.
47. LeVeque FG, et al. A multicenter, randomized, double-blind, placebo-controlled, dose-titration study of oral pilocarpine for treatment of radiation-induced xerostomia in head and neck cancer patients. J Clin Oncol. 1993;11(6):1124–31.
48. Frankart AJ, et al. Osteoradionecrosis: exposing the evidence not the bone. Int J Radiat Oncol Biol Phys. 2021;109(5):1206–18.
49. Melissant HC, et al. Body image distress in head and neck cancer patients: what are we looking at? Support Care Cancer. 2021;29(4):2161–9.
50. Manier KK, et al. The impact and incidence of altered body image in patients with head and neck tumors: a systematic review. Neurooncol Pract. 2018;5(4):204–13.
51. van Deudekom FJ, et al. Functional and cognitive impairment, social environment, frailty and adverse health outcomes in older patients with head and neck cancer, a systematic review. Oral Oncol. 2017;64:27–36.
52. Akmansu M, et al. Influence of locoregional radiation treatment on tumor necrosis factor-alpha and interleukin-6 in the serum of patients with head and neck cancer. Cytokine. 2005;31(1):41–5.
53. Wulff-Burchfield E, et al. Late systemic symptoms in head and neck cancer survivors. Support Care Cancer. 2019;27(8):2893–902.
54. Rogers LQ, et al. Factors associated with fatigue, sleep, and cognitive function among patients with head and neck cancer. Head Neck. 2008;30(10):1310–7.
55. Maes M, Twisk FN, Johnson C. Myalgic Encephalomyelitis (ME), Chronic Fatigue Syndrome (CFS), and Chronic Fatigue (CF) are distinguished accurately: results of supervised learning techniques applied on clinical and inflammatory data. Psychiatry Res. 2012;200(2–3):754–60.
56. Dantzer R, et al. The neuroimmune basis of fatigue. Trends Neurosci. 2014;37(1):39–46.
57. Macfarlane TV, et al. Head and neck cancer pain: systematic review of prevalence and associated factors. J Oral Maxillofac Res. 2012;3(1):e1.
58. Cramer JD, Johnson JT, Nilsen ML. Pain in head and neck cancer survivors: prevalence, predictors, and quality-of-life impact. Otolaryngol Head Neck Surg. 2018;159(5):853–8.
59. Borsook D, Erpelding N, Becerra L. Losses and gains: chronic pain and altered brain morphology. Expert Rev Neurother. 2013;13(11):1221–34.
60. Bond SM, Dietrich MS, Murphy BA. Neurocognitive function in head and neck cancer patients prior to treatment. Support Care Cancer. 2012;20(1):149–57.
61. Meyers CA, et al. Neurocognitive effects of therapeutic irradiation for base of skull tumors. Int J Radiat Oncol Biol Phys. 2000;46(1):51–5.
62. Haisfield-Wolfe ME, et al. Prevalence and correlates of depression among patients with head and neck cancer: a systematic review of implications for research. Oncol Nurs Forum. 2009;36(3):E107–25.
63. Gilbert J, et al. Depression in patients with head and neck cancer and a functional genetic polymorphism of the serotonin transporter gene. Head Neck. 2012;34(3):359–64.
64. Felger JC, Lotrich FE. Inflammatory cytokines in depression: neurobiological mechanisms and therapeutic implications. Neuroscience. 2013;246:199–229.
65. Argiles JM, et al. The role of cytokines in cancer cachexia. Curr Opin Support Palliat Care. 2009;3(4):263–8.
66. Jager-Wittenaar H, et al. High prevalence of cachexia in newly diagnosed head and neck cancer patients: an exploratory study. Nutrition. 2017;35:114–8.
67. Grossberg AJ, et al. Association of body composition with survival and locoregional control of radiotherapy-treated head and neck squamous cell carcinoma. JAMA Oncol. 2016;2(6):782–9.

13 Rare Head and Neck Cancers

Margaret B. Mitchell, Amy Juliano, and Jeremy Richmon

Key Points

- Rare head and neck cancers can be divided into those arising from epithelial (e.g., esthesioneuroblastoma, mucosal melanoma), nervous (e.g., malignant nerve sheath tumors), and soft and connective tissue (e.g., sarcomas) origin.
- Most of these rare cancers affect older adults, although rhabdomyosarcoma and Ewing's sarcoma are typically pediatric malignancies, and esthesioneuroblastoma can present across age groups.
- These pathologies have varying natural histories: while mucosal melanoma, neuroendocrine carcinomas, and certain sarcomas are aggressive and have poor survival rates, others like chondrosarcoma and Ewing's sarcomas have excellent prognosis.
- Many of these tumors are optimally treated with surgical resection with negative margins with or without adjuvant therapy; however, patients with extensive tumors involving critical structures and patients with metastatic spread are often treated non-operatively.
- The lack of large scale data detailing these rare pathologies poses a challenge in terms of standardized staging systems and treatment algorithms, especially as less invasive and potentially more efficacious treatment options like immunotherapy, targeted therapy, and stereotactic radiosurgery are increasingly introduced into practice.

M. B. Mitchell · J. Richmon (✉)
Department of Otolaryngology-Head & Neck Surgery, Harvard Medical School/Massachusetts Eye and Ear, Boston, MA, USA
e-mail: Margaret_Mitchell@meei.harvard.edu; Jeremy_Richmon@meei.harvard.edu

A. Juliano
Department of Radiology, Harvard Medical School/Massachusetts Eye and Ear, Boston, MA, USA
e-mail: Amy_Juliano@meei.harvard.edu

R. A. Chandra, R. J. Li (eds.), *Multidisciplinary Management of Head and Neck Cancer*,
https://doi.org/10.1007/978-3-031-05973-5_13

Exemplar Case

A 19-year-old man with 1.5 pack-year smoking history presented with progressive discomfort of the right jaw. Over a period of about 6 months he noticed the development of a 5 cm subcutaneous, fixed, mildly tender right-sided facial mass. A biopsy was consistent with synovial cell sarcoma. He had full facial nerve function bilaterally at the time of presentation. Imaging with CT, MRI, and a full-body PET scan at the time of diagnosis showed a large, destructive lesion of his right face extending over the mandible and invading the masticator space, parapharyngeal space, right maxilla, infratemporal fossa, and pterygoid space.

Medical Oncology and Radiation Oncology recommended neoadjuvant chemoradiation before surgical resection. He tolerated his neoadjuvant therapy well and then underwent segmental mandibulectomy, resection of right masseteric space, parapharyngeal space, infratemporal fossa, and skull base with parotidectomy and right facial nerve dissection, and limited right neck dissection as well as reconstruction of the mandible with a free fibular flap. Significant manipulation of the facial nerve branches were required to free the tumor, although at the end of the case all branches were structurally intact. His pathology did show positive margins.

He then tolerated adjuvant chemoradiation well and had partial return of facial movement on the right side of his face postoperatively although he had expected numbness in the V3 distribution.

Approximately 3.5 years after initial diagnosis he was found to have a new pulmonary nodule on surveillance chest imaging. This nodule was resected and found to be metastatic synovial cell carcinoma. He has not had any further evidence of metastasis since and continues to undergo serial surveillance imaging.

Introduction

The specialized tissues in the head and neck allow for an array of mechanical and sensory functions; however, this cellular diversity also provides a wide spectrum of potential malignancies. While the majority of head and neck cancers arise from epithelial cells, specifically the squamous cell mucosa of the aerodigestive tract [1], less common malignancies arise from specialized epithelia, as well as connective, muscle, nervous, and endocrine tissue.

In this chapter, we cover these rare head and neck malignancies, organized by the cell type of origin. We combine malignancies arising from muscle and connective tissue given the nonspecific derivation of sarcomas arising broadly from soft tissue origin, and include malignancies which affect pediatric populations given head and neck surgeons may be involved in care for these patients in addition to pediatric otolaryngologists. We focus specifically on natural history, diagnosis, and management for these pathologies and include pertinent elements of epidemiology; management strategies are summarized in Table 13.1. As precision medicine is increasingly integrated into head and neck cancer treatment along with less invasive radiation-based treatment options, we give special attention to emerging literature in these areas, allowing potentially both for greater treatment efficacy and less morbidity. Given the extensive breadth of rare head and neck malignancies, this chapter is not meant to be all-inclusive and covers those pathologies more frequently encountered within this group.

Table 13.1 Management modalities by pathology: Given the rarity of these pathologies and the lack of standardized treatment regimens, treatment modality supported by highest level of evidence and most commonly described in the available literature is included in the table. "+/–" is meant to reflect both the variability in whether adjuvant therapy may be warranted in patients (e.g., only in patients with positive surgical margins) and the overall uncertainty of benefit of adjuvant therapies in these pathologies

Pathology	Treatment
Esthesioneuroblastoma	Early stage: Radiation alone
	Late stage: Surgical resection +/– adjuvant radiation
Mucosal melanoma	Surgical resection + adjuvant radiation
Neuroendocrine carcinoma	Surgical resection +/– adjuvant chemotherapy and/or radiation versus chemoradiation alone
Chondrosarcoma	Surgical resection or debulking
Osteogenic sarcoma	Surgical resection +/– adjuvant radiation with or without chemotherapy
Synovial cell sarcoma	Surgical resection + adjuvant chemotherapy and/or radiation
Ewing's sarcoma	Surgical resection +/– adjuvant radiation
Rhabdomyosarcoma	Chemotherapy + surgical resection or chemoradiation alone if unresectable
Leiomyosarcoma	Surgical resection +/– adjuvant radiation
Angiosarcoma	Surgical resection + chemoradiation or chemoradiation alone if unresectable
Liposarcoma	Surgical resection +/– adjuvant radiation and/or chemotherapy
Kaposi sarcoma	Antiretroviral therapy + chemotherapy; surgical resection reserved for symptomatic localized or minimally aggressive cases
Teratocarcinosarcoma	Surgical resection +/– adjuvant radiation and/or chemotherapy
Malignant nerve sheath tumors	Surgical resection +/– adjuvant radiation or chemotherapy
Malignant paraganglioma	Surgical resection +/– adjuvant radiation

Epithelial Tissue

Head and neck cancers arising from epithelial cell lines are by far the most common malignancies in this region, dominated by squamous cell carcinoma [1]. The epithelium in the head and neck not only is exposed to the potential toxins both from the environment (e.g., inhaled irritants, human papilloma virus) and the body itself (e.g., refluxed gastric acid, trauma from teeth), but is forced to regenerate upon injury or chronic inflammation, providing an opportunity for uncontrolled growth if cellular mechanisms to prevent this process are disrupted. Integration of sensory organs and neurons as well as endocrine glands increases the complexity of the epithelial layers in this region.

Esthesioneuroblastoma

Arising from the olfactory neuroepithelium, esthesioneuroblastoma occurs in 1 per 2.5 per million people and generally presents in a bimodal fashion in either the second or sixth decade of life, although rare pediatric presentations have been reported as well [2]. These tumors can fill the nasal cavity and extend into nearby structures, such as the orbits and intracranially ([3], [Fig. 13.1]), and most commonly present as an obstructing nasal mass causing headaches, congestion, and epistaxis although metastatic neck masses are possible at the time of presentation. Histological examination of this tumor is notable for neurosecretory granules with neurofilaments and microtubules along with clas-

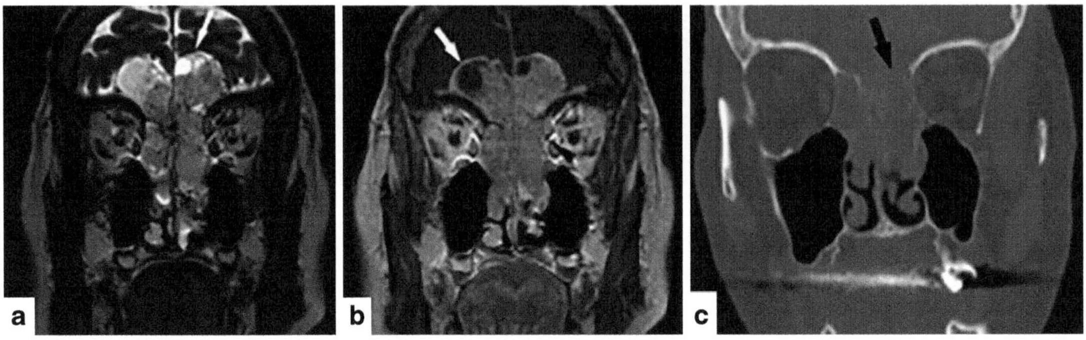

Fig. 13.1 Esthesioneuroblastoma. A 70-year-old female who presented with nasal congestion. (**a**) Coronal T2-weighted MR image shows a bulky lobulated mass of intermediate T2 signal in bilateral nasal cavities with intracranial extension. Note the cysts at the leading edge of the intracranial component (arrow), a characteristic feature of esthesioneuroblastoma when extending intracranially. (**b**) Coronal contrast-enhanced T1-weighted MR image shows homogeneous avid enhancement of the mass except in the cysts at the leading edge (white arrow). Note extension of the mass to the left orbit abutting the left superior oblique muscle (black arrowhead), suggesting invasion of the periorbita. (**c**) Coronal CT image in bone algorithm shows bony erosion of the ethmoid septations and ethmoid roof (arrow)

sic pathological features of high nucleus to cytoplasm ratio and scant, poorly defined cytoplasm [4]. Esthesioneuroblastomas are assessed based on their histological grade with the Hyams grading system which correlates with prognosis and disease-free survival [5].

Multiple meta-analyses and systemic reviews focus on treatment options given these tumors are chemo- and radio-sensitive as well as potentially amenable to surgical resection [6–8]. Current data suggests that early stage tumors can be treated with radiotherapy alone, while more extensive tumors require surgery. In terms of surgical management, subtotal endoscopic resection followed by adjuvant radiation is in many treatment centers replacing attempted total resection; for example, endoscopic endonasal surgery is recommended for low grade tumors due to its lower morbidity. Other more invasive approaches can also be used for resection, including a transcranial approach with lumbar drain placement and bifrontal craniotomy, or for more extensive tumors crossing midline, a Draf III frontal sinus approach. And while both pre- and post-operative radiation have been shown to improve surgical outcomes, there is again no standard management algorithm to clarify optimal timing of different treatment modalities. In patients who present with cervical disease, upfront neck dissection and adjuvant radiation treatment seem to be beneficial, although adjuvant chemotherapy may also play a role in preventing nodal recurrence after surgery. Newer treatment modalities like stereotactic radiosurgery as well as proton-beam therapy appear promising. Given the recent finding of frequent expression of somatostatin receptor 2a in these tumors, there is also potential for diagnosis with radioreceptor techniques as well as treatment with somatostatin analogues [4].

Recurrence rates vary widely (from 0% to 86%) across studies, and also vary based on initial tumor stage as well as treatment modality [4]. Post-treatment, these tumors have an average time of recurrence of 2–6 years, although tumors can recur up to 19 years after treatment [9]. Given these late recurrences, serial MRIs [3] as well as yearly chest radiographs [10] should be done to monitor metastases. Survival is similarly dependent both on tumor grade and extent as well as treatment modality. Five-year overall survival for patients undergoing combined chemoradiation alone and surgical resection alone may be similar—51% [6] versus 48–78% [4]. Survival in this time interval may be improved with the addition of adjuvant treatment to surgery, as surgical resection with adjuvant radiation has a reported 5-year survival rate of 65–75% [4].

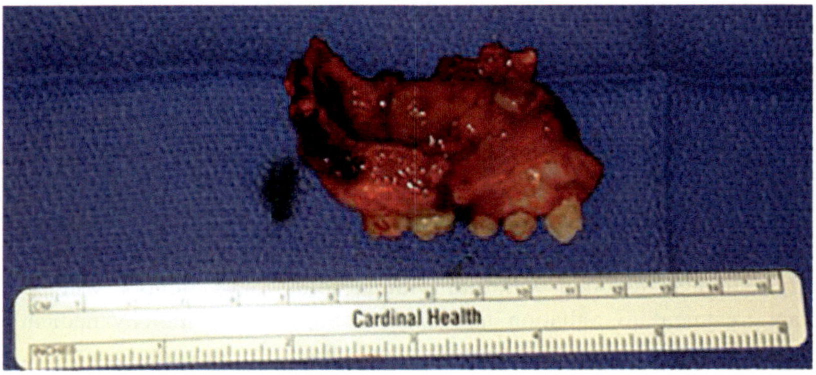

Fig. 13.2 Sinonasal mucosal melanoma: Specimen from a partial maxillectomy with mucosal melanoma arising from the floor of the right maxillary sinus

Mucosal Melanomas

Arising from melanocytes and most commonly in the nasal cavity, paranasal sinus, and oral cavity [11], mucosal melanoma (MM) is an exceedingly rare and aggressive tumor, with a reported incidence in Europe of 1 per 660,000 [12]. Smoking and exposure to formaldehyde may be risk factors [13, 14], as well as advanced age and male gender [12]. Presentation varies with the primary tumor site, as patients may present with signs and symptoms of nasal obstruction, epistaxis, facial pain and may have a polypoid, fleshy and pigmented lesion (Fig. 13.2) with possible surrounding satellite lesions up to centimeters away from the main tumor. Similar to other pathologies described in this chapter, no specific staging system exists for mucosal melanoma [15].

Multiple meta-analyses note that surgery with adjuvant radiation is optimal treatment for locoregional control, as the addition of radiation to surgery decreases local recurrence rates and has moderate survival benefit [16–18]. Radiation is recommended for patients with unresectable tumors, and immunotherapy may be a potential option for control of distant metastases [17]. While the use of targeted therapies is widespread in cutaneous melanoma, the pathophysiology of mucosal melanoma is, in contrast, poorly understood. MM has been excluded from many clinical trials in cutaneous melanoma, and thus data is limited for the use of molecular targeted therapies in MM. Additionally, response of MM to immunotherapy agents like ipilimumab is largely unknown [15]. Unfortunately, prognosis for mucosal melanoma is poor: 1-year, 3-year, and 5-year survival rates are 63%, 30%, and 20%, respectively [12].

Neuroendocrine Carcinomas

Because neuroendocrine tumors have dual tissue origin, they have historically been divided between those of epithelial versus neuronal origin (e.g. paragangliomas or olfactory neuroblastomas) [19]. We cover the former in this section. Per a proposal from the International Agency for Research on Cancer (IARC) and World Health Organization (WHO), these epithelial-derived tumors are classified as well-differentiated (typical carcinoids), moderately differentiated (atypical carcinoids), or poorly differentiated tumors. These latter tumors are referred to collectively as neuroendocrine carcinomas (NECs), and can be further divided into small or large cell tumors histologically [20]. Analysis with immunohistochemical markers is the hallmark for diagnosis [21]. Given the rarity of NECs in the head and neck, the available data to describe their presentation and outcomes is generally from small case series which either differentiate between small and large cell tumors or collectively group all NECs together.

Small cell NECs have an array of presentations, as they have been documented arising in an array of head and neck sites, including the oral cavity, nasopharynx, oropharynx, hypopharynx, and paranasal sinuses [22, 23]. Large cell

NECs again present broadly and have been reported in the oropharynx, sinonasal tract, and larynx [24].

Because there is not a standardized treatment algorithm for NECs as a whole, and because NEC presents in multiple sites with varying ability to completely resect, patients in these studies have undergone an array of treatment regimens, from surgical resection with and without adjuvant treatment to radiation and chemotherapy alone. Even within patients with identical tumor sites, treatment modalities vary across and within studies. However, there is growing evidence that treatment of NECs must be more aggressive and involve chemoradiotherapy as opposed to their lower grade counterparts, typical and atypical carcinoids, which may be addressed with surgery alone [25]. One recent meta-analysis, in fact, suggested that combined chemoradiotherapy yielded the best disease-specific survival for patients with laryngeal small cell NEC instead of surgery, radiation, or chemotherapy alone [26]. In contrast, another meta-analysis recommended sinonasal NECs be treated with surgery alone and with supplemental radiotherapy only in poorly differentiated tumors, as chemotherapy did not appear to influence survival [27].

Both small and large cell carcinomas of the head and neck are aggressive cancers and characterized by widespread metastases [22, 24]; in fact, one meta-analysis found almost 70% of patients with small and large cell carcinomas of the larynx had metastases at the time of presentation [26]. Survival rates for small cell NEC have been reported as 56% and 36% at 1 year and 3 years, respectively [22]. Five-year survival has been reported as low as 10% [23]. Large cell carcinoma similarly has a reported 5-year disease-free survival of 15% [28].

While current survival rates are poor, the understanding of the pathophysiology and treatment for these tumors is evolving—interestingly, tumors which show evidence of human papilloma virus (HPV) infection have been shown in a small study to be associated with better prognosis amongst a group of patients with poorly differently head and neck NEC [29]; additionally, there is increasing interest in the expression of PD-1 in these pathologies and potential use of immunotherapies as treatment [30].

Connective and Muscle Tissue

Malignant tumors arising from connective tissue and muscle (sarcomas) are both rare and aggressive tumors with poor survival and warrant evaluation by medical oncologists who specialize in sarcoma treatment.

Chondrosarcoma

Chondrosarcoma, which arises from chondroid matrix in cartilage, soft tissue, or bone, represents only 0.1% of head and neck neoplasms [31]. The most common primary sites include mandible, larynx, nasal cavity, sinus, and maxilla [31, 32]. Chondrosarcoma arising from cartilaginous or soft tissue origin is more common in males over 50 years of age and in women less than 50. Patients may present with an array of symptoms related to their primary site: for example, patients with paranasal sinus primaries presenting with ocular symptoms, or patients with jugular foramen primary tumors presenting with facial swelling. Invasion into surrounding neurological structures may be seen at the time of presentation—for example, tumors arising from the mastoid may present with ipsilateral facial paralysis. Patients with laryngeal involvement may present solely with a neck mass [33] or with voice changes (Fig. 13.3). On histological assessment, chondrosarcomas can be difficult to diagnose given they have varying histological appearance from benign chondroid tumor to undifferentiated neoplasm [34]. A histological grading system developed by Evans et al. can be used which may correlate with recurrence rates [33].

While there is not high-level evidence comparing treatment regimens for chondrosarcoma of the head and neck as a whole, a review of the American College of Surgeons' National Cancer Database in 2000 described surgery as the mainstay of treatment for this malignancy, with some patients additionally undergoing adjuvant radia-

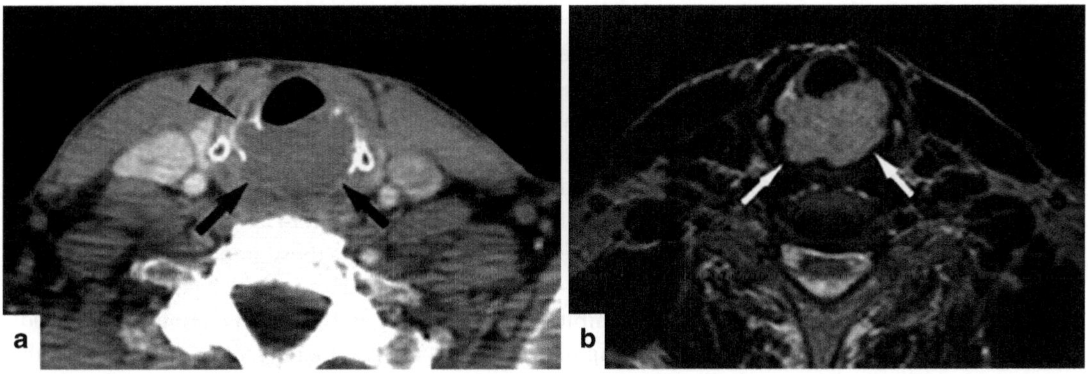

Fig. 13.3 Cricoid chondrosarcoma. A 64-year-old male with hoarseness. (**a**) Axial contrast-enhanced CT image in soft tissue algorithm shows an expansile relatively hypodense lesion involving and destroying the posterior cricoid ring (arrows). Note the sharp, well-defined, scalloped margin (arrowhead) and normal cricoid cartilage anteriorly (**b**) Axial T2-weighted MR image shows high T2 signal of the mass, characteristic of a chondroid lesion. Surgical pathology revealed a grade 1 chondrosarcoma

tion [31]. Technique of surgical resection is dependent on primary tumor site and size—for example, laser and robotic resections may be appropriate in patients with epiglottic lesions. Because chondrosarcoma is slower growing and less likely to metastasize than most cancers, subtotal debulking may be utilized, especially in areas with high morbidity like the larynx. In fact, a more recent systematic review of laryngeal chondrosarcoma reflected the important role of surgical treatment, with less than 1% of patients studied undergoing either radiation or chemotherapy [35]. Survival rates for head and neck chondrosarcoma are generally favorable, with literature quoting survival rates of 70–87.5% (with variable follow-up durations) [32, 33].

Osteogenic Sarcoma

Sarcomas arising from bone origin include both osteosarcoma, arising from mesenchymal cells producing osteoid and immature bone, and fibrosarcoma, arising from fibroblasts [36]. These tumors can occur both spontaneously and after prior head and neck radiation—this radiation-induced sarcoma of the head and neck reportedly occurs in 0.15% of patients with a mean interval between radiation and occurrence of 11 years. The most common radiation-induced sarcomas are osteosarcomas in the nasopharynx [37]. Osteosarcomas a whole account for approximately 1% of all head and neck cancers, and generally arise in the maxilla or mandible in patients in their fourth decade.

Optimal treatment for osteo- and fibrosarcomas is complete resection with consideration for adjuvant radiotherapy in patients with close or positive margins. While in a prior meta-analysis the role of adjuvant chemotherapy appeared unclear in osteosarcomas of the head and neck [38], a recent study suggested there may be a benefit for patients with maxillary tumors, positive surgical margins, or high-grade tumors [39]. The 5-year disease-specific and overall survival rates are approximately 60–70%, and if patients do have a recurrence, most are within 5 years [40]. Although data is limited, fibrosarcoma appears to have similar recurrence and survival rates as osteosarcoma [36].

Synovial Cell Sarcoma

Also arising from mesenchymal cells, and named after their histologic resemblance to the synovium, synovial sarcomas (SS) usually arise in the extremities, although they can be found in head and neck and compose about 3% of head and neck sarcomas [41]. These head and neck SS arise most commonly in men during middle age, and have been reported in a variety of locations, including face, oral cavity, pharynx, larynx, oral cavity, and other soft tissues of the neck, and can present with hoarseness, dysphagia, odynophagia, and bleeding [42]. Rarely orbital involvement has been reported [43]. These tumors are divided into monophasic and biphasic tumors depending on the presence of an epithelial component in addition to a spindle cell component (versus a spindle cell component alone), and almost all (90%) tumors exhibit chromosomal translocation t(X; 18), resulting in a fusion gene product between the two genes, synaptotagmin1 gene, *SYT1* and either SSX family member 1 or 2 genes, *SSX1* and *SSX2*, respectively [41].

These tumors are treated with surgical excision with adjuvant treatment with radiation or chemotherapy [42, 44]. More studies and higher level evidence is needed to clarify which patients may benefit from adjuvant therapy. One study suggests chemotherapy as a treatment option in high-risk surgical candidates given the chemosensitivity of these tumors [41]. Synovial cell sarcomas have high rates of metastasis to the lungs and have survival rates at 10 years of less than 50% [45], although head and neck synovial cell sarcomas may have better survival rates up to 82% [46].

Rhabdomyosarcoma

Rhabdomyosarcoma (RMS) is the most common soft tissue sarcoma in children and is derived from mesenchymal cells not yet differentiated into skeletal muscle; about 40% of RMS present in the head and neck [47]. These tumors are generally divided into those in parameningeal (nasopharynx, nasal cavity, paranasal sinus, infratemporal and pterygopalatine fossa, middle ear) and nonparameningeal locations (all other sites) and present with a variety of symptoms. Given that parameningeal tumors can spread along the dura and encase vessels, many patients have cranial nerve deficits and skull base erosion and orbital involvement at the time of presentation (Fig. 13.4). MRI is an essential in characterizing dural involvement, involvement of orbital structures, and perineural spread in these tumors, and FDG PET can elucidate the extent of metastatic disease [48].

Treatment includes a combination of chemotherapy as well as local control with surgery or radiation [49]. Nonparameningeal tumors may be entirely resectable, but because of the extent of disease at presentation, many patients with parameningeal tumors may not have resectable disease [48]. While there are no meta-analyses

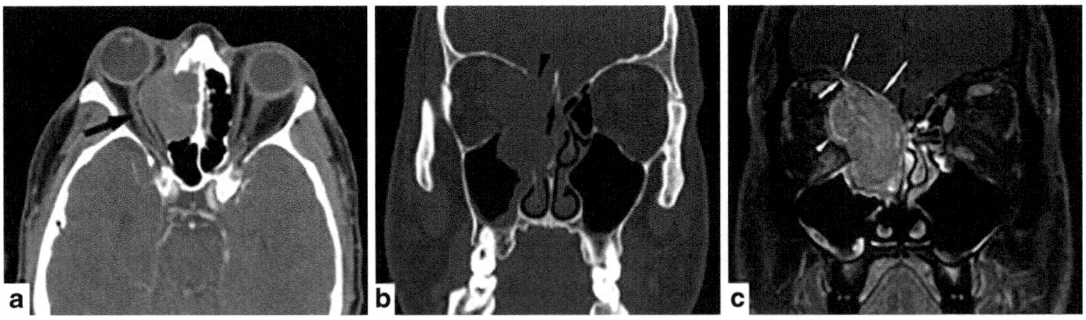

Fig. 13.4 Sinonasal rhabdomyosarcoma. A 34-year-old male who presented to the emergency department with right proptosis and nasal congestion. (**a**) Axial contrast-enhanced CT image in soft tissue algorithm shows a bulky lobulated mass in the right ethmoid region bulging laterally through eroded bone into the right orbit, displacing the medial rectus and optic nerve (arrow) laterally and causing right proptosis. (**b**) Coronal CT image in bone algorithm demonstrates bony destruction around the margins of the mass, including the right lamina papyracea, right orbital floor, floor of the right anterior cranial fossa (arrowhead), and the right aspect of the crista galli (arrow). (**c**) Coronal contrast-enhanced T1-weighted MR image shows avid homogeneous enhancement of the mass. There is right orbital invasion, with tumor extending along the orbital roof (small arrow) and displacing the medial rectus muscle laterally (small white arrowhead). It crosses midline to the left (black arrowhead), and there is intracranial involvement as evidenced by thickened dural enhancement (long arrows)

defining optimal treatment regimens for head and neck RMS, a recent pooled analysis in parameningeal RMS noted patients who received radiation as part of their initial treatment regimen exhibited better survival at 5 and 10 years than patients without radiation early in their treatment course [50]. Prognosis varies broadly with histologic, genetic, as well as anatomical factors of tumors along with extent of disease at presentation. For example, patients with nonparameningeal disease without metastases have reported survival rates up to 100% at 3 years [48] while the overall survival for patients with metastatic disease is 39% [51].

Ewing Sarcoma

In contrast to osteosarcoma, Ewing's sarcoma, a tumor of small round cells arising from soft tissue, most frequently affects children and adolescents and can present in the skull, mandible, and maxilla. Although it is the second most common primary bone malignancy in these age groups [52], only 1–15% of Ewing's sarcomas are found in the head and neck [53]. Treatment generally consists of adjuvant chemotherapy followed by local treatment with surgical resection or radiation combined with consolidation or maintenance chemotherapy. While data for head and neck Ewing's tumor is limited, a recent meta-analysis of patients with Ewing's sarcoma across all sites (all of whom underwent neoadjuvant chemotherapy) showed improved outcomes in terms of local recurrence and survival rates in patients who underwent surgery alone and surgery with adjuvant radiation than in patients who underwent radiation alone [54]. Risk factors for recurrence include metastases at the time of diagnosis, poor histological response to therapy, and large tumor size. In contrast to osteosarcoma, the survival rates of Ewing's sarcoma in the head and neck are favorable—up to 90% at 5 years. However, long-term sequelae from treatment is common (occurring in up to 88% of patients) including functional deficits, growth abnormalities with facial asymmetry, aesthetic defects, psychosocial impairments, and endocrine and neurological disorders [55].

Leiomyosarcoma

In contrast to the prior two pediatric pathologies, leiomyosarcoma is a smooth-muscle derived malignancy that typically presents in the sixth or seventh decade of life. These tumors present generally as solitary nodules and can be divided based on their site of presentation into cutaneous (arising from erector pili muscles of hair follicles) and noncutaneous (arising from article tunica media) [56]. Noncutaneous tumors have been reported in a variety of head and neck sites, including oral cavity, nasopharynx, nasal cavity and sinuses, larynx and hypopharynx, as well as parotid and thyroid in rare cases. A review of the National Cancer Database in 2018 noted surgical therapy with tumor excision is the most common treatment modality used in the majority of patients, although radiation has been used both in the adjuvant and neoadjuvant setting for patients with positive surgical margins [57]. Chemotherapy may play a role in treatment, as improved local and distal recurrence has been reported in patients who underwent adjuvant chemotherapy prior to other treatment modalities [58]. Although neck metastasis is rare with this disease, prognosis overall is again poor—about half of patients survive 5 years after their diagnosis [56].

Angiosarcoma

Arising from endothelial cells lining vasculature, angiosarcoma is a rare tumor generally affecting older patients over 60 years of age [59, 60]. In the head and neck region, these tumors are found in the scalp or face although they have been reported in the larynx, hypopharynx, and sinuses as well [61]. These tumors appear as bruise-like macules with possible ulceration and palpable nodules [60]. While there is not high-level evidence specifically comparing different treatment modalities for head and neck angiosarcoma, patients with these tumors generally undergo wide surgical resection with negative margins, although this may not be possible given the microscopic subcutaneous spread of these tumors. Adjuvant therapy with chemotherapy, radiation, or combined chemoradiation treatments have all been reported

[62], with data clearly supporting the use of adjuvant radiation to improve survival [63]. Patients with unresectable tumors generally undergo chemoradiation [62]. Local recurrence rates after surgery have been reported as high as 90–94% [60, 64] and occur usually within a year of treatment, although patients with negative margins may present later with recurrences [61]. Survival rates are poor with studies reporting both 2- and 5-year overall survival rates of 20–30% [61, 64].

Liposarcoma

Liposarcomas arising from adipose tissue are malignant tumors including well-differentiated liposarcomas, as well as de-differentiated, myxoid, and pleomorphic liposarcomas [65]. Like the prior pathologies already discussed, surgery with negative margins is the ideal treatment, although adjuvant radiation as well as systemic chemotherapy have been shown to improve overall survival in high-risk patients with positive margins, high-grade, deep, and larger tumors and are recommended for these subsets of patients by the European Society for Medical Oncology (ESMO) and European Rare Cancer (EURACAN) practice guidelines from 2018 [66]. Liposarcomas of the head and neck, in comparison to lipomas in the rest of their body, are more likely present early stage, and low grade, as well as to be treated with surgery alone, and had better overall survival (approximately 80% vs. 70% overall survival at 5 years [67]).

Kaposi Sarcoma

Kaposi sarcoma is another malignancy arising in adults from mesenchymal cells lining lymphatic channels or vessels and is associated with human herpes virus 8 (HHV8) infection. While the majority of reported cases have been in patient with human immunodeficiency virus (HIV), there are also reported cases in HIV-negative patients [68]. Tumors arise in a variety of locations including the skin, oral and oropharyngeal mucosa, nasal cavity and paranasal sinuses, cervical lymph nodes, and ear (pinna and external auditory canal). While patients with tumors in the nasal cavity may present with symptoms secondary to blockage, patients with disease in the neck may present with painless neck masses with skin color changes. Mucosal and skin lesions appear as slightly raised colored spots or small purple, brown, or red lumps [69]. Although data is limited to case reports and series, HIV status may be related to tumor location, as HIV-negative patients may be more likely to have tonsillar or ear tumors, while HIV-positive may have tumors presenting more often in the oral cavity, sinuses, or neck [70]. Many patients may have multifocal disease with tumors in non-head and neck locations as well [68]. This diagnosis should be distinguished from bacillary angiomatosis which also presents as a similar vasoproliferative lesion in HIV positive patients with *Bartonella henselae* infection; immunohistochemistry, Warthin–Starry stain, and/or molecular testing can distinguish these diagnoses. Kaposiform hemangioendothelioma may also present similarly, although it presents in patients without evidence of immunodeficiency or HHV8 infection.

As described in two Cochrane reviews, the standard of treatment for adults and children is systemic therapy with chemotherapy in combination with antiretrovirals in patients with HIV. Local treatments such as surgical excision, freezing, or radiation are reserved only for symptomatic localized or minimally aggressive cases [71, 72]. Although immune-modulating agents like interferon (IFN) have been used in the past with variable tolerance [73], data is now growing regarding the use of immune checkpoint inhibitors like nivolumab and pembrolizumab, which may have efficacy in achieving partial response in patients [74]. Prognosis of this tumor is difficult to discern from existing data as many patients in these studies have a concurrent AIDS diagnosis and thus poor survival prior to tumor diagnosis [69].

Teratocarcinosarcoma

Teratocarincosarcoma is a rare and aggressive tumor found primarily in the sinonasal region.

These tumors are morphologically heterogeneous and have multiple components on histological examination including benign respiratory-type epithelium, malignant glands resembling adenocarcinoma, squamous differentiation including fetal-type, primitive neuroepithelium, and mesenchymal components [75]. They present most commonly in middle aged men in the sinonasal region with nasal obstruction and epistaxis [75, 76], although they have also been reported as a congenital tumor presenting in the nasopharynx of an infant with the congenital absence of the ipsilateral Eustachian tube [77].

Much of the data on this malignancy is from small case series and reports, describing treatment most commonly with a combination of surgery and radiation although with limited success [75, 76, 78]. A recent meta-analysis in fact found up to 67% of patients treated with surgery and 80% of patients treated initially with radiation had recurrence, metastasis, or were unresponsive to treatment [79]; this study noted, however, that of the patients who did survive to 1 year post-diagnosis, the majority (70%) had undergone combined surgery and adjuvant therapies, supporting the use of aggressive treatment regimens. A recent small study explored the use of neoadjuvant chemotherapy in patients with large and unresectable tumors at presentation [80]; although both patients described in this study were noted to be asymptomatic post-treatment, follow-up ranged only 2–3 months, so more studies with longer-term follow-up are needed to confirm its efficacy. Given the ability of these tumors to spread rapidly intracranially and have high rates of recurrence, less than half of patients survive to 3 years post diagnosis even with aggressive treatment [81].

Nervous Tissue

Nervous tissue in the head and neck consists of autonomic nerve fibers as well as sensory and motor nerves and their surrounding supportive cells. We cover malignancies derived from these tissues below, as we include neuroendocrine and neuroepithelial malignancies in the first section.

Malignant Nerve Sheath Tumors

Malignant nerve sheath tumors arise in the peripheral nervous system although they are considered a type of soft tissue sarcoma. They are most commonly associated with neurofibromatosis type 1 (in 50–60% of patients) and less commonly occur after prior radiation exposure or sporadically [82] Patients are diagnosed generally between the third and seventh decade of life. Traditionally these tumors have been treated with surgical excision, as the role of adjuvant radiation and chemotherapy is unclear [83, 84]. A recent meta-analysis investigating prognostic factors for malignant peripheral nerve sheath tumors found that chemotherapy and radiation therapy individually were associated with patients with better survival [85]; however, more studies are needed to define this benefit and clarify which subsets of patients with these tumors may benefit from adjuvant treatment. Survival overall for these tumors is poor, with 5-year rates between 35% and 65%. Current staging system is likely inadequate for prognostication. Interestingly, patients of extremes of age (patients younger than 40 or older than 60) and patients with larger tumor size are at higher risk of worse outcomes [86].

Paraganglioma

Paragangliomas are tumors arising from the ganglia of the paraganglion system throughout the body, and most are benign and occur in the adrenal paraganglia. However, 3% of these tumors occur in the head and neck [87], arising most commonly in the carotid body, jugular foramen (Fig. 13.5) and along the vagus nerve, although they can occur in the larynx, nasal cavity, orbit, trachea, aortic body, and mediastinum [88]. About 9–19% of head and neck paragangliomas are malignant [89, 90], as certain mutations in the gene encoding the enzyme succinate dehydrogenase (SDH) may deem these tumors higher risk for malignant transformation [91]. We focus on this important distinction between benign and malignant paragangliomas later in this section.

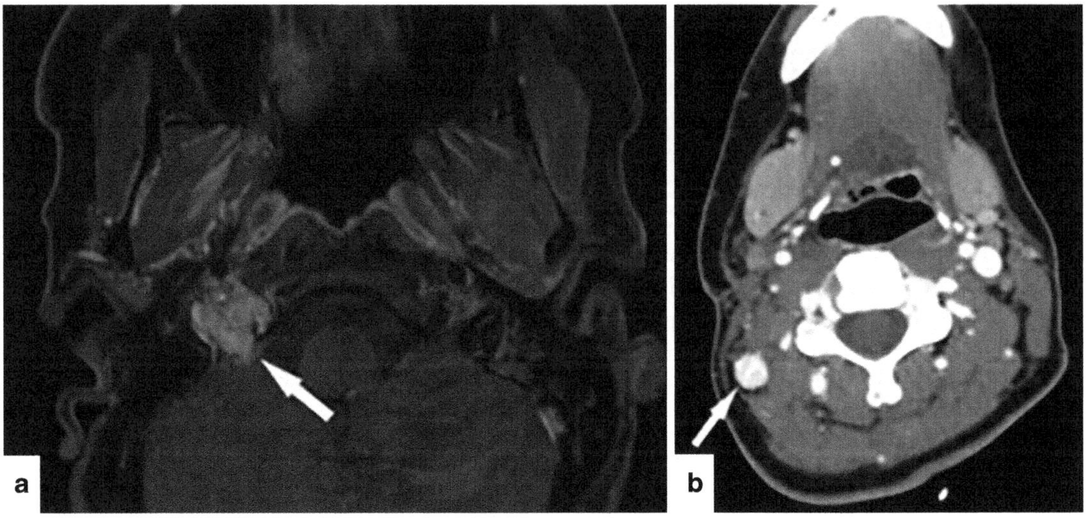

Fig. 13.5 Metastatic paraganglioma. A 33-year-old female with a right jugular foramen paraganglioma (glomus jugulare). (**a**) Axial contrast-enhanced T1-weighted MR image shows an avidly enhancing mass in the right jugular foramen (arrow) representing a paraganglioma. (**b**) Axial contrast-enhanced CT image in soft tissue algorithm shows an avidly enhancing soft tissue nodule (arrow). Surgical resection of this nodule was performed, with pathology revealing paraganglioma involving a lymph node and considered highly indicative of metastatic disease

Patients with paragangliomas, both benign and malignant, may present with neck masses or cranial neuropathies depending on the location of the tumor—vagal tumors may present with difficulty speaking or swallowing, while jugular foramen tumors may present with hearing loss, shoulder weakness and pain, or facial paralysis. Workup of these tumors includes a detailed history and physical exam including laryngoscopy to assess vocal fold motion, as well as imaging with a CT, MRI, or MRA to assess tumor relationship to surrounding vasculature. Patients may additionally undergo testing with plasma or urine catecholamines, metanephrines, chromogranin A, and vanillylmandelic acid (VMA) to assess functional status of these tumors, as up to a third of both benign and malignant paragangliomas can secrete catecholamines [89, 92]. Chromogranin A, often secreted with catecholamines, can be monitored throughout a patient's course as a marker of secretory function [93]. Patients with malignant paragangliomas may also present with evidence of regional or distant metastases.

Treatment for both malignant and secretory paragangliomas is surgical excision [89]. However, given the extensive vascularity of these tumors and their proximity to cranial nerves, these surgeries often have significant associated morbidity both intraoperatively in terms of blood loss and postoperatively in terms of patient quality of life. Thus special preoperative considerations may be taken into account: for example, if carotid artery resection or reconstruction is considered, preoperative carotid artery balloon occlusive testing with single-photo emission CT may be used to evaluate cerebral perfusion [91]. Prior to resection of jugular foramen and high vagal tumors preoperative embolization is recommended—while this has been shown to reduce blood loss intraoperatively, strokes and cranial neuropathies have also been reported with its use [94].

Importantly, benign and malignant paragangliomas cannot be differentiated preoperatively based on cytological, histological, immunohistochemical, or molecular criteria given they are indistinguishable in these assays; however, lymph node tissue obtained intraoperatively can provide this diagnosis [89]. Thus traditionally selective neck dissection has been recommended for patients with evidence of metastasis noted [95].

However, a recent review suggests selective neck dissection in all patients in order to avoid delay in treatment for patients who do have a malignant tumor [89, 96]. There may be additional benefit to adjuvant radiation in these patients with malignant paragangliomas [97, 98], as well as a potential role for chemotherapy in patients with metastatic or unresectable disease [99].

For patients with malignant paraganglioma, survival at 5 years has been reported to be 88.1% for patients who underwent surgical excision alone and 66.5% in patients undergoing adjuvant radiation (a nonsignificant difference in this study). Patients with carotid body tumors and younger patients (less than 50 years of age) have a survival advantage when compared to patients with other primary sites and advanced age, respectively [90].

Conclusions

The rare cancers discussed above arise from several unique and specialized tissue types in the head and neck and show significant variation in their epidemiology. Given their rarity, randomized clinical trials identifying the optimal treatment for each entity are unlikely. Best level evidence tends to be retrospective cohort studies with all of the inherent limitations thereof. Multidisciplinary treatment approaches are of paramount importance to successfully managing these rare cancers, especially given rapid innovations in targeted therapeutics and immunotherapies. Inter-institutional collaboration as well as nationwide and international databases may allow for better delineation of optimal treatment modalities, natural history, and long-term prognosis.

References

1. Vigneswaran N, Williams MD. Epidemiologic trends in head and neck cancer and aids in diagnosis. Oral Maxillofac Surg Clin North Am. 2014;26(2):123–41.
2. Papacharalampous GX, Vlastarakos PV, Chrysovergis A, Saravakos PK, Kotsis GP, Davilis DI. Olfactory neuroblastoma (esthesioneuroblastoma): towards minimally invasive surgery and multi-modality treatment strategies - an updated critical review of the current literature. JBUON. 2013;18(3):557–63.
3. Dublin AB, Bobinski M. Imaging characteristics of olfactory neuroblastoma (esthesioneuroblastoma). J Neurol Surg B Skull Base. 2016;77(1):1–5.
4. Fiani B, Quadri SA, Cathel A, Farooqui M, Ramachandran A, Siddiqi I, et al. Esthesioneuroblastoma: a comprehensive review of diagnosis, management, and current treatment options. World Neurosurg. 2019;126:194–211.
5. Van Gompel JJ, Giannini C, Olsen KD, Moore E, Piccirilli M, Foote RL, et al. Long-term outcome of esthesioneuroblastoma: hyams grade predicts patient survival. J Neurol Surg B Skull Base. 2012;73(5):331–6.
6. Dulguerov P, Allal AS, Calcaterra TC. Esthesioneuroblastoma: a meta-analysis and review. Lancet Oncol. 2001;2(11):683–90.
7. Marinelli JP, Janus JR, Van Gompel JJ, Link MJ, Foote RL, Lohse CM, et al. Esthesioneuroblastoma with distant metastases: systematic review & meta-analysis. Head Neck. 2018;40(10):2295–303.
8. Gore MR, Zanation AM. Salvage treatment of local recurrence in esthesioneuroblastoma: a meta-analysis. Skull Base. 2011;21(1):1–6.
9. Sheehan J, Payne R. Esthesioneuroblastomas. Youmans and Winn neurological surgery. New York, NY: Elsevier Health Sciences; 2016. p. 1284–92.
10. Slevin NJ, Irwin CJ, Banerjee SS, Gupta NK, Farrington WT. Olfactory neural tumours--the role of external beam radiotherapy. J Laryngol Otol. 1996;110(11):1012–6.
11. Mikkelsen LH, Larsen AC, von Buchwald C, Drzewiecki KT, Prause JU, Heegaard S. Mucosal malignant melanoma - a clinical, oncological, pathological and genetic survey. APMIS. 2016;124(6):475–86.
12. RARECARENet: Information Network on Rare Cancers. Executive agency for health and consumers of the European Commission. n.d.. www.Rarecarenet.istitutotumori.mi.it.
13. Axéll T, Hedin CA. Epidemiologic study of excessive oral melanin pigmentation with special reference to the influence of tobacco habits. Scand J Dent Res. 1982;90(6):434–42.
14. Holmstrom M, Lund VJ. Malignant melanomas of the nasal cavity after occupational exposure to formaldehyde. Br J Ind Med. 1991;48(1):9–11.
15. Ascierto PA, Accorona R, Botti G, Farina D, Fossati P, Gatta G, et al. Mucosal melanoma of the head and neck. Crit Rev Oncol Hematol. 2017;112:136–52.
16. Jarrom D, Paleri V, Kerawala C, Roques T, Bhide S, Newman L, et al. Mucosal melanoma of the upper airways tract mucosal melanoma: a systematic review with meta-analyses of treatment. Head Neck. 2017;39(4):819–25.
17. Li W, Yu Y, Wang H, Yan A, Jiang X. Evaluation of the prognostic impact of postoperative adjuvant radio-

therapy on head and neck mucosal melanoma: a meta-analysis. BMC Cancer. 2015;15:758.
18. Grant-Freemantle MC, Lane O'Neill B, Clover AJP. The effectiveness of radiotherapy in the treatment of head and neck mucosal melanoma: systematic review and meta-analysis. Head Neck. 2021;43(1):323–33.
19. Subedi N, Prestwich R, Chowdhury F, Patel C, Scarsbrook A. Neuroendocrine tumours of the head and neck: anatomical, functional and molecular imaging and contemporary management. Cancer Imaging. 2013;13(3):407–22.
20. Rindi G, Klimstra D, Abedo-Ardekani B, Asa SL, Bosman FT, Brambilla EA, et al. A common classification framework for neuroendocrine neoplasms: an International Agency for Research on Cancer and World Health Organization expert consensus proposal. Mod Pathol. 2018;31:1770–86.
21. Lewis JS Jr, Chernock RD, Bishop JA. Squamous and neuroendocrine specific immunohistochemical markers in head and neck squamous cell carcinoma: a tissue microarray study. Head Neck Pathol. 2018;12(1):62–70.
22. Wakasaki T, Yasumatsu R, Masuda M, Matsuo M, Tamae A, Kubo K, et al. Small cell carcinoma in the head and neck. Ann Otol Rhinol Laryngol. 2019;128(11):1006–12.
23. Hosokawa S, Takahashi G, Baba S, Mineta H. Small cell neuroendocrine carcinomas arising in the head and neck region. J Oral Maxillofac Surg. 2016;74(5):1091–5.
24. Thompson ED, Stelow EB, Mills SE, Westra WH, Bishop JA. Large cell neuroendocrine carcinoma of the head and neck: a clinicopathologic series of 10 cases with an emphasis on HPV status. Am J Surg Pathol. 2016;40(4):471–8.
25. López F, Hunt JL, Nixon IJ, Rinaldo A, Williams MD, Cardesa A, Ferlito A. How phenotype guides management of the neuroendocrine carcinomas of the larynx. J Laryngol Otol. 2018;132(7):568–74.
26. van der Laan TP, Plaat BE, van der Laan BF, Halmos GB. Clinical recommendations on the treatment of neuroendocrine carcinoma of the larynx: a meta-analysis of 436 reported cases. Head Neck. 2015;37(5):707–15.
27. van der Laan TP, Iepsma R, Witjes MJ, van der Laan BF, Plaat BE, Halmos GB. Meta-analysis of 701 published cases of sinonasal neuroendocrine carcinoma: the importance of differentiation grade in determining treatment strategy. Oral Oncol. 2016;63:1–9.
28. Halmos GB, van der Laan TP, van Hemel BM, et al. Is human papillomavirus involved in laryngeal neuroendocrine carcinoma? Eur Arch Otorhinolaryngol. 2013;270:719–25.
29. Benzerdjeb N, Traverse-Glehen A, Philouze P, Bishop J, Devouassoux-Shisheboran M. Poorly differentiated neuroendocrine carcinoma of the head and neck: human papillomavirus tumour status/p16 status and impact on overall survival. Histopathology. 2020;76(4):581–91.
30. Bahr K, Zimmer S, Springer E, Fottner C, Becker S, Ernst BP, et al. High-grade neuroendocrine carcinoma of the head and neck: human papillomavirus status and PD-L1 expression. ORL J Otorhinolaryngol Relat Spec. 2019;81(5–6):309–16.
31. Koch BB, Karnell LH, Hoffman HT, Apostolakis LW, Robinson RA, Zhen W, et al. National cancer database report on chondrosarcoma of the head and neck. Head Neck. 2000;22(4):408–25.
32. Burkey BB, Hoffman HT, Baker SR, Thornton AF, McClatchey KD. Chondrosarcoma of the head and neck. Laryngoscope. 1990;100:1301–5.
33. Lee SY, Lim YC, Song MH, Seok JY, Lee WS, Choi EC. Chondrosarcoma of the head and neck. Yonsei Med J. 2005;46(2):228–32.
34. Batsakis JG, Solomon AR, Rice DH. The pathology of head and neck tumors: neoplasm of cartilage, bone, and the notochord, part 7. Head Neck Surg. 1980;3:43–57.
35. Chin OY, Dubal PM, Sheikh AB, Unsal AA, Park RC, Baredes S, et al. Laryngeal chondrosarcoma: a systematic review of 592 cases. Laryngoscope. 2017;127(2):430–9.
36. Han S, Yin X, Xu W, Wang Y, Han W. The management of head and neck sarcoma. J Craniofac Surg. 2020;31(2):e189–92.
37. Coca-Pelaz A, Mäkitie AA, Strojan P, Corry J, Eisbruch A, Beitler JJ, et al. Radiation-induced sarcomas of the head and neck: a systematic review. Adv Ther. 2021;38(1):90–108.
38. Kassir RR, Rassekh CH, Kinsella JB, Segas J, Carrau RL, Hokanson JA. Osteosarcoma of the head and neck: meta-analysis of nonrandomized studies. Laryngoscope. 1997;107(1):56–61.
39. Liang L, Zhang T, You Y, He Q, Fan Y, Liao G. An individual patient data meta-analysis on the effect of chemotherapy on survival in patients with craniofacial osteosarcoma. Head Neck. 2019;41(6):2016–23.
40. Mendenhall WM, Fernandes R, Werning JW, Vaysberg M, Malyapa RS, Mendenhall NP. Head and neck osteosarcoma. Am J Otolaryngol. 2011;32(6):597–600.
41. Harb WJ, Luna MA, Patel SR, Ballo MT, Roberts DB, Sturgis EM. Survival in patients with synovial sarcoma of the head and neck: association with tumor location, size, and extension. Head Neck. 2007;29(8):731–40.
42. Gopalakrishnan V, Amini B, Wagner MJ, Nowell EN, Lazar AJ, Lin PP, et al. Synovial sarcoma of the head and neck: a single institution review. Sarcoma. 2017;2017:2016752.
43. Kusuma S, Skarupa DJ, Ely KA, Cmelak AJ, Burkey BB. Synovial sarcoma of the head and neck: a review of its diagnosis and management and a report of a rare case of orbital involvement. Ear Nose Throat J. 2010;89(6):280–3.
44. Stanbouly D, Litman E, Lee KC, Philipone E. Synovial sarcoma of the head & neck: a review of reported cases in the literature. J Stomatol Oral Maxillofac Surg. 2021;122:505.
45. Oda Y, Hashimoto H, Tsuneyoshi M, Takeshita S. Survival in synovial sarcoma: a multivariate study

of prognostic factors with special emphasis on the comparison between early death and long-term survival. Am J Surg Pathol. 1993;17(1):35–44.
46. Mallen-St Clair J, Arshi A, Abemayor E, St John M. Factors associated with survival in patients with synovial cell sarcoma of the head and neck: an analysis of 167 cases using the SEER (Surveillance, Epidemiology, and End Results) database. JAMA Otolaryngol Head Neck Surg. 2016;142(6):576–83.
47. Huh WW, Skapek SX. Childhood rhabdomyosarcoma: new insight on biology and treatment. Curr Oncol Rep. 2010;12(6):402–10.
48. Reilly BK, Kim A, Peña MT, Dong TA, Rossi C, Murnick JG, Choi SS. Rhabdomyosarcoma of the head and neck in children: review and update. Int J Pediatr Otorhinolaryngol. 2015;79(9):1477–83.
49. Malempati S, Hawkins DS. Rhabdomyosarcoma: review of the Children's Oncology Group (COG) Soft-Tissue Sarcoma Committee experience and rationale for current COG studies. Pediatr Blood Cancer. 2012;59(1):5–10.
50. Merks JH, De Salvo GL, Bergeron C, Bisogno G, De Paoli A, Ferrari A, et al. Parameningeal rhabdomyosarcoma in pediatric age: results of a pooled analysis from North American and European cooperative groups. Ann Oncol. 2014;25(1):231–6. https://doi.org/10.1093/annonc/mdt426. Erratum in: Ann Oncol. 2014;25(3):756. Erratum in: Ann Oncol. 2014;25(3):756. Erratum in: Ann Oncol. 2014;25(3):756.
51. Breneman JC, Lyden E, Pappo AS, Link MP, Anderson JR, Parham DM, et al. Prognostic factors and clinical outcomes in children and adolescents with metastatic rhabdomyosarcoma--a report from the Intergroup Rhabdomyosarcoma Study IV. J Clin Oncol. 2003;21(1):78–84.
52. Gatta G, Botta L, Rossi S, Aareleid T, Bielska-Lasota M, Clavel J, et al. Childhood cancer survival in Europe 1999–2007: results of EUROCARE-5–a population-based study. Lancet Oncol. 2014;15(1):35–47.
53. Daw NC, Mahmoud HH, Meyer WH, Jenkins JJ, Kaste SC, Poquette CA, et al. Bone sarcomas of the head and neck in children: the St Jude Children's Research Hospital experience. Cancer. 2000;88(9):2172–80.
54. Zhu H, Li Y, Xie X, Zhang S, Xue Y, Fan T. Efficacy of local control strategies for Ewing sarcoma after neoadjuvant chemotherapy: a network meta-analysis. Indian Pediatr. 2020;57(6):527–32.
55. Bouaoud J, Temam S, Cozic N, Galmiche-Rolland L, Belhous K, Kolb F, et al. Ewing's sarcoma of the head and neck: margins are not just for surgeons. Cancer Med. 2018;7(12):5879–88.
56. Oztürk K, Keleş B, Arbag H, Yöndemll F, Avunduk MC. Postauricular subcutaneous leiomyosarcoma. Auris Nasus Larynx. 2004;31(3):323–8.
57. Workman AD, Farquhar DR, Brody RM, Parasher AK, Carey RM, Purkey MT, et al. Leiomyosarcoma of the head and neck: a 17-year single institution experience and review of the National Cancer Data Base. Head Neck. 2018;40(4):756–62. https://doi.org/10.1002/hed.25054.
58. Pervaiz N, Colterjohn N, Farrokhyar F, Tozer R, Figueredo A, Ghert M. A systematic meta-analysis of randomized controlled trials of adjuvant chemotherapy for localized resectable softtissue sarcoma. Cancer. 2008;113:573–81.
59. Fury MG, Antonescu CR, Van Zee KJ, Brennan MF, Maki RG. A 14- year retrospective review of angiosarcoma: clinical characteristics, prognostic factors, and treatment outcomes with surgery and chemotherapy. Cancer J. 2005;11(3):241–7.
60. Ogawa K, Takahashi K, Asato Y, Yamamoto Y, Taira K, Matori S, et al. Treatment and prognosis of angiosarcoma of the scalp and face: a retrospective analysis of 48 patients. Br J Radiol. 2012;85(1019):e1127–33.
61. Mullins B, Hackman T. Angiosarcoma of the head and neck. Int Arch Otorhinolaryngol. 2015;19(3):191–5.
62. Shin JY, Roh SG, Lee NH, Yang KM. Predisposing factors for poor prognosis of angiosarcoma of the scalp and face: systematic review and meta-analysis. Head Neck. 2017;39(2):380–6.
63. Bi S, Chen S, Wu B, Cen Y, Chen J. The effectiveness of different treatment modalities of cutaneous angiosarcoma: results from meta-analysis and observational data from SEER database. Front Oncol. 2021;11:627113.
64. Mark RJ, Tran LM, Sercarz J, Fu YS, Calcaterra TC, Juillard GF. Angiosarcoma of the head and neck. The UCLA experience 1955 through 1990. Arch Otolaryngol Head Neck Surg. 1993;119(9):973–8.
65. Barisella M, Giannini L, Piazza C. From head and neck lipoma to liposarcoma: a wide spectrum of differential diagnoses and their therapeutic implications. Curr Opin Otolaryngol Head Neck Surg. 2020;28(2):136–43.
66. Casali PG, Abecassis N, Aro HT, Bauer S, Biagini R, Bielack S, et al. ESMO Guidelines Committee and EURACAN. Soft tissue and visceral sarcomas: ESMO-EURACAN clinical practice guidelines for diagnosis, treatment and follow-up. Ann Oncol. 2018;29(Suppl 4):iv51–67.
67. Gerry D, Fox NF, Spruill LS, Lentsch EJ. Liposarcoma of the head and neck: analysis of 318 cases with comparison to non-head and neck sites. Head Neck. 2014;36(3):393–400.
68. Agaimy A, Mueller SK, Harrer T, Bauer S, Thompson LDR. Head and neck kaposi sarcoma: clinicopathological analysis of 11 cases. Head Neck Pathol. 2018;12(4):511–6.
69. Yu DX, Pi SJ, Zhang WS. Clinical manifestation of Kaposi sarcoma in otorhinolaryngology head and neck surgery. Zhonghua Er Bi Yan Hou Tou Jing Wai Ke Za Zhi. 2013;48(3):241–3.
70. Sethia R, Hari-Raj A, Koenigs M, Ozer E. Non-HIV oral kaposi sarcoma: a case report and review of the literature. Ear Nose Throat J. 2021;100:NP214.
71. Gbabe OF, Okwundu CI, Dedicoat M, Freeman EE. Treatment of severe or progressive Kaposi's sar-

coma in HIV-infected adults. Cochrane Database Syst Rev. 2014;(9):CD003256.
72. Anglemyer A, Agrawal AK, Rutherford GW. Treatment of Kaposi sarcoma in children with HIV-1 infection. Cochrane Database Syst Rev. 2014;(1):CD009826.
73. Dupin N. Update on oncogenesis and therapy for Kaposi sarcoma. Curr Opin Oncol. 2020;32(2):122–8.
74. Galanina N, Goodman AM, Cohen PR, Frampton GM, Kurzrock R. Successful treatment of HIV-associated kaposi sarcoma with immune checkpoint blockade. Cancer Immunol Res. 2018;6(10):1129–35.
75. Miller M, Newberry CI, Witt B, Oakley GM. Sinonasal teratocarcinosarcoma-a rare and highly aggressive neoplasm. JAMA Otolaryngol Head Neck Surg. 2021;147:106.
76. Smith SL, Hessel AC, Luna MA, Malpica A, Rosenthal DI, El-Naggar AK. Sinonasal teratocarcinosarcoma of the head and neck: a report of 10 patients treated at a single institution and comparison with reported series. Arch Otolaryngol Head Neck Surg. 2008;134(6):592–5.
77. Rotenberg B, El-Hakim H, Lodha A, MacCormick A, Ngan BY, Forte V. Nasopharyngeal teratocarcinosarcoma. Int J Pediatr Otorhinolaryngol. 2002;62(2):159–64.
78. Wahid FI, Javaid M, Khan Q, Khan IA. Sinonasal teratocarcinosarcoma. J Coll Physicians Surg Pak. 2012;22(5):335–7.
79. Wei S, Carroll W, Lazenby A, Bell W, Lopez R, Said-Al-Naief N. Sinonasal teratocarcinosarcoma: report of a case with review of literature and treatment outcome. Ann Diagn Pathol. 2008;12(6):415–25.
80. Joshi A, Noronha V, Sharma M, Dhumal S, Juvekar S, Patil VM, et al. Neoadjuvant chemotherapy in advanced sinonasal teratocarcinosarcoma with intracranial extension: report of two cases with literature review. J Cancer Res Ther. 2015;11(4):1003–5.
81. Pai SA, Naresh KN, Masih K, Ramarao C, Borges AM. Teratocarcinosarcoma of the paranasal sinuses: a clinicopathologic and immunohistochemical study. Hum Pathol. 1998;29(7):718–22.
82. Wanebo JE, Malik JM, VandenBerg SR, Wanebo HJ, Driesen N, Persing JA. Malignant peripheral nerve sheath tumors. A clinicopathologic study of 28 cases. Cancer. 1993;71(4):1247–53.
83. Kolberg M, Høland M, Agesen TH, Brekke HR, Liestøl K, Hall KS, et al. Survival meta-analyses for >1800 malignant peripheral nerve sheath tumor patients with and without neurofibromatosis type 1. Neuro-Oncology. 2013;15(2):135–47.
84. Perrin RG, Guha A. Malignant peripheral nerve sheath tumors. Neurosurg Clin N Am. 2004;15(2):203–16.
85. Cai Z, Tang X, Liang H, Yang R, Yan T, Guo W. Prognosis and risk factors for malignant peripheral nerve sheath tumor: a systematic review and meta-analysis. World J Surg Oncol. 2020;18(1):257.
86. Mowery A, Clayburgh D. Malignant peripheral nerve sheath tumors: analysis of the national cancer database. Oral Oncol. 2019;98:13–9.
87. Wasserman PG, Savargaonkar P. Paragangliomas: classification, pathology, and differential diagnosis. Otolaryngol Clin N Am. 2001;34:845–62.
88. Myssiorek D. Head and neck paragangliomas: an overview. Otolaryngol Clin N Am. 2001;34:829–36.
89. McCrary HC, Babajanian E, Calquin M, Carpenter P, Casazza G, Naumer A, et al. Characterization of malignant head and neck paragangliomas at a single institution across multiple decades. JAMA Otolaryngol Head Neck Surg. 2019;145(7):641–6.
90. Sethi RV, Sethi RK, Herr MW, Deschler DG. Malignant head and neck paragangliomas: treatment efficacy and prognostic indicators. Am J Otolaryngol. 2013;34(5):431–8.
91. Moore MG, Netterville JL, Mendenhall WM, Isaacson B, Nussenbaum B. Head and neck paragangliomas: an update on evaluation and management. Otolaryngol Head Neck Surg. 2016;154(4):597–605.
92. van Duinen N, Corssmit EP, de Jong WH, Brookman D, Kema IP, Romijn JA. Plasma levels of free metanephrines and 3-methoxytyramine indicate a higher number of biochemically active HNPGL than 24-h urinary excretion rates of catecholamines and metabolites. Eur J Endocrinol. 2013;169(3):377–82.
93. Martucci VL, Pacak K. Pheochromocytoma and paraganglioma: diagnosis, genetics, management, and treatment. Curr Probl Cancer. 2014;38(1):7–41.
94. Gaynor BG, Elhammady MS, Jethanamest D, Angeli SI, AzizSultan MA. Incidence of cranial nerve palsy after preoperative embolization of glomus jugulare tumors using Onyx. J Neurosurg. 2014;120:377–81.
95. American College of Surgeons Commission on Cancer; American Cancer Society. National Cancer Data Base report on malignant paragangliomas of the head and neck. Cancer. 2002;94(3):730–7.
96. Taïeb D, Kaliski A, Boedeker CC, Martucci V, Fojo T, Adler JR Jr, Pacak K. Current approaches and recent developments in the management of head and neck paragangliomas. Endocr Rev. 2014;35(5):795–819.
97. Hinerman RW, Amdur RJ, Morris CG, Kirwan J, Mendenhall WM. Definitive radiotherapy in the management of paragangliomas arising in the head and neck: a 35-year experience. Head Neck. 2008;30(11):1431–8.
98. Moskovic DJ, Smolarz JR, Stanley D, Jimenez C, Williams MD, Hanna EY, Kupferman ME. Malignant head and neck paragangliomas: is there an optimal treatment strategy? Head Neck Oncol. 2010;2:23.
99. Patel SR, Winchester DJ, Benjamin RS. A 15-year experience with chemotherapy of patients with paraganglioma. Cancer. 1995;76(8):1476–80.

14 Speech, Voice, and Swallowing Rehabilitation for Patients with Head and Neck Cancers

Heather M. Starmer and Jocelen Hamilton

Key Points

- The speech–language pathologist plays a critical role in the management of communication and swallowing issues in patients with head and neck cancer.
- Preoperative assessment of communication and swallowing issues provides the patient with realistic expectations of posttreatment function and the oncology treatment team with information that may influence treatment decisions.
- Education prior to surgery and nonsurgical therapy is associated with increased patient satisfaction and decreased patient anxiety.
- Communication and swallowing issues are common following treatment for head and neck cancer and depend on the site of disease, stage of disease, and treatments employed.
- A proactive approach to rehabilitation should be adopted whenever possible, particularly in patients undergoing radiation therapy.

H. M. Starmer (✉)
Department of Otolaryngology, Head and Neck Surgery, Stanford University, Palo Alto, CA, USA
e-mail: hstarmer@stanford.edu

J. Hamilton
Head and Neck Cancer Care Program, Stanford Cancer Center, Palo Alto, CA, USA
e-mail: jrhamilton@stanfordhealthcare.org

Introduction

Patients with head and neck malignancies commonly encounter challenges with speech, voice, and/or swallow function prior to, during, or following oncologic management. Speech and swallowing function are critical aspects of being human, and degradation of these creates significant impacts on quality of life. Being able to swallow after oncologic treatment is rated as one of the highest priorities by patients [1, 2]. Furthermore, changes in one's ability at the physical level (i.e., speech, voice, swallow function) often disrupt one's social interactions and participation leading to degradation of physiological well-being which may already be under insult due to distress related to the head neck cancer diagnosis and treatment [3].

The speech–language pathologist (SLP) plays a pivotal role on the multidisciplinary team, pro-

R. A. Chandra, R. J. Li (eds.), *Multidisciplinary Management of Head and Neck Cancer*,
https://doi.org/10.1007/978-3-031-05973-5_14

viding skilled assessment and treatment of such functional impairments. Inclusion of the SLP from the onset of care may help to educate the treatment team about functional status to assist with decisions regarding the most appropriate oncologic treatment modalities, mitigate functional impairments associated with cancer treatment, and provide the patient and their caregivers with realistic expectations for function across the treatment continuum. In this chapter, we will outline how different treatment approaches may impact speech, voice, and swallowing as well as the SLP's role in assessing and managing these difficulties.

Value of Baseline Assessment

Swallowing function prior to head and neck cancer treatment is influenced by disease site (more prevalent in hypopharyngeal and laryngeal primary tumors), tumor stage (more common in T3-4 disease), prior surgical treatment, and other patient factors such as pain [4, 5]. As tumors are commonly slow growing, patients often adapt to the functional changes related to the tumor and thus under-estimate the presence and degree of dysphagia. Patient report and clinician impressions (such as those gained through patient interviews or instruments like the Performance Status Scale Head and Neck (PSS-HN) [6] and the Common Terminology Criteria for Adverse Events (CTCAE)) typically under-report dysphagia compared to what is noted on instrumental, objective measures of swallowing such as videofluoroscopic swallow studies (VFSS) or flexible endoscopic evaluation of swallowing (FEES) [7, 8]. Given that there is often an expectation of worsening function in response to treatment as well as potential immunocompromise for those patients undergoing chemoradiotherapy, it is of great value to assess baseline swallow function prior to treatment to mitigate negative consequences of dysphagia during treatment. Further, as baseline function may correlate with long-term function, pretreatment assessment allows the treatment team to appropriately prepare and advise the patient regarding realistic posttreatment expectations [9].

Patient reported changes prior to treatment may include a sense of food sticking in the throat, coughing/choking during eating and drinking, odynophagia, a need to alter their diet, and increased time to consume meals [10]. Presence of baseline dysphagia has been shown to correlate with the need for feeding tube placement during radiation therapy [11, 12]. As it is increasingly understood that prophylactic feeding tubes may contribute to posttreatment dysphagia, risk stratification regarding which patients may truly need a feeding tube is critical. Other factors which have been associated with increased feeding tube use during radiation-based treatment include advanced T stage (increased risk) [13], advanced/level II nodal stage (increased risk) [13], posterior hypopharyngeal tumors (increased risk) [14], baseline malnutrition (increased risk) [13], and prophylactic use of gabapentin (reduced risk) [15]. Thus, the baseline swallowing evaluation plays an important role in determining the need for proactive feeding tube placement [16].

The baseline swallowing evaluation should include both measures of patient perception of dysfunction and measurement of physiologic function and is recommended by various international agencies as standard of care for patients with head and neck cancer [7, 9]. Common patient-reported outcome tools include the MD Anderson Dysphagia Inventory (MDADI) [17], the Eating Assessment Tool (EAT-10) [18], and the Sydney Swallow Questionnaire (SSQ) [19] among others. Clinician-rated measures such as the PSS-HN [6], the Functional Oral Intake Scale (FOIS) [20], and the CTCAE dysphagia rating may compliment the information provided by patient-reported outcome measures. Additionally, a formal cranial nerve examination, measurement of maximal mouth opening (MMO), tongue mobility using a standardized scale [21], and tongue strength using a device such as the Iowa Oral Performance Instrument (IOPI) may provide additional valuable information. Finally, instrumental assessment of swallowing is warranted given the potential for silent swallowing dysfunction. Both the VFSS and FEES examinations are appropriate options for instrumental assessment and may provide complimentary clinical information. The Dynamic Imaging Grade of Swallowing Toxicity (DIGEST) [22] and DIGEST-FEES [23] provide overall ratings of safety, efficiency, and overall swallowing impairment during instrumental swallowing evaluations using the nomenclature of the CTCAE.

In addition to issues with swallowing prior to treatment, many patients may also experience baseline changes to speech or voice. As with swallowing, changes in communication function are strongly associated with tumor site and stage with oral cavity cancers more likely to cause changes in speech function and laryngeal cancers more likely to cause changes in voice [24, 25]. These difficulties have been associated with reduced communication interactions, overall quality of life, and depression [26]. As with swallowing, patients presenting with baseline communication impairments should have a formal assessment of function including a full cranial nerve examination, completion of patient reported outcome measures such as the Voice Handicap Index (VHI) [27], the Speech Handicap Index (SHI) [28], and the Communication Participation Item Bank (CPIB) [29], formal measures of speech or voice such as acoustic and aerodynamic measures of voice, vocal rating scales such as the GRBAS scale [30] and the Consensus on Auditory Perceptual Evaluation of Voice (CAPE-V) [31], and formal assessment of speech precision and intelligibility using tools such as the Goldman Fristoe Test of Articulation [32] and Tikofsky's 50-word intelligibility test [33].

It is particularly critical to complete a comprehensive baseline assessment in those patients undergoing salvage treatment or treatment occurring after recurrence of the head and neck cancer. Because these individuals have had prior treatment and potentially have chronic sequela from that treatment, it is critical to understand their current level of function in order to provide a realistic prognosis as well as for making treatment decisions such as whether a patient requires a G-tube versus an NG tube.

Preoperative Assessment and Counseling

SLPs assist patients in maximizing speech and swallowing outcomes along the continuum of oncologic care and pre-oncologic treatment evaluation, and counseling is key to setting the stage for a productive rehabilitation journey [34]. While the specifics of the counseling session are dependent on the tumor characteristics and surgical plans, some general themes apply across patients.

Providing presurgical counseling for anticipated changes in function of speech and swallowing after surgery has been shown to support QOL outcomes later [35]. Having realistic expectations regarding posttreatment function allows the patient and family begin to address how they might adjust their lives and prepare for the changes. SLPs, as communication specialists, can monitor comprehension of information and provide additional teaching regarding normal and altered anatomy and physiology to maximize patient/family comprehension of information. Patients with a good understanding of their oncologic care are more likely to follow-through with therapeutic recommendations. Presurgical counseling is also important for the development of a strong client–clinician alliance which is key to facilitating patient adherence and buy in.

Baseline evaluations of communication and swallowing are typically completed first (as outlined above), and results are reviewed with the patient. Recorded images from a VFSS or FEES can be used to assist with education regarding normal and altered anatomy and physiology. Key components of the presurgical counseling are to review the general surgical plan and likely changes to communication and swallowing function, as well as timelines of recovery. It can be beneficial to first have the patient share their understanding of the planned surgery and its likely impact on speech/voice/swallow function. This then allows the counseling session to be tailored to maximize patient understanding. Discussion and careful listening can elucidate misperceptions, commonly that a surgery will "fix" an existent speech or swallowing deficit. At this point, the overall prognosis for communication and swallowing function is discussed. It is beneficial to outline the anticipated status of communication and swallowing function at the acute (immediately following surgery to discharge), subacute (2 weeks to 2–3 months after surgery), and long-term (3+ months) time-points and to provide an overview of key strategies and therapy approaches that will be utilized to maximize function. The subacute time frame of recovery is influenced by whether adjuvant chemoradiation is needed. Briefly outlining functional speech and swallowing changes during and after radiation can be helpful before moving on to review potential for long-term, chronic functional changes.

For many patients, counseling regarding the immediate postoperative recovery period (inpatient stay) can help prepare them for major, yet sometimes temporary, changes if they are aphonic due to a tracheostomy tube or NPO with feeding tube. The patient is introduced to the role of inpatient SLP services to assess and treat swallowing and communication function and, in the case of possible or likely tracheostomy placement, the rationale for a tracheostomy speaking valve is introduced using a visual model to show placement of a trach tube and resulting changes in airflow impacting breathing and voicing. Additional communication strategies, known as Augmentative and Alternative Communication (AAC) tools, are reviewed such as use of a whiteboard or tablet to write or use of a text-to-speech app. If a language barrier is present or if the patient has reading or writing difficulties, it is beneficial to develop other communication tools prior to surgery such as a low-tech AAC board with picture supports. Additionally, the SLP introduces the concept of normal postsurgical edema and how the rehabilitation process coincides with the process of healing and resolution of swelling. Patients undergoing free flap reconstruction are prepared to anticipate a large bulky flap resulting in difficulty with oral management of secretions as well as exaggerated postoperative speech disturbances.

A general timeline for inpatient SLP services is provided along with potential status of diet, likely use of swallowing strategies and status of communication function at the time of discharge. The rationale for postoperative assessment and treatment is outlined along with anticipated frequency of therapy and an overview of therapy targets.

Determine the Status of Related Factors: Health Literacy, Support, Resources

Presurgical evaluations are also important for determining the status of factors that may influence the patient's ability to achieve maximal recovery. SLPs may detect memory or reasoning deficits that could influence poor follow-through. Additionally, identification of gaps in patient understanding may be identified due to low health literacy skills. As communication specialists, SLPs can adjust education materials to better meet the patient's understanding and/or refer to other team members or schedule additional sessions to maximize understanding. They may also alert the broader treatment team to the ongoing need for additional support. Referrals to registered dietitians are common as compromised nutrition may be identified at baseline or anticipated as part of the treatment process. Patients and family are also counseled regarding additional supplies and resources needed during the rehabilitation process and options for their obtainment. These include supplies and devices such as heat and moisture exchangers and an electrolarynx if having a laryngectomy. Through conversation, barriers to participation in therapy, whether transportation or access to a device for teletherapy, can begin to be addressed and a suggestion for a social work referral can be brought to the multidisciplinary team. Last, but not least, inclusion of the patient's caregiver(s) throughout the rehabilitation process is emphasized, and the patient and family/caregiver's mental and emotional needs are supported with referrals for additional support through supportive cancer care programs, social work, and/or oncology psychology.

Considerations for Specific Surgical Procedures

In the following sections, we will outline information regarding swallowing and communication management for various anatomic structures and regions. Speech and swallowing outcomes following surgery will depend on the location of the resection, the size of the resection [36], and the nature of reconstruction [37]. Knowing the physiological correlates for anatomical sites of the aerodigestive tract allows SLPs to grossly predict speech and swallowing changes following surgery. It is important to keep in mind that each case needs careful consideration as surgical resections are often complex, particularly when tumors extend to multiple sites. Besides anatomical changes due to ablation and reconstruction, nerve manipulation during surgery or nerve sacrifice can alter motor and/or sensory function to the area and contribute to speech and swallowing difficulties. In the subacute phase, scar formation and lymphedema may also develop, leading to reduced range of motion and flexibility which require specific management.

Glossectomy

Surgical resection of the tongue may have functional implications for both communication and eating. In respect to swallowing, the tongue plays an important role in both the oral and pharyngeal phases. The tongue tip, tongue body, and tongue base are the major functional regions of the tongue. A resection involving the tongue tip is likely to lead to reduced bolus hold, impaired bolus movement for adequate mastication, and difficulty with posterior bolus movement. Surgeries involving the body of the tongue impair bolus collection and posterior propulsion, whereas resection of the lateral tongue often results in impaired bolus lateralization during mastication and reduced bolus clearance on the resected side. Reduced bolus control may impact swallowing safety if premature spillage into the pharynx occurs while the airway is open. For larger and more posterior tumors, the pharyngeal swallow may also be compromised due to reduced propulsive forces on the bolus, leading to more stasis within the pharynx. Further, if floor of mouth musculature is resected, a reduction of hyolaryngeal complex elevation may result, leading to reduced laryngeal vestibule closure and reduced cricopharyngeal opening [38, 39]. Additional disruptions that patients experience also include impaired sensation and coordination which may lead to tongue biting, reduced bolus control and monitoring of bolus size, as well as impaired oral hygiene due to inability to extract particles from sulci or even to lick one's lips. Most patients can tolerate an oral diet following a partial glossectomy, including patients who undergo large resections of the oral tongue with intact floor of mouth and tongue base musculature. Again, depending on the location and extent of resection, patients may need routine use of compensatory strategies and food texture alterations or avoidances.

Larger surgical resections such as a subtotal [>50% but <90%] or total glossectomy [>90%] result in major swallowing and speech deficits. Baseline swallowing is often compromised due to the large tumor volume and, postsurgically, these patients typically experience a high rate of aspiration after the swallow due to poor bolus clearance and impaired movement of the hyolaryngeal complex as a result of removal of the suprahyoid musculature [40]. In these cases where the suprahyoid musculature is removed, surgeons may attempt to facilitate improved airway protection by suspending the larynx in a raised position by fixing the hyoid bone to the mandible [41]. A systematic review by Dziegielewski et al. found ~24% of patients were G-tube dependent 12 months after total glossectomy with free flap reconstruction [42].

Structural and functional integrity of the tongue is key for the production of different speech sounds with reduced speech precision and intelligibility noted as common consequences following tongue resection. If a lesion and subsequent surgery involve the tongue tip, articulation is greatly impaired as this portion of the tongue is key for articulating many sounds including the dental (TH), alveolar (T, D, S, Z, N, L), and postalveolar sounds (SH, CH, J, R). Resection of lateral tongue lesions often impairs tongue elevation, resulting in a lateral lisp for S, Z, SH, CH, and J. Tongue resections involving the posterior tongue will influence the production of velar consonants K and G. Vowel production is influenced by the vertical and horizontal positioning of the tongue within the oral cavity, and vowel distortions are common after surgery.

Consideration of the surgical approach and type of reconstruction is important for anticipating postsurgical function and developing the rehabilitation plan. These are influenced by several factors including the type and extent of tumor, whether there is involvement of multiple subsites, and patient factors such as overall health and prior radiation therapy [43]. Generally, it follows that the smaller the tumor, the less invasive the surgery and the smaller the impact on function. The surgical approach itself may contribute to the nature of the dysphagia. Bhattacharya et al. found that use of an extra-oral surgical approach for glossectomy, such as a mandibulotomy, was correlated with worse swallowing outcomes on VFSS [36]. Patients with small tongue lesions typically can be resected using a transoral approach with primary closure and are expected to experience milder forms of articulation and oral bolus manipulation disruption than those with a larger tumor necessitating a flap reconstruction. However, if the primary closure for a small tongue lesion leads to a tethering of a portion of the tongue, articulation could be significantly disrupted.

Larger tongue resections may require use of skin grafts (containing skin and underlying subcutaneous tissue) or myocutaneous flaps that include underlying muscle for larger defects. In the case of glossectomy reconstruction, the aim of reconstruction methods is to provide enough bulk to enable contact between the flap and the palate to assist with speech and swallowing function while minimizing any tethering of the native tongue, if present. Tissue transfer may use a locoregional flap which maintains a pedicle to the original site such as the facial artery myomucosal flap (FAMM) or, for larger defects, a pectoralis major flap. For tongue reconstruction, locoregional flaps are typically used if the defect size comprises less than 1/3 of the oral tongue [43]. Free flap use involves transfer of tissue from a distant location and uses microvascular surgery to complete anastomosis for blood supply. Use of free flaps, such as the radial forearm free flap (RFFF) or anterolateral thigh free flap (ALT) enables the surgeon to be more selective in designing the flap dimensions to support maximal speech and swallowing outcomes with the aim to avoid both inadequate and excessive bulk. Free flaps may be sensate by way of anastomosis of nerves in the flap to regional nerves to allow for sensory innervation or insensate. Many studies have looked at whether use of sensate flaps results in superior speech and/or swallowing function and findings indicate that use of innervated free flaps supports more favorable speech and swallowing outcomes when used for reconstruction of hemi- and subtotal glossectomies [44].

Floor of Mouth and Composite Resections

Preparation for speech and swallowing changes following an oral composite resection, resections that involve removal of more than one structure within the oral cavity, will depend on the specifics of the tumor location and the surgery plan. Floor of mouth resections may result in tethering of the anterior portion of the tongue and lead to difficulties with bolus hold, bolus manipulation, and oral clearance as well as impaired articulation for T, D, L, R, and N. Resections including the alveolar ridge typically involve removal of dentition affecting mastication as well as articulation. It is important to document current MMO and review the risk for long-term trismus following surgery, especially if the masseter, lateral pterygoid, and/or medial pterygoids are within the surgical site.

Mandibulectomy

Mandibular resections often lead to challenges with mastication, trismus (especially if the temporomandibular joint is resected), and impaired occlusion. Furthermore, reduced overall jaw stability may negatively affect swallowing efficiency, as well as oral containment and laryngeal elevation [45]. Reconstruction with osseocutaneous flaps has been shown to improve speech as well as mastication and swallowing outcomes [46]. Patients should be prepared to experience trismus and difficulty masticating foods. Often, surgeons place life-long restrictions on mastication of hard foods such as nuts following mandibular reconstruction. Speech changes may be expected for anterior surgeries due to removal of dentition and postsurgical lip weakness may also contribute to speech distortions as well as oral incompetence. As indicated, patients are prepared for potential presence of a temporary feeding tube and/or tracheostomy tube.

Maxillectomy

Resections of the maxilla may involve resection of the hard and soft palate leading to communication between the oral and nasal cavities. Depending on the type and staging of reconstruction, patients may have reconstruction of the defect at the time of the resection, utilize a prosthesis life-long, or, in some cases, return for a free flap reconstruction later. Patients typically do not present with speech or swallowing complaints in the setting of maxillary tumors unless an oronasal fistula is already present. Patients are counseled regarding changes to swallowing which may include nasal regurgitation through an oronasal fistula, hypernasality, and reduced speech intelligibility. Mastication and speech may also be affected if the resection involves removal of dentition. For patients who do not undergo primary reconstruction, collaboration between the maxillofacial prosthodontist and the SLP for palatal obturator optimization will improve speech and swallowing outcomes. Alternatively, for those patients undergoing free flap reconstruction, patients should be prepared to experience the presence of a bulky flap which can lead to reduced bolus containment and altered speech sound production as it creates a lower "ceiling" within the oral cavity.

Oropharyngeal Resections

Transoral robotic surgery (TORS) is increasingly used for resection of oropharyngeal tumors and has been shown to significantly reduce swallowing-related morbidity compared with the traditional approach which required splitting the mandible. Despite these advancements, patients still experience dysphagia following TORS. Research shows that swallowing function improves by 3–6 months [47, 48], but some degree of dysphagia does typically persist, even as long as 1 year following surgery [49]. As is noted with other surgical sites, the larger the tumor and necessary resection, the greater the impact on swallowing function. Hutcheson and colleagues [48] found that moderate-severe dysphagia per VFSS post-TORS was significantly associated with T stage and baseline primary tumor volume.

Patients may experience swallowing difficulties related to pain for 1–2 weeks following oropharyngeal resections. Delayed initiation of the pharyngeal swallow may occur in anticipation of pain. Scarring/contracture of the tonsillar region may also lead to reduced lateral pharyngeal wall contraction. Speech and swallowing outcomes will be negatively affected if the resection extends to the soft palate leading to impaired velopharyngeal closure. Hypernasality of speech and velopharyngeal incompetence both negatively impact speech clarity and frequently result in reduced loudness of speech due to loss of power through the nasal passage. Nasal regurgitation may occur. Impaired velopharyngeal closure may lead to reduced intraoral/pharyngeal pressure generation and subsequent reduced driving pressure on the bolus leading to increased pharyngeal residue. Base of tongue resections may result in reduced bolus propulsion and reduced bolus clearance through the pharynx. Most patients demonstrate a functional swallow and are able to begin a soft, moist diet within a day of surgery. Use of small bites, chewing on the nonsurgical side (if applicable) and use of liquid wash are often beneficial. In addition to reviewing likely postsurgical swallowing changes, patients undergoing TORS for oropharyngeal tumors are informed that pain management is very important in the weeks following surgery to keep pain from preventing function during rehabilitation. If extensive resection of the soft palate is needed, use of a palatal lift prosthesis can improve velopharyngeal closure, especially if the lateral and pharyngeal wall are preserved to assist with closure [50].

Thyroidectomy

Swallowing and voice disorders following thyroid surgery are most commonly caused by disruption of the recurrent or superior laryngeal nerves. Injury to the recurrent laryngeal nerve may lead to vocal fold paralysis and impaired glottic closure during swallowing. Injury to the external branch of the superior laryngeal nerve, responsible for supraglottic sensation, may lead to reduced patient sensitivity for penetration/aspiration as well as some limitations to vocal range. Of note, some patients may present with dysphagia and/or dysphonia at baseline if the thyroid tumor has already compromised the laryngeal nerve function. Large tumors may also press against the esophagus resulting in dysphagia.

Laryngectomy

Patients undergoing partial or total laryngectomy will benefit from a comprehensive understanding of the impact of surgery on voice and swallowing. Both supraglottic and supracricoid laryngectomy are associated with compromised airway protection; therefore, there is particular value in baseline instrumental swallowing assessment in this population, including training in compensatory maneuvers that will be used during rehabilitation. Realistic timelines for recovery and the role of the SLP in rehabilitation should be discussed [51]. Acute and long-term changes in communication should also be reviewed.

For patients undergoing total laryngectomy, a comprehensive preoperative counseling session is required and associated with reduced length of stay and readmission rates [52–54]. Removal of the larynx prevents phonation, thereby creating a need for an alternative means of sound generation or communication. Patients should be introduced to different options including the electrolarynx, tracheoesophageal voice restoration, and esophageal speech. The SLP should review the pros and cons of each option. It is often helpful for the patient to see and hear examples of each method in use so that they are able to determine which option will be best suited for their communicative needs. In addition to communication changes, there may be reduction of swallowing efficiency as well as enjoyment of oral intake due to diminished taste and smell [55, 56]. Patients should be prepared for a period of NPO status during acute healing as well as a realistic timeline of return to oral intake, depending on the extent of their surgery and reconstruction. Finally, the diversion of the upper airway influences respiratory function, and the patient should be prepared for the permanent changes to their respiratory anatomy and the need for additional safety considerations. The changes associated with total laryngectomy can lead to dramatic changes in patient perceived quality of life, lifestyle, and interpersonal relations and patients benefit from an honest conversation to prepare them as best as is possible [57].

Postsurgical Management

Postsurgical management will depend on the type and extent of the surgical procedure. It is critical for SLPs to review operative reports prior to assessing patients and to clarify information with the surgical team in order to fully grasp the extent of the resection and impact on cranial nerves. The goal of postsurgical rehabilitation is to re-establish, to the greatest extent possible, functional communication and swallowing abilities.

Following surgery, inpatient rehabilitation primarily targets patient education, establishing functional communication, and establishing oral intake when appropriate by providing the patient with strategies to facilitate safe and efficient swallowing. Clinical and/or instrumental swallowing evaluations (VFSS, FEES) are completed as indicated. In many cases, a clinical swallowing assessment is typically performed during the inpatient stay instead of a VFSS or FEES as dysphagia is an expected outcome that is typically well-managed with compensatory strategies.

Often instrumental evaluations are deferred to the outpatient setting where acute pain and edema are less confounding. One exception to this is the patient who has undergone partial laryngectomy. Due to the high incidence of silent aspiration in this patient population, an instrumental swallowing assessment is warranted prior to transition to full PO diet [58, 59]. During early recovery, the SLP should work with the patient to habituate compensatory strategies targeting improved airway closure and to engage the swallowing musculature to prevent disuse atrophy.

Patients may need continued alternative methods of nutrition to support proper healing despite initiation of oral intake. For those with extensive resections resulting in significant dysphagia with either an NG tube or G-tube, SLP intervention may focus primarily on management of secretions. In regard to swallowing, the timing of initiation of swallowing trials will vary depending upon the extent of surgical resection.

If a tracheostomy tube is present, the SLP plays an important role in maximizing the restoration of airflow through the larynx and upper airway through evaluation and treatment for use of a tracheostomy speaking valve. In addition to helping to restore verbal communication (further addressed below), use of a speaking valve enables improved generation of subglottal pressures for increased cough strength for secretion management and postsurgical swallowing strategies and maneuvers. Multiple studies have shown that a speaking valve may improve aspiration status [60–62]. Furthermore, tolerance of a speaking valve and subsequent improved management of secretions can facilitate timely decannulation which is ultimately beneficial for restoration of upper airway breathing and improved subglottal pressures in the absence of glottal incompetence (i.e., vocal fold paresis). Note, even if patients are unable to tolerate extended speaking valve use, therapy targeting use of brief digital occlusion for speech and brief digital occlusion while swallowing or expectorating secretions may still be useful.

Initial postsurgical assessments of communication focus on establishing the best methods for functional communication. Speech assessments following an oral or oropharyngeal resection are typically informal as postsurgical swelling precludes a detailed speech evaluation for several weeks until swelling has largely subsided. Patients are taught to slurp and swallow secretions which can reduce further distortion of their speech due to pooled secretions. Inpatient SLP support addresses use of AAC tools such as writing, mouthing, and gestures. AAC may also include low-technology supports such as a picture or word board or high-tech options such as a text-to-speech app on a smartphone or tablet. Compensatory communication strategies are also introduced to the patient and may include use of over-articulation, gaining the attention of the listener prior to speaking and explicit identification of the topic for the listener. Family members and staff can be taught supportive communication strategies such as repeating back the message they heard to allow the patient to confirm and/or clarify. Patients may experience fatigue and increasing pain with extended talking so may need to use writing or other preferred forms of AAC throughout each day.

Regardless of tracheostomy status, if unexpected dysphonia is appreciated, further evaluation for vocal fold motion impairment can be completed and treatment planned, such as an injection laryngoplasty, so that voicing can be improved.

The inpatient admission is an opportune time to establish functional communication following total laryngectomy. The SLP will introduce the patient to the electrolarynx and initiate training in the use of the device. This will include training in proper device use/maintenance, optimal placement of the device for sound generation, and the importance of over-articulation and slowed rate to facilitate understanding. Structured speech practice should be provided on a daily basis as is feasible, and the family/caregivers should be engaged in this process. In addition to electrolaryngeal training, the SLP should review information initially introduced during the preoperative counseling session regarding changes in the respiratory system and safety considerations following laryngectomy [53]. While swallowing trials commonly are deferred until after hospital discharge, particularly in the instance of salvage laryngectomy, the SLP can reinforce teaching around swallowing changes and expectations.

In addition to initiation of assessment and treatment for specific speech, voice, and swallowing deficits (detailed below), inpatient SLPs continue to provide counseling and education regarding expected timelines of recovery and reinforce the importance of continued outpatient rehabilitation.

Outpatient Rehabilitation

Intensive outpatient speech and swallowing rehabilitation typically begins about 2 weeks after surgery as most of the postsurgical swelling is typically resolved by that time [63]. Postsurgical swallowing evaluations often begin with an instrumental evaluation such as VFSS or FEES with a heavy focus on the assessment of swallowing safety and efficiency. Goals include determination of whether the patient can consume anything safely, whether the patient can consume enough to sustain nutrition/hydration and to outline swallowing physiology for treatment planning. Even if swallowing safety and efficiency are significantly impaired, SLPs work to find the best strategies for the patient to utilize so they may begin swallowing for the purposes of rehabilitation (i.e., therapeutic PO trials) while minimizing risks for aspiration-induced illness.

VFSS and FEES are both valuable in assessing dysphagia and provide complimentary information. VFSS allows for a dynamic view of bolus flow and physiology of swallowing structures through all stages of the swallow, whereas FEES does not assess the oral or cervical esophageal phases. In contrast, FEES includes direct visualization of postsurgical anatomical changes (including edema), secretion management, and evaluation of laryngeal function. Additionally, FEES is useful for completion of extended therapeutic trials and biofeedback and does not expose the patient to radiation. Selection of assessment type is influenced by many factors including tumor location, time postsurgery, and severity of dysphagia. For example, VFSS is ideal following surgical treatment of the hypopharynx or cervical esophagus, whereas FEES may be useful for initial postoperative assessment in the setting of severe dysphagia to assess secretion management and to allow for extended therapeutic trials. Additionally, each institution may have its own practice patterns influenced by access to equipment and personnel for each test.

Dysphagia management following surgery typically employs compensatory strategies, dietary modifications as well as range of motion and strengthening exercises. The goal of therapy is to facilitate a safe and efficient swallow with the least restrictive diet texture possible and the fewest compensations. Rehabilitation may include exercises to facilitate airway closure and pharyngeal clearance, bolus-driven therapies, and cough strength/efficiency using interventions such as the expiratory muscle strength trainer (EMST) [64]. The specific plan of care should be established based upon the findings of the instrumental swallowing evaluation.

Table 14.1 lists common impairments, strategies, therapeutic maneuvers, and exercises for different surgical sites. Exercises should be selected to strengthen intact musculature, whereas stretches should be utilized to facilitate maximal range of motion for the swallowing structures. Biofeedback using devices such as the Iowa Oral Performance Instrument (IOPI) or devices using submental EMG may help to optimize treatment outcomes.

Completion of range of motion stretches for the first 3 months following surgery results in improved long-term outcomes [65]. Manual therapy, including active stretching and soft tissue mobilization, helps improve tissue extensibility and prevent scar tension. For patients with severe chronic dysphagia after surgery, swallowing intervention is still very much needed to address secretion management as well as oral, pharyngeal, and laryngeal hygiene. Intervention also addresses the development of some degree of oral intake, not only for hygiene, but also for quality of life.

In regard to swallowing rehabilitation following total laryngectomy, many of the causes of

Table 14.1 Common impairments, strategies, maneuvers, and exercises for dysphagia across different surgical sites

Location	Impairment	Strategies and therapeutic maneuvers	Exercises
Glossectomy	Reduced bolus control (pocketing/pooling)	Place bolus to stronger side	Lingual range of motion stretches and manual therapy to surgical region
	Impaired bolus manipulation	Use of a straw on the nonsurgical side	Lingual strengthening exercises
	Impaired bolus transport to oropharynx	Tilt head to nonsurgical side or tilt head posteriorly[a]	
	Premature spillage into pharynx	Slurp and swallow[a]	
	Drooling	Liquid wash	
		Gavage[a]	
		Palatal drop prothesis[a]	
		Thicker viscosities for impaired bolus control	
		Thinner viscosities for impaired bolus propulsion[a]	
		Supraglottic swallow[a]	
Floor of mouth resection	Tongue tethering	Place bolus to stronger side	Lingual and jaw range of motion stretches and manual therapy to surgical region
	Impaired bolus manipulation	Tilt head to nonsurgical side	Lingual strengthening exercises
	Impaired bolus containment	Use of straws	
	Trismus	Liquid wash	
	Aspiration due to reduced hyolaryngeal excursion	Mendelsohn maneuver	
Mandibulectomy	Loss of dentition	Bolus placement to nonsurgical side	Jaw range of motion exercises
	Impaired mastication	Liquid wash	Manual therapy for trismus
	Trismus	Soft consistency foods	
	Reduced laryngeal elevation	Mendelsohn maneuver	
Maxillectomy	Nasal regurgitation	Tilt head to nonsurgical side or tilt head posteriorly	Not applicable
		Obturator	
Soft palate resection	Nasal regurgitation	Posterior head tilt	Exercises to improve VP closure using biofeedback
	Reduced bolus propulsion	Manual nares occlusion	EMST
		Effortful swallow	
		Obturator	
Tonsillectomy	Premature spillage	Turn head to surgical side if unilateral	Not applicable
		Tilt head to nonsurgical side	
		Chin tuck	

(continued)

Table 14.1 (continued)

Location	Impairment	Strategies and therapeutic maneuvers	Exercises
Base of tongue resection	Impaired bolus propulsion	Posterior head tilt	Masako
	Vallecular stasis	Turn head to surgical side if unilateral	Mendelsohn
	Reduced laryngeal elevation	Chin tuck with head turn	Effortful swallow
		Effortful swallow	
		Multiple swallows	
		Liquid wash	
Partial laryngectomy	Reduced laryngeal vestibular closure	Supraglottic swallow	Super supraglottic swallow
	Laryngeal penetration/ aspiration	Super Supraglottic swallow	Bio-FEES-back
	Reduced sensation	Chin tuck	Airway protected swallow
		Head rotation to the more impaired side	
		Thicker liquids	
Total laryngectomy	Reduced pharyngeal clearing forces	Liquid wash	Effortful swallow
	Pseudovallecula	Upright positioning	
	Reduced UES opening	Thinner foods/liquids	

[a]Indicates frequent use for subtotal/total glossectomy

swallowing difficulty are less amenable to behavioral intervention. In patients with swallowing complaints, a modified barium swallow study should be pursued to establish the cause of dysphagia. Causes may include reduced upper esophageal sphincter (UES) opening, presence of pseudoepiglottis/pseudovallecula, and reduced neo-pharyngeal clearing forces due to changes in underlying swallowing physiology [66]. When the cause of dysphagia is reduced UES opening or pseudoepiglottis, the patient should be referred back to the surgical team for consideration of surgical interventions. When dysphagia is related to poor pharyngeal clearance in the absence of the aforementioned issues, swallowing therapy should be offered targeting tongue strengthening, use of compensatory strategies, and exploration of appropriate food options. Pharyngeal manometry may provide beneficial information when trying to establish the relative contribution of outflow obstruction at the UES versus reduced pharyngeal propulsive forces [67, 68].

Speech and voice evaluations may be completed during the initial outpatient assessment. Depending on the case specifics and degree of postsurgical swelling, these communication features may initially undergo informal assessments and treatment to allow more time for swelling to subside before a formal evaluation is completed. Formal speech evaluations previously discussed in the preoperative evaluation section can be repeated at different intervals following surgery. Intervention includes maximizing residual function and/or teaching alternate articulatory contacts to approximate the target sound. As mentioned above, compensatory strategy use is also addressed as well as family education to utilize supportive communication strategies. Assessment and support for ongoing use of AAC is also addressed across time to adjust the tools to meet the current communication needs.

Rehabilitation following total laryngectomy will depend upon the chosen alaryngeal speech modality. For patients who have selected the electrolarynx as their primary communication tool, the outpatient SLP will continue providing structured speech practice, gradually increasing the complexity and challenge of tasks. For patients who have undergone primary tracheoesophageal puncture, the SLP will provide training to the patient and their caregivers regarding how to care for the device and how to manage any emergency issues that may arise. Patients will need to understand different patterns of prosthetic leakage and how to handle each. The SLP will typically initiate TE voicing trials ~2 weeks post-op assuming medical clearance to do so. Speech therapy may include facilitation of proper stomal occlusion, improving coordination of respiration and phonation, and avoidance of strain and tension. When patients have difficulty with voicing or with prosthesis-related complications, the modified barium swallow is a valuable tool to help assess for structural and functional issues [69]. If patients have chosen a non-indwelling prosthesis which is managed at home rather than in the clinic, the SLP will need to train the patient in how to change the prosthesis using safe and appropriate techniques, and ensuring patient competence before allowing the patient to manage the prosthesis independently. Additional interventions for stoma management using laryngectomy tubes and buttons and pulmonary rehabilitation using heat–moisture exchange units (HMEs) will be provided by the SLP. Initiation of hands-free speech is generally reserved for later stages of rehabilitation once the patient has demonstrated fluent TE speech, and the clinician has identified the best interface option for the hands-free valve. For patients who have elected to pursue esophageal speech, the SLP will provide training in different methods of esophageal air insufflation and help the patient determine which yields the most fluent esophageal speech. As with electrolaryngeal training, the focus then turns to gradually increasing challenges with speech drills.

Comprehensive speech pathology services also address the impact the communication and/or swallowing problem has on one's participation in social events and personal feelings. Some patients only feel comfortable eating alone or in the presence of immediate family. Other patients may avoid phone communication and social interactions with strangers while running errands due to their speech changes. These personal and participation factors, in addition to impaired speech and swallowing, greatly influence quality of life. Thus, intervention may include encouragement to seek out head and neck cancer support groups and counseling services. Ongoing intervention also addresses the changing needs of the patient as they participate in different events that involve eating and/or communication.

Lymphedema

Lymphedema is a common consequence of surgical resection in the head and neck region and results in stasis of protein-rich lymphatic fluid in the interstitial spaces. While acute edema is anticipated in early postoperative recovery, lymphedema may develop as early as 4–6 weeks after surgery. Lymphedema has been associated with poorer functional outcomes, and thus the management of lymphedema is an important part of head and neck rehabilitation [70]. The management of head and neck lymphedema may be the responsibility of SLPs, physical therapists, and occupational therapists with specific training in head and neck lymphedema. Which rehabilitation professional treats head and neck lymphedema is often a care center/hospital preference based on the availability of trained personnel.

Surgical Case Presentation: Mr. T

Mr. T is a 59-year-old man diagnosed with verrucous carcinoma of the left mandible involving the left floor of mouth with extension into left masticator space. Additionally, the patient had chronic hypersensitive pneumonitis which necessitated supplemental oxygen with activity. Prior to his surgery, Mr. T met with the SLP for a swallowing evaluation and presurgical counseling during which his history of a cervical esophageal web was revealed. His last dilation was completed 2 months prior to his presurgical workup. A clinical swallowing evaluation was completed. Mr. T was consuming a general texture diet and denied symptoms of pharyngeal dysphagia, but he had to take small bites due to moderate-severe trismus with maximal mouth opening (MMO) noted to vary between 5 and 15 mm over the prior 2 months. MMO noted to be 15 mm on this date. His cranial nerve examination was otherwise unremarkable with clear voicing and clear speech production. Mr. T completed the MD Anderson Dysphagia Inventory (MDADI) with an overall score of 74/100, indicating a mild self-perceived swallowing handicap.

The surgical plan was reviewed: composite resection of L mandible with fibula free flap reconstruction, left neck dissection, tracheostomy, and NG tube placement. Expected postsurgical speech and swallowing changes were reviewed, and Mr. T was informed that, to start, he would not be able to produce voice due to the tracheostomy. He was encouraged to use a marker board/notebook and/or computer to supplement oral communication after surgery until resolution of postsurgical swelling allowed him to begin digital trach occlusion for voicing. The role of the inpatient SLP team was introduced to potentially evaluate him for a tracheostomy speaking valve and to complete a swallowing evaluation. The need for outpatient therapy was reviewed in order to maximize the Mr. T's speech and swallowing function following surgery and prior to starting adjuvant radiation. He was informed that his speech would be distorted initially due to post-surgical swelling and that he would experience postoperative dysphagia of the oral stage. Mr. T was made aware that the degree of speech and swallowing changes depend on the extent of the surgical resection with some degree of pharyngeal dysphagia likely given the surgical approach which disrupts the floor of mouth musculature which is key for laryngeal vestibule closure. Mr. T was also prepared to experience postoperative trismus. An overview of likely targets for future therapy was provided: completion of FEES, introduction of swallowing strategies/modifications (e.g., place food to nonsurgical side, use small bites, use liquid wash), as well as jaw and lingual ROM and swallow strengthening exercises. His surgeon was informed of the history of the esophageal web, and esophageal dilation was then included on the surgical plan, especially given Mr. T's likely need for postoperative radiation which would preclude any intervention for some time.

The above-mentioned surgery was completed with the operative note detailing that the medial pterygoid and a portion of the lateral pterygoid were resected as well as the lingual nerve. The cervical esophagus was dilated serially from 40 to 54 Fr. Mr. T was decannulated on POD 7 and the inpatient SLP completed a clinical swallowing assessment noting a breathy voice and overt signs/symptoms of pharyngeal dysphagia with thin and puree consistencies and recommended continuation of NG tube feeds and small sips of water. At the 2-week postoperative visit with the surgical team, Mr. T was found to have a left true vocal fold paralysis. His voice was severely dysphonic (breathy) with no speech distortions appreciated. Oral motor examination was notable for left lower lip weakness and severe trismus, MMO was 15 mm. His MDADI score was 58/100 indicating a moderate-severe self-perceived swallowing handicap. A FEES was completed which allowed visualization of the left TVF paralysis. Thin, nectar, and puree consistency trials were completed. The primary problems were reduced laryngeal vestibule closure, which led to penetration of all consistencies and aspiration of thin liquids. A strategy that helped was left head turn with chin tuck for liquid swallows. Mr. T's swallowing was deemed to be not safe nor efficient for maintenance of oral diet and a conservative plan was recommended (water sips after oral care using the left head turn) given his baseline compromised pulmonary status.

Mr. T was able to undergo an injection laryngoplasty later that same day and was seen 7 days later for a repeat swallowing evaluation. His MMO was unchanged and his voice had improved from severe dysphonia to moderate dysphonia with a strong cough noted. FEES was completed, showing the ability to achieve full glottic closure as well as improved safety and efficiency with thin, puree, and soft chew bolus trials with use strategies. A left head turn with chin tuck was effective for eliminating penetration, and multiple swallows with liquid wash was effective for clearing vallecular stasis. Initiation of a soft diet with thin liquids was recommended with use of the above-mentioned strategies. Mr. T felt he could maintain his nutrition and hydration needs, and his NG tube was removed with plans for the dietitian to see him shortly to support nutrition. Instruction for jaw stretches were provided.

Mr. T was seen for additional therapy sessions, 1 week, and then again 2 weeks later, at which time he demonstrated improved voicing (only mild roughness) and improved MMO of 30 mm. He was consuming a soft texture diet without need for the head turn with chin tuck with liquids. Mr. T did report occasional coughing on water (1×/day), mostly after eating food. He demonstrated anterior spillage of food on the left due to his left lower lip weakness. Mr. T was maintaining his weight, completing jaw stretches daily and was provided with oropharyngeal swallowing exercises and additional jaw and neck stretches in preparation for adjuvant radiation.

Transitioning from Surgery to Adjuvant Therapy

In many cases, patients will require adjuvant radiation or chemoradiation following surgical intervention. It is critical to restore function to the greatest extent possible in the time between surgery and start of adjuvant therapy. The SLP should work with the surgical team to optimize oral intake. Additionally, the SLP should partner with the registered dietician to optimize nutrition prior to the start of radiation. Rehabilitative exercises and familiarization with compensatory strategies may help to minimize the potential of complications during adjuvant treatment.

Nonoperative Treatment of Head and Neck Cancer

Patients undergoing nonoperative treatment of head and neck cancer are at risk for functional impairment of speech, swallowing, and voice both during and after treatment. While nonoperative approaches are commonly referred to as "organ sparing" treatments, it is increasingly evident that preservation of structure does not equate to preservation of function. The SLP plays an important role on the multidisciplinary team, providing assessment and management of these issues prior to, during, and after treatment. Involvement of the SLP from the time of diagnosis is preferred to a reactive approach as a proactive approach provides value in regard to patient education, setting realistic expectations, mitigating short- and long-term complications, and optimizing functional outcomes [71, 72].

Functional Changes Associated with Nonoperative Treatment

A substantial body of literature exists which demonstrates the potential for both short-term and long-term functional implications of nonoperative head and neck cancer therapy. Anatomic and physiologic changes associated with radiation therapy may impact swallowing, voice, speech, and mouth opening, all of which contribute to quality of life. Functional issues may be related to muscle atrophy, soft tissue fibrosis, lymphedema, or cranial neuropathies [73]. The SLP plays a key role in the prevention and remediation of these difficulties.

Importance of Education and Pre-habilitation

Prior to initiating radiation therapy, the SLP should meet with the patient, not only to conduct baseline assessment measures, as previously discussed, but also to provide the patient with information about radiation side effects and their potential impact on short-term and long-term functions. Qualitative studies of patients receiving head and neck radiation therapy have revealed that, in general, patients do not feel adequately prepared for the severity of difficulties encountered during treatment [74]. The patient should be educated about acute radiation toxicities which may make eating and speaking challenging. These include, but are not limited to, xerostomia, pain, taste changes, and edema. The SLP can provide the perspective of how these side effects may interfere with normal function and provide recommendations for how to prepare for and address these issues. The patient also needs to be educated about the potential for long-term or permanent functional restrictions related to the radiation. This is critical to the patient understanding the importance of engaging with prophylactic therapies and maintaining an oral diet.

It is increasingly evident that engaging patients in activities to maintain muscle function during radiation is important to long-term outcomes [75–78]. This may include a range of motion activities, exercise protocols, and/or maintaining oral intake throughout treatment. The goal of these interventions is to maintain the mobility and strength of the structures critical for safe and efficient swallow function. Some studies have failed to demonstrate a relationship between prophylactic swallowing exercises and improved function; however, adherence to treatment in such studies was poor, suggesting that those in the exercise groups actually did not have adequate exercise to assess its efficacy [79]. Importantly, it has been established that while adherence is challenging during treatment, there are strategies that can be employed to enhance adherence [80–82].

Maintenance of oral intake during radiation therapy has also been associated with more favorable swallowing outcomes. Patients with complete tube feeding dependence showed poorer diet levels up to 1 year posttreatment in contrast to those on full or partial oral diets [83]. Gillespie et al. demonstrated that individuals with NPO status for greater than 2 weeks during radiation had poorer swallowing-related quality of life scores on the MD Anderson Dysphagia Inventory [84]. Further observational data demonstrated that patients maintaining at least some oral intake during radiation therapy had significantly shorter durations of feeding tube use and that those individuals who maintained a full oral diet throughout treatment had a two times greater odds ratio of returning to a normal diet following treatment [85]. Despite the benefits of maintaining oral intake during treatment, side effects often make this a difficult goal to achieve [86]. Management of acute toxicities such as pain may have a significant impact on patients' ability to maintain an oral diet as well as longer term swallowing outcomes [15, 87–89]. SLPs play an important role in encouraging continued oral intake by helping patients understand what foods are appropriate at different stages of treatment [75]. Between engaging patients in exercise protocols and encouraging continued oral intake, the SLP can play a critical role in maintaining long-term swallow function.

Less is understood about the role of preventative voice or speech intervention during radiation. From a voice perspective, it is reasonable to assume that patients receiving higher doses to the larynx may benefit from exercises to target mobility and strength as is done for swallowing therapy. While there is evidence of benefit from voice therapy after completion of radiation, none exists for preventative exercise at the time of this publication [90, 91]. Even less is known about the impact of preventative speech intervention during radiation, therapy, likely compounded by the fact that most individuals receiving high doses to articulatory structures also had prior surgical resection prior to radiation.

Management During Treatment

During radiation therapy, the SLP plays a critical role in keeping the patient engaged in activities to preserve muscular function. Adherence to these interventions is often very challenging, and it is incumbent on the treatment team to ensure measures are in place to facilitate adherence [92]. Clinician-guided treatment during radiation has been shown to facilitate adherence compared to a patient-guided treatment approach [78, 92]. The SLP should have multiple contacts with patients undergoing head and neck radiation to encourage continued performance of prophylactic exercises, oral intake, and management of treatment toxicities. It is important for SLPs working with this population to have an understanding of techniques that may facilitate adherence. A systematic review performed by Govender et al. [93] explored behavioral strategies cited in the literature as potential contributors to adherence in this patient population [93]. They found that a number of behavior change techniques were more commonly associated with more favorable outcomes, including practical social support, behavioral practice, self-monitoring of behavior, and a skilled clinician administering the intervention. Thus, the SLP should integrate such strategies into their treatment approach.

First-Year Posttreatment

In the first year following radiation-based treatment, functional status is expected to change significantly over time. Acute toxicities are expected to continue to interfere with oral functions for at least the first 6 weeks following treatment [94]. In the first 6 months posttreatment, acute treatment toxicities decrease significantly, making eating experiences easier for the majority of patients due to lessened odynophagia, dysgeusia, and xerostomia, with the majority of recovery expected by 12 months posttreatment [1, 95]. Despite the improvements in toxicities that impact eating, most patients do not report a return to baseline function by 1-year postradiation [96]. Additionally, physiologic changes related to lymphedema and fibrosis may worsen over time [97]. Given the anticipated changes over the first-year posttreatment, the SLP should remain an active part of the treatment team. Given what is known about resolution of acute toxicities and development of lymphedema and fibrosis, we would advocate for instrumental swallowing assessment of all patients undergoing head and neck radiation 2–3 months posttreatment as well as 12 months posttreatment, at a minimum. In those individuals with posttreatment dysphagia, early intervention has been shown to be more effective than late intervention, therefore swallowing therapy should be implemented as soon as difficulties are documented [98].

Similarly, thorough assessment of communication challenges should be completed for those patients experiencing voice difficulties after radiation therapy. Laryngeal edema is a common consequence of head and neck radiation which may contribute to vocal changes [99]. Videostroboscopic evaluation of laryngeal function provides valuable information about laryngeal function which is critical for treatment planning [100]. Further, use of a validated laryngeal edema scale such as the revised Patterson Edema Scale provides an opportunity to track changing laryngeal edema over time [101]. Several studies demonstrate the efficacy of voice therapy for patients with dysphonia following head and neck radiotherapy [90, 102, 103].

Lymphedema is increasingly implicated as a contributor to postradiation functional impairments. Lymphedema management, known collectively as Complete Decongestive Therapy (CDT), consists of manual lymphatic drainage, compression, exercise, and skin care [104]. While evidence regarding the efficacy of CDT for managing head and neck lymphedema is limited, due in part to a lack of consistent, reliable methods of measuring head and neck lymphedema, early evidence is suggestive of treatment benefit [105, 106]. While pneumatic compression is frequently used in the management of limb lymphedema, there is limited evidence regarding its role in the management of head and neck lymphedema [107]. While theoretically it is reasonable to postulate that reduction of head and neck lymphedema would equate to reduced functional impairment, which has not yet been established in the literature. Patients with head and neck lymphedema should be referred to an experienced head and neck lymphedema clinician who may be a SLP, occupational therapist, or physical therapist, depending upon institutional preferences.

Long-Term Management

The majority of patients undergoing radiation-based head and neck cancer treatment will see functional status plateau between the first- and second-year posttreatment. However, there is a small subset of patients who will develop gradually progressive difficulties with speech, voice, and swallowing in the years following treatment. Late radiation associated dysphagia (Late-RAD) was first described as a unique entity in the literature in 2012 and has been defined as dysphagia with onset >2 to 4+ years posttreatment after a period of normal function [108–110]. Alternatively, some patients will present with gradually progressive dysphagia due to fibrosis [111, 112]. A key difference between patients with Late-RAD compared to those with chronic progressive dysphagia is the presence of cranial neuropathies seen in the majority of patients with Late-RAD [109, 112]. In fact, recent evidence suggests that radiation dose to the hypoglossal nerve has been associated with Late-RAD [110]. Patients with Late-RAD may also experience laryngeal denervation which may lead to vocal fold paralysis and subsequent voice impairment [113, 114]. These individuals may benefit from laryngeal augmentation to reduce glottic insufficiency with a resultant improvement in voice and swallowing [113, 115]. Patients with Late-RAD, as well as those with gradually progressive dysphagia, are at risk for complications such as permanent feeding tube dependence and aspiration pneumonia [108, 116].

Unfortunately, our ability to predict which patients may experience late dysphagia is limited, and frequently Late-RAD is identified due to complications such as aspiration pneumonia which has been implicated in a very high number of fatalities [117]. As a result, we advocate for an annual functional surveillance examination with an SLP at which time a thorough cranial nerve examination, assessment of speech and voice, instrumental swallowing evaluation, and patient reported outcome measures can be obtained. Patients should also be educated annually about the need to monitor for early signs of radiation associated dysphagia, such as increased difficulty swallowing foods and liquids, changes in voice/speech, unexplained weight loss, and/or coughing with swallowing that should be brought to the attention of the treatment team to minimize the risk of negative consequences such as aspiration pneumonia. Early identification of late dysphagia may help to prevent negative health consequences such as pneumonia and diminished quality of life, though more research in this realm is needed. Though there is limited evidence to suggest that swallowing therapy for Late-RAD can reverse underlying dysphagia pathophysiology, there is some evidence of the benefit of therapy to help patients compensate for underlying deficits [118, 119]. Further, mitigation strategies such as oral care may be implemented to further minimize the risk of aspiration pneumonia [120, 121].

Nonsurgical Case Presentation: Mr. R

Mr. R was 57 years old when he noted a right sided neck mass when shaving. He was diagnosed with a T2N2 P16+ squamous cell carcinoma of the right base of tongue. Given that the tumor was more endophytic and neared midline, it was determined that he was most appropriate for a nonsurgical treatment approach. He was referred to an SLP prior to the start of treatment at which time he underwent a FEES exam which revealed mild reduction in pharyngeal clearing forces leading to mild vallecular residue. His swallow was deemed both safe and efficient for oral intake. He had no abnormalities of speech or voice and had a normal MMO of 42 mm. The SLP provided the patient with education about the side effects of radiation and the importance of maintaining muscle strength and mobility during treatment. He was provided with a series of prophylactic exercises and scheduled for follow-up on the fourth week of radiation.

When Mr. R was seen mid-radiation, he reported poor adherence to swallowing exercises due to pain with swallowing. Evaluation of the chart revealed he had a prescription for narcotic analgesics; however, he was resistant to taking these medications due to fear of potential addiction. In addition to not doing the exercises, he was also consuming only 1000–1200 cal/day and had lost 5 lb over the previous 2 weeks. The SLP counselled the patient on the importance of pain management to optimize adherence to therapeutic recommendations. As the patient had previously reported that eating was a major contributor to his quality of life, the SLP related the need to adhere to recommendations back to his goal of being able to resume normal eating after completion of treatment. The patient indicated better understanding of the need for pain management, and the SLP connected him back to the radiation oncology team to address his medication concerns. When the patient was seen back 2 weeks later, he reported continued issues with xerostomia; however, he had learned to be strategic in taking his pain medications 20–30 min before meals or swallowing exercises, and this strategy had allowed him to get back on track with swallowing recommendations.

The patient was seen back for a swallowing evaluation 3 months after the completion of radiation. While his pain issues had largely resolved, he was still struggling with changes in taste and saliva that made eating a challenge. His swallowing study revealed reduced tongue base retraction, pharyngeal constriction, and epiglottic tilt, leading to mild vallecular residue with liquids and moderate residue with solids. A liquid wash strategy was deemed effective in facilitating clearance of residue and the patient agreed that this small compensation was acceptable to him. He was provided with additional recommendations for managing his symptoms and encouraged to continue with his previously prescribed swallowing exercises. It was noted at that time that he had mild submental lymphedema so he was also provided with instruction in complete decongestive therapy. As he lived distant to the clinic, he opted for a home-based treatment strategy, and the SLP trained the patient's wife in how to perform manual lymphatic drainage. They were provided with written instructions as well as a tutorial video to support the home practice.

When Mr. R returned to clinic in 6 months, he had resolution of lymphedema and taste changes and moderate self-reported improvement in salivary flow. He reported excellent adherence to swallowing exercises on a daily basis as well as daily lymphedema treatment. His swallowing evaluation revealed only mild pharyngeal residue, improved from the last evaluation. Given excellent functional status, the patient was educated regarding functional changes that would warrant expedited follow-up as well as the value of an annual functional surveillance evaluation. He continued to follow up annually for a FEES evaluation as well as screening of speech, voice, trismus, and lymphedema.

Conclusions

Patients with head and neck cancer have a high potential of experiencing communication or swallowing issues prior to, during, or following oncologic treatment. The SLP is a critical member of the head and neck cancer care team and can provide assessment and management of these issues at all points of care. Integrating SLP care into the care pathway shortly after diagnosis may provide multiple benefits such as improved patient outcomes, reduced patient anxiety, and improved patient satisfaction. There is ample research demonstrating the value of services provided by SLPs to patients with head and neck cancer.

References

1. Roe JW, Drinnan MJ, Carding PN, Harrington KJ, Nutting CM. Patient-reported outcomes following parotid-sparing intensity-modulated radiotherapy for head and neck cancer. How important is dysphagia? Oral Oncol. 2014;50(12):1182–7.
2. Windon MJ, Fakhry C, Faraji F, et al. Priorities of human papillomavirus-associated oropharyngeal cancer patients at diagnosis and after treatment. Oral Oncol. 2019;95:11–5.
3. Doyle PC, MacDonald C. Well-being and quality of life in head and neck cancer. In: Doyle P, editor. Clinical care and rehabilitation in head and neck cancer. Cham: Springer; 2019.
4. Lango MN, Egleston B, Fang C, Burtness B, Galloway T, Liu J, Mehra R, Ebersole B, Moran K, Ridge JA. Baseline health perceptions, dysphagia, and survival in patients with head and neck cancer. Cancer. 2014;120(6):840–7.
5. Starmer H, Gourin C, Lua LL, Burkhead L. Pretreatment swallowing assessment in head and neck cancer patients. Laryngoscope. 2011;121(6):1208–11.
6. List MA, Ritter-Sterr C, Lansky SB. A performance status scale for head and neck cancer patients. Cancer. 1990;66(3):564–9.
7. Pedersen A, Wilson J, McColl E, Carding P, Patterson J. Swallowing outcome measures in head and neck cancer--how do they compare? Oral Oncol. 2016;52:104–8.
8. Arrese LC, Schieve HJ, Graham JM, Stephens JA, Carrau RL, Plowman EK. Relationship between oral intake, patient perceived swallowing impairment, and objective videofluoroscopic measures of swallowing in patients with head and neck cancer. Head Neck. 2019;41(4):1016–23.
9. Patterson J, Wilson JA. The clinical value of dysphagia preassessment in the management of head and neck cancer patients. Curr Opin Otolaryngol Head Neck Surg. 2011;19(3):177–81.
10. van der Molen L, van Rossum MA, Ackerstaff AH, et al. Pretreatment organ function in patients with advanced head and neck cancer: clinical outcome measures and patients' views. BMC Ear Nose Throat Disord. 2009;9:10.
11. Lango MN, Galloway TJ, Mehra R, et al. Impact of baseline patient-reported dysphagia on acute gastrostomy placement in patients with head and neck squamous cell carcinoma undergoing definitive radiation. Head Neck. 2016;38(Suppl 1):E1318–24.
12. Bhayani MK, Hutcheson KA, Barringer DA, et al. Gastrostomy tube placement in patients with oropharyngeal carcinoma treated with radiotherapy or chemoradiotherapy: factors affecting placement and dependence. Head Neck. 2013;35(11):1634–40.
13. Anderson NJ, Jackson JE, Smith JG, et al. Pretreatment risk stratification of feeding tube use in patients treated with intensity-modulated radiotherapy for head and neck cancer. Head Neck. 2018;40(10):2181–92.
14. Bhayani MK, Hutcheson KA, Barringer DA, Roberts DB, Lewin JS, Lai SY. Gastrostomy tube placement in patients with hypopharyngeal cancer treated with radiotherapy or chemoradiotherapy: factors affecting placement and dependence. Head Neck. 2013;35(11):1641–6.
15. Yang W, McNutt TR, Dudley SA, et al. Predictive factors for prophylactic percutaneous endoscopic gastrostomy (PEG) tube placement and use in head and neck patients following intensity-modulated radiation therapy (IMRT) treatment: concordance, discrepancies, and the role of gabapentin. Dysphagia. 2016;31(2):206–13.
16. Brown TE, Spurgin AL, Ross L, et al. Validated swallowing and nutrition guidelines for patients with head and neck cancer: identification of high-risk patients for proactive gastrostomy. Head Neck. 2013;35(10):1385–91.
17. Chen AY, Frankowski R, Bishop-Leone J, et al. The development and validation of a dysphagia-specific quality-of-life questionnaire for patients with head and neck cancer: the M. D. Anderson dysphagia inventory. Arch Otolaryngol Head Neck Surg. 2001;127(7):870–6.
18. Belafsky PC, Mouadeb DA, Rees CJ, et al. Validity and reliability of the Eating Assessment Tool (EAT-10). Ann Otol Rhinol Laryngol. 2008;117(12):919–24.
19. Wallace KL, Middleton S, Cook IJ. Development and validation of a self-report symptom inventory to assess the severity of oral-pharyngeal dysphagia. Gastroenterology. 2000;118(4):678–87.
20. Crary MA, Mann GD, Groher ME. Initial psychometric assessment of a functional oral intake scale for dysphagia in stroke patients. Arch Phys Med Rehabil. 2005;86(8):1516–20.

21. Lazarus CL, Husaini H, Jacobson AS, et al. Development of a new lingual range-of-motion assessment scale: normative data in surgically treated oral cancer patients. Dysphagia. 2014;29(4):489–99.
22. Hutcheson KA, Barrow MP, Barringer DA, et al. Dynamic imaging grade of swallowing toxicity (DIGEST): scale development and validation. Cancer. 2017;123(1):62–70.
23. Starmer HS, Arrese L, Langmore S, et al. Adaptation and validation of the dynamic imaging grade of swallowing toxicity for flexible endoscopic evaluation of swallowing: DIGEST™-FEES. J Speech Lang Hear Res. 2021;64:1802.
24. Stelzle F, Oetter N, Goellner LT, et al. Speech intelligibility in patients with oral cancer: an objective baseline evaluation of pretreatment function and impairment. Head Neck. 2019;41(4):1063–9.
25. Starmer HM, Tippett DC, Webster KT. Effects of laryngeal cancer on voice and swallowing. Otolaryngol Clin N Am. 2008;41(4):793–vii.
26. Sauder C, Kapsner-Smith M, Baylor C, Yorkston K, Futran N, Eadie T. Communicative participation and quality of life in pretreatment oral and oropharyngeal head and neck cancer. Otolaryngol Head Neck Surg. 2021;164(3):616–23.
27. Jacobson BH, Johnson A, Grywalski C, Silbergleit A, Jacobson G. The voice handicap index: development and validation. Am J Speech Lang Pathol. 1997;6(3):66–70.
28. Rinkel RN, Verdonck-de Leeuw IM, van Reij EJ, Aaronson NK, Leemans CR. Speech Handicap Index in patients with oral and pharyngeal cancer: better understanding of patients' complaints. Head Neck. 2008;30(7):868–74.
29. Baylor CR, Yorkston KM, Eadie TL, Miller RM, Amtmann D. Developing the communicative participation item bank: Rasch analysis results from a spasmodic dysphonia sample. J Speech Lang Hear Res. 2009;52(5):1302–20.
30. De Bodt MS, Wuyts FL, Van de Heyning PH, Croux C. Test-retest study of the GRBAS scale: influence of experience and professional background on perceptual rating of voice quality. J Voice. 1997;11(1):74–80.
31. Kempster GB, Gerratt BR, Verdolini Abbott K, Barkmeier-Kraemer J, Hillman RE. Consensus auditory-perceptual evaluation of voice: development of a standardized clinical protocol. Am J Speech Lang Pathol. 2009;18(2):124–32.
32. Goldman R, Fristoe M. Goldman Fristoe test of articulation. Circle Pines, MN: American Guidance Service; 1986.
33. Tikofsky RS. A revised list for the estimation of dysarthric single word intelligibility. J Speech Hear Res. 1970;13:59.
34. Dwivedi RC, Kazi RA, Agrawal N, Nutting CM, Clarke PM, Kerawala CJ, Rhys-Evans PH, Harrington KJ. Evaluation of speech outcomes following treatment of oral and oropharyngeal cancers. Cancer Treat Rev. 2009;35(5):417–24.
35. Llewellyn CD, McGurk M, Weinman J. How satisfied are head and neck cancer (HNC) patients with the information they receive pre-treatment? Results from the satisfaction with cancer information profile (SCIP). Oral Oncol. 2006;42(7):726–34.
36. Bhattacharya S, Thankappan K, Joseph ST, et al. Volume and location of the defect as predictors of swallowing outcome after glossectomy: correlation with a classification. Dysphagia. 2021;36:974.
37. Pauloski BR. Rehabilitation of dysphagia following head and neck cancer. Phys Med Rehabil Clin N Am. 2008;19(4):889–928.
38. Arrese LC, Hutcheson KA. Framework for speech-language pathology services in patients with oral cavity and oropharyngeal cancers. Oral Maxillofac Surg Clin North Am. 2018;30(4):397–410.
39. Goldsmith T, Jacobson MC. Revisiting swallowing function following contemporary surgical interventions for oral/oropharyngeal cancer: key underlying issues. Perspect Swall Swall Dis. 2015;24:89–98.
40. Hirano M, Matsuoka H, Kuroiwa Y, Sato K, Tanaka S, Yoshida T. Dysphagia following various degrees of surgical resection for oral cancer. Ann Otol Rhinol Laryngol. 1992;101(2):138–41.
41. Navach V, Zurlo V, Calabrese L, Massaro MA, Bruschini R, Giugliano G, Ansarin M, Chiesa F. Total glossectomy with preservation of the larynx: oncological and functional results. Br J Oral Maxillofac Surg. 2013;51:217–23.
42. Dziegielewski PT, Ho ML, Rieger J, Singh P, Langille M, Harris JR, Seikaly H. Total glossectomy with laryngeal preservation and free flap reconstruction: objective functional outcomes and systematic review of the literature. Laryngoscope. 2013;123:140–5.
43. Vosler PS. Reconstruction of the oral tongue. In: Fakhry C, Pitman KT, Kiess AP, et al., editors. Oral cancer: evaluation, therapy, and rehabilitation. 1st ed. Leipzig: Thieme; 2020. p. 129–33.
44. Chang EI, Yu P, Skoracki RJ, Liu J, Hanasono MM. Comprehensive analysis of functional outcomes and survival after microvascular reconstruction of glossectomy defects. Ann Surg Oncol. 2015;22:3061–9.
45. Arrese LC, Lazarus CL. Special groups: head and neck cancer. Otolaryngol Clin N Am. 2013;46(6):1123–36.
46. Buchbinder D, Urken ML, Vickery C, Weinberg H, Sheiner A, Biller H. Functional mandibular reconstruction of patients with oral cancer. Oral Surg Oral Med Oral Pathol. 1989;68(4 Pt 2):499–503; discussion 503–4.
47. Albergotti WG, Jordan J, Anthony K, Abberbock S, Wasserman-Wincko T, Kim S, Ferris RL, Duvvuri U. A prospective evaluation of short-term dysphagia after transoral robotic surgery for squamous cell carcinoma of the oropharynx. Cancer. 2017;123:3132–40.
48. Hutcheson KA, Warneke CL, Yao CMKL, et al. Dysphagia after primary transoral robotic surgery

with neck dissection vs nonsurgical therapy in patients with low- to intermediate-risk oropharyngeal cancer. JAMA Otolaryngol Head Neck Surg. 2019;145(11):1053–63.
49. Charters E, Wu R, Milross C, et al. Swallowing and communication outcomes following primary transoral robotic surgery. Head Neck. 2021;43:1–11.
50. Lazarus CL. Management of swallowing disorders in head and neck cancer patients: optimal patterns of care. Semin Speech Lang. 2000;21(4):293–309.
51. Lips M, Speyer R, Zumach A, Kross KW, Kremer B. Supracricoid laryngectomy and dysphagia: a systematic literature review. Laryngoscope. 2015;125(9):2143–56.
52. Fitzgerald E, Perry A. Pre-operative counselling for laryngectomy patients: a systematic review. J Laryngol Otol. 2016;130(1):15–20.
53. Graboyes EM, Kallogjeri D, Zerega J, et al. Association of a perioperative education program with unplanned readmission following total laryngectomy. JAMA Otolaryngol Head Neck Surg. 2017;143(12):1200–6.
54. Shenson JA, Craig JN, Rohde SL. Effect of preoperative counseling on hospital length of stay and readmissions after total laryngectomy. Otolaryngol Head Neck Surg. 2017;156(2):289–98.
55. Maclean J, Cotton S, Perry A. Post-laryngectomy: it's hard to swallow: an Australian study of prevalence and self-reports of swallowing function after a total laryngectomy. Dysphagia. 2009;24(2):172–9.
56. Riva G, Sensini M, Corvino A, Pecorari G, Garzaro M. Smell and taste impairment after total laryngectomy. Ann Otol Rhinol Laryngol. 2017;126(7):548–54.
57. Wulff NB, Højager A, Wessel I, Dalton SO, Homøe P. Health-related quality of life following total laryngectomy: a systematic review. Laryngoscope. 2021;131(4):820–31.
58. Lewin JS, Hutcheson KA, Barringer DA, et al. Functional analysis of swallowing outcomes after supracricoid partial laryngectomy. Head Neck. 2008;30(5):559–66.
59. Webster KT, Samlan RA, Jones B, Bunton K, Tufano RP. Supracricoid partial laryngectomy: swallowing, voice, and speech outcomes. Ann Otol Rhinol Laryngol. 2010;119(1):10–6.
60. Elpern EH, Borkgren Okonek M, Bacon M, Gerstung C, Skrzynski M. Effect of the Passy-Muir tracheostomy speaking valve on pulmonary aspiration in adults. Heart Lung. 2000;29(4):287–93.
61. Gross RD, Mahlmann J, Grayhack JP. Physiologic effects of open and closed tracheostomy tubes on the pharyngeal swallow. Ann Otol Rhinol Laryngol. 2003;112(2):143–52.
62. Suiter DM, McCullough GH, Powell PW. Effects of cuff deflation and one-way tracheostomy speaking valve placement on swallow physiology. Dysphagia. 2003;18:284–92.
63. Colangelo LA, Logemann JA, Pauloski BR, et al. T stage and functional outcome in oral and oropharyngeal cancer patients. Head Neck. 1996;18:259–68.
64. Palmer AD, Bolognone RK, Thomsen S, Britton D, Schindler J, Graville DJ. The safety and efficacy of expiratory muscle strength training for rehabilitation after supracricoid partial laryngectomy: a pilot investigation. Ann Otol Rhinol Laryngol. 2019;128(3):169–76.
65. Logemann JA, Pauloski BR, Rademaker AW, Colangelo LA. Speech and swallowing rehabilitation for head and neck cancer patients. Oncology. 1997;11(5):651–6, 659; discussion 659, 663–4.
66. van der Kamp MF, Rinkel RNPM, Eerenstein SEJ. The influence of closure technique in total laryngectomy on the development of a pseudo-diverticulum and dysphagia. Eur Arch Otorhinolaryngol. 2017;274(4):1967–73.
67. Zhang T, Szczesniak M, Maclean J, et al. Biomechanics of pharyngeal deglutitive function following total laryngectomy. Otolaryngol Head Neck Surg. 2016;155(2):295–302.
68. Lippert D, Hoffman MR, Britt CJ, et al. Preliminary evaluation of functional swallow after total laryngectomy using high-resolution manometry. Ann Otol Rhinol Laryngol. 2016;125(7):541–9.
69. van As CJ, Op de Coul BM, van den Hoogen FJ, Koopmans-van Beinum FJ, Hilgers FJ. Quantitative videofluoroscopy: a new evaluation tool for tracheoesophageal voice production. Arch Otolaryngol Head Neck Surg. 2001;127(2):161–9.
70. Deng J, Murphy BA, Dietrich MS, et al. Impact of secondary lymphedema after head and neck cancer treatment on symptoms, functional status, and quality of life. Head Neck. 2013;35:1026–35.
71. Ajmani GS, Nocon CC, Brockstein BE, et al. Association of a proactive swallowing rehabilitation program with feeding tube placement in patients treated for pharyngeal cancer. JAMA Otolaryngol Head Neck Surg. 2018;144(6):483–8.
72. Starmer HM, Ayoub N, Byward C, et al. The impact of developing a speech and swallow rehab program: improving patient satisfaction and multidisciplinary care. Laryngoscope. 2017;127(11):2578–81.
73. King SN, Dunlap NE, Tennant PA, Pitts T. Pathophysiology of radiation-induced dysphagia in head and neck cancer. Dysphagia. 2016;31(3):339–51.
74. Nund RL, Ward EC, Scarinci NA, Cartmill B, Kuipers P, Porceddu SV. Survivors' experiences of dysphagia-related services following head and neck cancer: implications for clinical practice. Int J Lang Commun Disord. 2014;49(3):354–63.
75. Hutcheson KA, Gomes A, Rodriguez V, Barringer D, Khan M, Martino R. Eat all through radiation therapy (EAT-RT): structured therapy model to facilitate continued oral intake through head and neck radiotherapy-user acceptance and content validation. Head Neck. 2020;42(9):2390–6.
76. Messing BP, Ward EC, Lazarus CL, et al. Prophylactic swallow therapy for patients with head and neck cancer undergoing chemoradiotherapy: a randomized trial. Dysphagia. 2017;32(4):487–500.

77. Peng KA, Kuan EC, Unger L, Lorentz WC, Wang MB, Long JL. A swallow preservation protocol improves function for veterans receiving chemoradiation for head and neck cancer. Otolaryngol Head Neck Surg. 2015;152(5):863–7.
78. Carnaby-Mann G, Crary MA, Schmalfuss I, Amdur R. "Pharyngocise": randomized controlled trial of preventative exercises to maintain muscle structure and swallowing function during head-and-neck chemoradiotherapy. Int J Radiat Oncol Biol Phys. 2012;83(1):210–9.
79. Mortensen HR, Jensen K, Aksglæde K, Lambertsen K, Eriksen E, Grau C. Prophylactic swallowing exercises in head and neck cancer radiotherapy. Dysphagia. 2015;30(3):304–14.
80. Starmer HM, Abrams R, Webster K, et al. Feasibility of a mobile application to enhance swallowing therapy for patients undergoing radiation-based treatment for head and neck cancer. Dysphagia. 2018;33(2):227–33.
81. Govender R, Wood CE, Taylor SA, Smith CH, Barratt H, Gardner B. Patient experiences of swallowing exercises after head and neck cancer: a qualitative study examining barriers and facilitators using behaviour change theory. Dysphagia. 2017;32(4):559–69.
82. van der Molen L, van Rossum MA, Burkhead LM, et al. A randomized preventative rehabilitation trial in advanced head and neck cancer patients treated with chemoradiotherapy: feasibility, compliance, and short-term effects. Dysphagia. 2011;26:115–70.
83. Langmore S, Krisciunas GP, Miloro KV, et al. Does PEG cause dysphagia in head and neck cancer patients? Dysphagia. 2012;27(2):251–9.
84. Gillespie MB, Brodsky MB, Day TA, Lee FS, Martin-Harris B. Swallowing-related quality of life after head and neck cancer treatment. Laryngoscope. 2004;114(8):1362–7.
85. Hutcheson KA, Bhayani MK, Beadle BM, et al. Eat and exercise during radiotherapy or chemoradiotherapy for pharyngeal cancers: use it or lose it. JAMA Otolaryngol Head Neck Surg. 2013;139(11):1127–34.
86. Beadle BM, Liao KP, Giordano SH, et al. Reduced feeding tube duration with intensity-modulated radiation therapy for head and neck cancer: a surveillance, epidemiology, and end results-medicare analysis. Cancer. 2017;123(2):283–93.
87. Starmer HM, Yang W, Raval R, et al. Effect of gabapentin on swallowing during and after chemoradiation for oropharyngeal squamous cell cancer. Dysphagia. 2014;29(3):396–402.
88. Starmer HM, Yang W, Gourin CG, et al. One-year swallowing outcomes in patients treated with prophylactic gabapentin during radiation-based treatment for oropharyngeal cancer. Dysphagia. 2017;32(3):437–42.
89. Smith DK, Cmelak A, Niermann K, et al. Preventive use of gabapentin to decrease pain and systemic symptoms in patients with head and neck cancer undergoing chemoradiation. Head Neck. 2020;42(12):3497–505.
90. Angadi V, Dressler E, Kudrimoti M, et al. Efficacy of voice therapy in improving vocal function in adults irradiated for laryngeal cancers: a pilot study. J Voice. 2020;34(6):962.e9–962.e18.
91. Karlsson T, Johansson M, Andréll P, Finizia C. Effects of voice rehabilitation on health-related quality of life, communication and voice in laryngeal cancer patients treated with radiotherapy: a randomised controlled trial. Acta Oncol. 2015;54(7):1017–24.
92. Wall LR, Ward EC, Cartmill B, Hill AJ, Porceddu SV. Adherence to a prophylactic swallowing therapy program during (chemo) radiotherapy: impact of service-delivery model and patient factors. Dysphagia. 2017;32(2):279–92.
93. Govender R, Smith CH, Taylor SA, Barratt H, Gardner B. Swallowing interventions for the treatment of dysphagia after head and neck cancer: a systematic review of behavioural strategies used to promote patient adherence to swallowing exercises. BMC Cancer. 2017;17(1):43.
94. Ihara Y, Crary MA, Madhavan A, et al. Dysphagia and oral morbidities in chemoradiation-treated head and neck cancer patients. Dysphagia. 2018;33(6):739–48.
95. Hunter KU, Schipper M, Feng FY, et al. Toxicities affecting quality of life after chemo-IMRT of oropharyngeal cancer: prospective study of patient-reported, observer-rated, and objective outcomes. Int J Radiat Oncol Biol Phys. 2013;85(4):935–40.
96. Patterson JM, McColl E, Wilson J, Carding P, Rapley T. Head and neck cancer patients' perceptions of swallowing following chemoradiotherapy. Support Care Cancer. 2015;23(12):3531–8.
97. Eisbruch A, Lyden T, Bradford CR, et al. Objective assessment of swallowing dysfunction and aspiration after radiation concurrent with chemotherapy for head-and-neck cancer. Int J Radiat Oncol Biol Phys. 2002;53(1):23–8.
98. Van Daele DJ, Langmore SE, Krisciunas GP, et al. The impact of time after radiation treatment on dysphagia in patients with head and neck cancer enrolled in a swallowing therapy program. Head Neck. 2019;41(3):606–14.
99. Sanguineti G, Adapala P, Endres EJ, Brack C, Fiorino C, Sormani MP, Parker B. Dosimetric predictors of laryngeal edema. Int J Radiat Oncol Biol Phys. 2007;68(3):741–9.
100. Marciscano AE, Charu V, Starmer HM, Best SR, Quon H, Hillel AT, Akst LM, Kiess AP. Evaluating post-radiotherapy laryngeal function with laryngeal videostroboscopy in early stage glottic cancer. Front Oncol. 2017;7:124.
101. Starmer HM, Drinnan M, Bhabra M, Watson LJ, Patterson J. Development and reliability of the revised Patterson Edema Scale. Clin Otolaryngol. 2021;46:752.

102. van Gogh CD, Verdonck-de Leeuw IM, Boon-Kamma BA, et al. The efficacy of voice therapy in patients after treatment for early glottic carcinoma. Cancer. 2006;106(1):95–105.
103. Karlsson T, Tuomi L, Andréll P, Johansson M, Finizia C. Effects of voice rehabilitation after radiotherapy for laryngeal cancer: a longitudinal study. Logoped Phoniatr Vocol. 2017;42(4):167–77.
104. Smith BG, Hutcheson KA, Little LG, et al. Lymphedema outcomes in patients with head and neck cancer. Otolaryngol Head Neck Surg. 2015;152(2):284–91.
105. Tyker A, Franco J, Massa ST, Desai SC, Walen SG. Treatment for lymphedema following head and neck cancer therapy: a systematic review. Am J Otolaryngol. 2019;40(5):761–9.
106. Yao T, Beadle B, Holsinger CF, Starmer HM. Effectiveness of a home-based head and neck lymphedema management program: a pilot study. Laryngoscope. 2020;130(12):E858–62.
107. Gutiérrez C, Mayrovitz HN, Naqvi SHS, Karni RJ. Longitudinal effects of a novel advanced pneumatic compression device on patient-reported outcomes in the management of cancer-related head and neck lymphedema: a preliminary report. Head Neck. 2020;42(8):1791–9.
108. Hutcheson KA, Lewin JS, Barringer DA, et al. Late dysphagia after radiotherapy-based treatment of head and neck cancer. Cancer. 2012;118(23):5793–9.
109. Awan MJ, Mohamed AS, Lewin JS, et al. Late radiation-associated dysphagia (late-RAD) with lower cranial neuropathy after oropharyngeal radiotherapy: a preliminary dosimetric comparison. Oral Oncol. 2014;50(8):746–52.
110. Gharzai LA, Li P, Schipper MJ, et al. Characterization of very late dysphagia after chemoradiation for oropharyngeal squamous cell carcinoma. Oral Oncol. 2020;111:104853.
111. Christianen MEMC, Verdonck-de Leeuw IM, Doornaert P, Chouvalova O, Steenbakkers RJ, Koken PW, et al. Patterns of long-term swallowing dysfunction after definitive radiotherapy or chemoradiation. Radiother Oncol. 2015;117:139–44.
112. Strojan P, Hutcheson KA, Eisbruch A, et al. Treatment of late sequelae after radiotherapy for head and neck cancer. Cancer Treat Rev. 2017;59:79–92.
113. Crawley BK, Sulica L. Vocal fold paralysis as a delayed consequence of neck and chest radiotherapy. Otolaryngol Head Neck Surg. 2015;153(2):239–43.
114. Hamdan AL, Abou Rizk S, Ghanem A, El Natout T. Irradiation-induced vocal fold paralysis: a delayed complication. Ear Nose Throat J. 2019;100:NP274.
115. Siu J, Tam S, Fung K. A comparison of outcomes in interventions for unilateral vocal fold paralysis: a systematic review. Laryngoscope. 2016;126(7):1616–24.
116. Aggarwal P, Zaveri JS, Goepfert RP, et al. Swallowing-related outcomes associated with late lower cranial neuropathy in long-term oropharyngeal cancer survivors: cross-sectional survey analysis. Head Neck. 2019;41(11):3880–94.
117. Wang JJ, Jiang RS, Yen TT, Liang KL. Risk factors for recurrent pneumonia in post-irradiated patients with nasopharyngeal carcinoma. J Chin Med Assoc. 2017;80:558–62.
118. Cousins N, MacAulay F, Lang H, MacGillivray S, Wells M. A systematic review of interventions for eating and drinking problems following treatment for head and neck cancer suggests a need to look beyond swallowing and trismus. Oral Oncol. 2013;49:387–400.
119. Martin-Harris B, McFarland D, Hill EG, Strange CB, Focht KL, Wan Z, et al. Respiratory-swallow training in patients with head and neck cancer. Arch Phys Med Rehabil. 2015;96:885–93.
120. Tada A, Miura H. Prevention of aspiration pneumonia (AP) with oral care. Arch Gerontol Geriatr. 2012;55(1):16–21.
121. Scannapieco FA. Poor oral health in the etiology and prevention of aspiration pneumonia. Dent Clin N Am. 2021;65(2):307–21.

Index

R. A. Chandra, R. J. Li (eds.), *Multidisciplinary Management of Head and Neck Cancer*, https://doi.org/10.1007/978-3-031-05973-5

GPSR Compliance

The European Union's (EU) General Product Safety Regulation (GPSR) is a set of rules that requires consumer products to be safe and our obligations to ensure this.

If you have any concerns about our products, you can contact us on ProductSafety@springernature.com

In case Publisher is established outside the EU, the EU authorized representative is:

Springer Nature Customer Service Center GmbH
Europaplatz 3
69115 Heidelberg, Germany

Batch number: 10732809

Printed by Printforce, the Netherlands